The
Idea
Factory

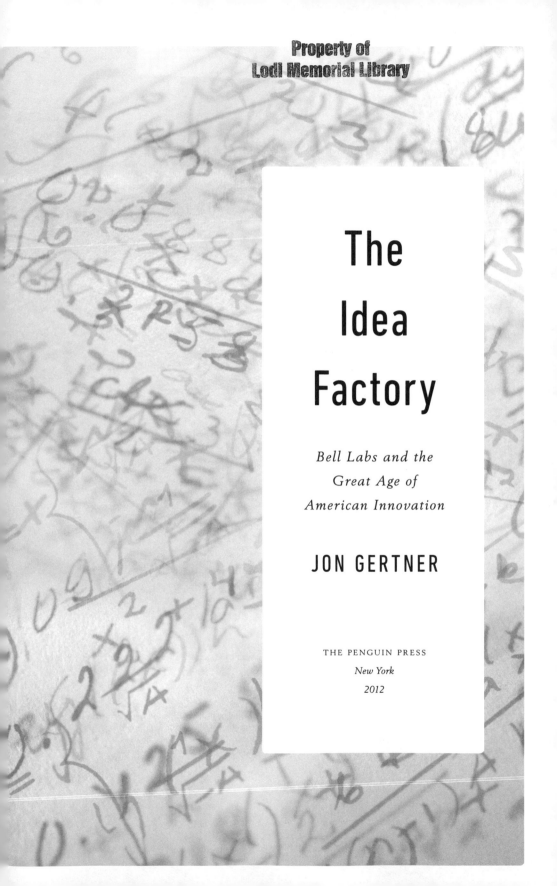

The
Idea
Factory

*Bell Labs and the
Great Age of
American Innovation*

JON GERTNER

THE PENGUIN PRESS

New York

2012

THE PENGUIN PRESS
Published by the Penguin Group
Penguin Group (USA) Inc., 375 Hudson Street, New York, New York 10014, U.S.A.
• Penguin Group (Canada), 90 Eglinton Avenue East, Suite 700, Toronto, Ontario,
Canada M4P 2Y3 (a division of Pearson Penguin Canada Inc.) • Penguin Books Ltd,
80 Strand, London WC2R 0RL, England • Penguin Ireland, 25 St Stephen's Green,
Dublin 2, Ireland (a division of Penguin Books Ltd) • Penguin Books Australia Ltd,
250 Camberwell Road, Camberwell, Victoria 3124, Australia (a division of Pearson
Australia Group Pty Ltd) • Penguin Books India Pvt Ltd, 11 Community Centre,
Panchsheel Park, New Delhi – 110 017, India • Penguin Group (NZ), 67 Apollo
Drive, Rosedale, Auckland 0632, New Zealand (a division of Pearson New Zealand
Ltd) • Penguin Books (South Africa) (Pty) Ltd, 24 Sturdee Avenue, Rosebank,
Johannesburg 2196, South Africa

Penguin Books Ltd, Registered Offices:
80 Strand, London WC2R 0RL, England

First published in 2012 by The Penguin Press,
a member of Penguin Group (USA) Inc.

Library of Congress Cataloging-in-Publication Data

Gernter, Jon.
The idea factory : the Bell Labs and
the great age of American innovation / Jon Gernter.
p.   cm.
Includes bibliographical references and index.
ISBN 978-1-59420-328-2
1. Bell Telephone Laboratories—History—20th century.
2. Telecommunication—United States—History—20th century.
3. Technological innovations—United States—History—20th century.
4. Creative ability—United States—History—20th century.   5. Inventors—United
States—History—20th century.   I. Title.
TK5102.3.U6G47 2012
384—dc23
2011040207

Printed in the United States of America
1   3   5   7   9   10   8   6   4   2

DESIGNED BY AMANDA DEWEY

ALWAYS LEARNING   PEARSON

For Liz, Emmy, and Ben

CONTENTS

Where is the knowledge we have lost in information?

—T. S. Eliot, *The Rock*

Introduction

WICKED PROBLEMS

This book is about the origins of modern communications as seen through the adventures of several men who spent their careers working at Bell Telephone Laboratories. Even more, though, this book is about innovation—about how it happens, why it happens, and who makes it happen. It is likewise about why innovation matters, not just to scientists, engineers, and corporate executives but to all of us. That the story is about Bell Labs, and even more specifically about life at the Labs between the late 1930s and the mid-1970s, isn't a coincidence. In the decades before the country's best minds began migrating west to California's Silicon Valley, many of them came east to New Jersey, where they worked in capacious brick-and-glass buildings located on grassy campuses where deer would graze at twilight. At the peak of its reputation in the late 1960s, Bell Labs employed about fifteen thousand people, including some twelve hundred PhDs. Its ranks included the world's most brilliant (and eccentric) men and women. In a time before Google, the Labs sufficed as the country's intellectual utopia. It was where the future, which is what we now happen to call the present, was conceived and designed.

For a long stretch of the twentieth century, Bell Labs was the most innovative scientific organization in the world. It was arguably among the world's most important commercial organizations as well, with countless entrepreneurs building their businesses upon the Labs' foundational inventions, which were often shared for a modest fee. Strictly speaking, this wasn't Bell Labs' intended function. Rather, its role was to support the research and development efforts of the country's then-monopolistic telephone company, American Telephone & Telegraph (AT&T), which was seeking to create and maintain a system—the word "network" wasn't yet common—that could connect any person on the globe to any other at any time. AT&T's dream of "universal" connectivity was set down in the early 1900s. Yet it took more than three-quarters of a century for this idea to mature, thanks largely to the work done at Bell Labs, into a fantastically complex skein of copper cables and microwave links and glass fibers that tied together not only all of the planet's voices but its images and data, too. In those evolutionary years, the world's business, as well as its technological progress, began to depend on information and the conduits through which it moved. Indeed, the phrase used to describe the era that the Bell scientists helped create, the age of information, suggested we had left the material world behind. A new commodity— weightless, invisible, fleet as light itself—defined the times.

A new age makes large demands. At Bell Labs, it required the efforts of tens of thousands of scientists and engineers over many decades— millions of "man-hours," in the parlance of AT&T, which made a habit of calculating its employees' toil to a degree that made its workers proud while also keeping the U.S. government (which closely monitored the company's business practices and long-distance phone monopoly) at bay. For reasons that are conceptual as well as practical, this book does not focus on those tens of thousands of Bell Laboratories workers. Instead, it looks primarily at the lives of a select and representative few: Mervin Kelly, Jim Fisk, William Shockley, Claude Shannon, John Pierce, and William Baker. Some of these names are notorious—Shockley, for instance, who won the Nobel Prize in Physics in 1956 and in his later years steadfastly pursued a scientific link between race and intelligence. Others, such as Shannon, are familiar to those within a certain area of interest (in Shannon's case, mathematics and artificial intelligence) while remaining largely unknown to the general public. Pierce, a nearly forgotten figure, was the father of satellite communications and an instigator of more ideas than can be properly accounted for here. Kelly, Fisk, and Baker were presidents of the Labs, and served as stewards during the institution's golden age. All these men knew one another, and some were extremely close. With the exception of Mervin Kelly, the eldest of the group, they were sometimes considered members of a band of Bell Labs revolutionaries known as the Young Turks. What bound them was a shared belief in the nearly sacred mission of Bell Laboratories and the importance of technological innovation.

The men preferred to think they worked not in a laboratory but in what Kelly once called "an institute of creative technology." This description aimed to inform the world that the line between the art and science of what Bell scientists did wasn't always distinct. Moreover, while many of Kelly's colleagues might have been eccentrics, few were dreamers in the less flattering sense of the word. They were paid for their imaginative abilities. But they were also paid for working within a culture, and within an institution, where the very point of new ideas was to make them into new things.

SHOULD WE CARE ABOUT how new ideas begin? Practically speaking, if our cell phones ring and our computer networks function we don't need to recall how two men sat together in a suburban New Jersey laboratory during the autumn of 1947 and invented the transistor, which is the essential building block of all digital products and contemporary life. Nor should we need to know that in 1971 a team of engineers drove around Philadelphia night after night in a trailer home stocked with sensitive radio equipment, trying to set up the first working cell phone system. In other words, we don't have to understand the details of the twentieth century in order to live in the twenty-first. And there's a good reason we don't have to. The history of technology tends to remain stuffed in attic trunks and the minds of aging scientists. Those breakthrough products of past decades—the earliest silicon solar cells, for example, which were invented at Bell Labs in the 1950s and now reside in a filing cabinet in a forlorn warehouse in central New Jersey—seem barely functional by today's standards. So rapid is the evolutionary development of technological ideas that the journey from state-of-the-art to artifact can occur in a mere few years.

Still, good arguments urge us to contemplate scientific history. Bill Gates once said of the invention of the transistor, "My first stop on any time-travel expedition would be Bell Labs in December 1947."[1] It's a perceptive wish, I think. Bell Labs was admittedly imperfect. Like any elite organization, it suffered at times from personality clashes, institutional arrogance, and—especially in its later years—strategic missteps. Yet understanding the circumstances that led up to that unusual winter of 1947 at Bell Labs, and what happened there in the years afterward, promises a number of insights into how societies progress. With this in mind, one might think of a host of reasons to look back at these old inventions, these forgotten engineers, these lost worlds.

While our engineering prowess has advanced a great deal over the past sixty years, the principles of innovation largely have not. Indeed, the techniques forged at Bell Labs—that knack for apprehending a vexing problem, gathering ideas that might lead to a solution, and then pushing toward the development of a product that could be deployed on a massive scale—are still worth considering today, where we confront a host of challenges (information overloads, infectious disease, and climate change, among others) that seem very nearly intractable. Some observers have taken to calling them "wicked problems." As it happens, the past offers the example of one seemingly wicked problem that was overcome by an innovative effort that rivals the Apollo program and Manhattan Project in size, scope, expense, and duration. That was to connect all of us, and all of our new machines, together.

"At first sight," the writer Arthur C. Clarke noted in the late 1950s, "when one comes upon it in its surprisingly rural setting, the Bell Telephone Laboratories' main New Jersey site looks like a large and up-to-date factory, which in a sense it is. But it is a factory for ideas, and so its production lines are invisible."[2] Some contemporary thinkers would lead us to believe that twenty-first-century innovation can only be accomplished by small groups of nimble, profit-seeking entrepreneurs working amid the frenzy of market competition. Those idea factories of the past—and perhaps their most gifted employees—have no lessons for those of us enmeshed in today's complex world. This is too simplistic. To consider what occurred at Bell Labs, to glimpse the inner workings of its invisible and now vanished "production lines," is to consider the possibilities of what large human organizations might accomplish.

Part 1

One

OIL DROPS

The first thing they tended to notice about Mervin Kelly was his restlessness. Anyone in the town of Gallatin, Missouri, could see it. The boy was antsy, impatient—barely able to contain himself in anticipation of some future event that could not possibly arrive quickly enough. You might think he'd been born with electricity running through his veins. He was serious about his schoolwork, but his excess of energy led him to a multitude of other jobs, too. At a very young age, he made extra money assisting in his father's store and leading cows to pasture for local farmers. At ten he began building a paper route business, and soon became an employer of other boys who did the work, rather than the one who made the deliveries. By his teenage years he was also helping his father keep the books at the shop downtown. His high school class was small—just eighteen students—but he was a striver, becoming both class president and valedictorian. His classmates called him "our Irish king." People in Gallatin noticed that, too. The young man was intent on being in charge. And in a place where people neither walked fast nor talked fast, young Mervin Kelly did both.

His father—kindly and bookish, and not nearly the go-getter his son was turning out to be—was named Joseph Fennimore Kelly. As a young man, Joe Kelly had taught high school history and English, but by 1900, when the Kelly family was counted for the first time in the Gallatin census, he was managing a hardware store on the east side of the town square. Despite being seventy-five miles from Kansas City, far enough away to be considered a backwater, Gallatin's downtown bustled. The clear reason was its location at the intersection of two train lines, the Rock Island and the Wabash, both of which stopped in town to take in and disgorge passengers. As a result, Gallatin, with a population of just 1,700, boasted three hotels and several restaurants. The town had two newspapers, two banks, five dentists, four druggists, two jewelers, and nine physicians. There were two cigar factories, four blacksmiths, and several saloons. In Gallatin, the Kelly family had settled in a prosperous place that was perched on the cusp of modernity.

All around was the simplicity of small-town life. The days were mostly free of noisy machinery or any kind of electric distractions. You butchered your own hogs and collected eggs from your own hens. Farmers and merchants alike visited with acquaintances around the crowded town square on Saturday nights. The Old West—the Wild West—had not quite receded, and so you listened quite regularly to reminiscences about the trial of Frank James, Jesse's outlaw brother, which Gallatin had hosted a few decades before. On hot days in the summer you walked or rode a horse a half mile from town to the banks of the Grand River, where you would go for a swim; and on some summer evenings, if you were a teenager (and if you were lucky), you danced with a girl at an ice cream social. There were no radio stations yet—the device was mostly a new toy for hobbyists—so instead there might be a primitive Edison phonograph or a string band at the party, some friends who could play fiddle and mandolin.

In the meantime, there was little doubt that Gallatin was moving ahead with the rest of the world. And the disruptions of technology, at least to a young man, must have seemed thrilling. It wasn't only the railroads. As Mervin Kelly attended high school, automobiles began arriving in Gallatin. Thanks to a diesel generator, the town now enjoyed a few hours of electricity each evening. A local telephone exchange—a small switchboard connecting the hundred or so phone subscribers in Gallatin—opened its office near the town square, in the same brick building as the Kelly hardware store. To see the switchboard in action, Kelly would only have had to step outside his father's store, turn right, and walk around the side of the building to the front door of the exchange. In a sense, his future was right around the corner.

At sixteen, he was awarded a scholarship to the Missouri School of Mines, located in the town of Rolla, 250 miles away. To someone from Gallatin, such a distance was almost unimaginably far, yet Kelly seemed to have no reservations about leaving. "I was really pretty lucky," he later said. Few people in his town made it through high school; fewer still made it to college. When he departed, the young man thought he might ultimately work as a geologist or mining engineer. That way, he would travel to the far reaches of the earth. He seemed well aware that the course of his life might be determined by his energetic impulses. "My zeal," Kelly noted in the Gallatin High School yearbook, "has condemned me."

IN 1910, when Kelly set off for mining school, few Americans recognized the differences between a scientist, an engineer, and an inventor. The public was far more impressed by new technology than the knowledge that created the technology. Thus it was almost certainly the case that the inventor of machinery seemed more vital to the modern age than someone—a trained physicist, for example—who might explain how and why the machine worked.

There seemed no better example of this than Thomas Edison. By the time Kelly was born, in 1894, Edison was a national hero, a beau ideal of American ingenuity and entrepreneurship. Uniquely intuitive, Edison had isolated himself with a group of dedicated and equally obsessive men at a small industrial laboratory in New Jersey. Edison usually worked eighteen hours a day or longer, pushing for weeks on end, ignoring family obligations, taking meals at his desk, refusing to pause for sleep or showers. He disliked bathing and usually smelled powerfully of sweat and chemical solvents.[1] When fatigue overcame him he would crawl under his table for a catnap or stretch out on any available space (though eventually his wife placed a bed in the library of his West Orange, New Jersey, laboratory). For his inventing, Edison used a dogged and systematic exploratory process. He tried to isolate useful materials—his stockroom was replete with everything from copper wire to horses' hoofs and rams' horns—until he happened upon a patentable, and marketable, combination.[2]

Though Edison became rich and famous for his phonograph and his filament for the electric light bulb, some of his less heralded inventions were arguably as influential on the course of modern life. One of those was a new use for a compressed carbon "button," which he discovered in 1877 could be placed inside the mouthpiece of a telephone to dramatically improve the quality and power of voice transmission. (He had first tried lead, copper, manganese, graphite, osmium, ruthenium, silicon, boron, iridium, platinum, and a wide variety of other liquids and fibers.) A decade later Edison improved upon the carbon button by proposing instead the use of tiny roasted carbon granules, derived from coal, in the vocal transmitter.[3] These discoveries made the telephone a truly marketable invention.

Edison's genius lay in making new inventions work, or in making existing inventions work better than anyone had thought possible. But *how* they worked was to Edison less important. It was not true, as his onetime protégé Nikola Tesla insisted, that Edison disdained literature or ideas. He read compulsively, for instance—classics as well as newspapers. Edison often said that an early encounter with the writings of Thomas Paine had set his course in life. He maintained a vast library in his laboratory and pored over chemistry texts as he pursued his inventions. At the same time, however, he scorned talk about scientific theory, and even admitted that he knew little about electricity. He boasted that he had never made it past algebra in school. When necessary, Edison relied on assistants trained in math and science to investigate the principles of his inventions, since theoretical underpinnings were often beyond his interest. "I can always hire mathematicians," he once said at the height of his fame, "but they can't hire me."[4]

And it was true. In the boom times of the Industrial Revolution, in the words of one science historian, inventing products such as the sewing machine or barbed wire "required mainly mechanical skill and ingenuity, not scientific knowledge and training." Engineers in the fields of mining, rubber, and energy on occasion consulted with academic geologists, chemists, and physicists. "But on the whole, the industrial machine throbbed ahead without scientists and research laboratories, without even many college-trained engineers. The advance of technology relied on the cut-and-try methods of ingenious tinkerers, unschooled save possibly for courses at mechanics institutes."[5] Indeed, by the time Mervin Kelly began his studies at the Missouri School of Mines around 1910, any sensible American boy with an eye on the future might be thinking of engineering; the new industrial age mostly needed men who could make bigger and better machines.

And yet the notion that scientists trained in subjects like physics could do intriguing and important work was gaining legitimacy. Americans still knew almost nothing about the sciences, but they were beginning to hear about a stream of revelations, all European in origin, regarding the hidden but fundamental structure of the visible world. Words like "radioactivity," "X-rays," and, especially, "quanta"—a new term for what transpired within the tiny world of molecules—started filtering into American universities and newspapers. These ideas almost certainly made their way to Missouri, where Kelly was paying his rent in Rolla—a room on the third floor of the metallurgical building—by working with the State Geological Survey for $18 a week numbering mineral specimens. During one of his summer breaks he took a job at a copper mine in Utah, an experience that repelled him permanently from a career as a mining engineer and pushed him closer to pure science. After graduating he took a one-year job teaching physics to undergraduates at the University of Kentucky. The school also gave him a master's degree in that subject. After that, he headed north to Chicago,

FOR DECADES, any serious American science student had to complete his education in Europe, most often at schools in Berlin and Gottingen, Germany, where he could sit at the feet of the masters as they lectured or carried on laboratory research. (The language of science was German, too.) But early in the twentieth century a handful of American schools, notably Johns Hopkins, Cornell, and the University of Chicago, began turning out accomplished graduates in physics and chemistry. In 1916, Robert Millikan at the University of Chicago was establishing himself as a leading physicist and teacher of the subject. Then in his forties, he would go on to win the Nobel Prize in Physics in 1923, and grace the cover of *Time* magazine in 1927. Ultimately, he would build the California Institute of Technology into one of the country's great scientific institutions, and throughout his career he would guide many of his brightest students to jobs with AT&T. To a student like Kelly, Millikan would have seemed heroic. His textbooks on physics were becoming the standard for college instruction, and his work on measuring the exact charge of an electron, an experiment that was continuing when Kelly arrived in Chicago to study with him, had made him famous in the small community of academic physicists.

Rather like Kelly himself, there was something authentically, irresistibly American about Millikan. Though he'd received a year's worth of instruction in Paris, Berlin, and Gottingen, he was nevertheless the son of an Iowa preacher, cheerful, earnest, conservative, boyishly handsome, and almost always neatly dressed in a collared shirt and bow tie. Also like Kelly, Millikan was a man of action. He worked himself not quite to Edison's extreme, but close, which suggested the bootstrap ethic could apply to physicists as well as inventors. As a younger man, the professor had almost missed his own wedding because he was so busy reviewing a scientific manuscript in his office.

By the early twentieth century, physicists were already dividing into camps: those who theorized and those who experimented. Millikan was an experimentalist. He shrewdly devised laboratory tests that validated theoretical work but also built upon the work of other experimentalists, "discovering the weak points that could be improved upon," as his student Paul Epstein described it. Millikan's first great claim to fame was something known as the oil-drop experiment, which was representative of those early-twentieth-century forays into laboratory physics. The experiment was both creative and demanding—creative in how it attempted to reveal the elements of the cosmos by way of a small device constructed from everyday materials, and demanding in how it required years of follow-up work (even after the results were first shared in 1910) before it could be deemed precise. It was also, not incidentally, Mervin Kelly's first real encounter with deep, fundamental research.[6]

The oil-drop experiment would, in Millikan's own words, serve as "the most direct and unambiguous proof of the existence of the electron." More precisely, it would attempt to put an exact value on e, which is the charge of the electron, and which in turn would make a range of precise calculations about subatomic physics possible. Other researchers had already tried to measure e by observing the behavior of a fine mist of water that had been subjected to an electric charge. The experimenter would spray a mist between two horizontal metal plates spaced less than an inch apart. One plate carried a negative charge and the other a positive charge. The electric field between the two plates would slow the fall of some droplets. The idea, or rather the hope, was to suspend a droplet of water between the plates; then, by measuring the speed of the falling droplet and the intensity of the electric field required to slow the droplet, you could calculate its electric charge. There was a problem, however: The water in the droplet evaporated so fast that it would only remain visible for a couple of seconds. It was proving difficult to get anything beyond a rough estimate of the charge. The experiment was going nowhere.

One of Millikan's great ideas—he would claim it came to him on a train traveling through the plains of Manitoba—was to change the measured substance from water to oil, because oil wouldn't evaporate, and measurements would thus improve. (It was more likely that a graduate student of Millikan's named Harvey Fletcher actually suggested the switch from water to oil and helped him create the testing apparatus.)[7]

In time, the experiment came to work something like this: A researcher would stand in front of a boxlike chamber and spray a fine mist of oil from a tool called an atomizer; he would look through a close-range telescope at the droplets, which were illuminated by a beam of light; he would then turn on the electric plates and measure (stopwatch in hand) how the oil drops behaved—how long it took for them to move down or up in their suspended state—and write down the observations.

When Millikan's student Harvey Fletcher first tried the experiment—when he looked through the telescope at the tiny oil drops suspended in air that sparkled like "stars in constant agitation"—he felt the urge to scream with excitement. To do the experiment for hour after hour, day after day, counting how long it took for a certain-sized drop to rise or fall a certain distance when a certain amount of current was applied, was a painstaking process. Fletcher was well matched for such work. But for someone in a hurry, for someone whose very constitution was unsuited to the practice of quiet and diligent observation, the time spent in the Millikan lab must have seemed like a kind of torture. Eventually, Fletcher's role in the lab was taken over by a younger graduate student—Mervin Kelly. On some evenings, Kelly asked his new wife, Katherine, a pretty girl from Rolla whom he had met as an undergraduate and had married after a brief courtship, to come to the lab with him. On Chicago's south side, late into the night, she would help him measure the drops.

LONG BEFORE Mervin Kelly came to Millikan's lab in 1915, a chain reaction began that would ultimately shape his own career and Bell Labs' singular trajectory. To understand how that chain of events started, it's helpful to pause for a moment on the image of the young physicist in the lab, counting oil drops late into the night, and go back in time a few years, to 1902. That year, Robert Millikan was married. What was significant about Millikan's wedding was not the ceremony itself. Rather, it was his best man: a slight, balding, cigar-smoking physicist named Frank Baldwin Jewett.

At Chicago, Jewett was pursuing a PhD when he met Millikan, a new faculty member who was nine years older. The two men lived in the same boardinghouse. Unlike Millikan, Jewett had grown up in the lap of privilege. He was the son of a railroad and electric utility executive, and his family had originally owned large tracts of land that became part of Pasadena and Greater Los Angeles. And yet Jewett wasn't exactly a snob; he was agile-minded and glib; he could talk with and befriend almost anyone. He was especially adept at earning the trust of older men. When he graduated from Chicago, Jewett considered returning west to join the ranks of California industrialists, like his father. But first he decided to teach at the Massachusetts Institute of Technology instead. Midway through his year as a physics instructor, he had a chance meeting with one of the engineers at American Telephone & Telegraph, who was quickly charmed and impressed by him. When Jewett was offered a job with the company in 1904, he accepted. His pay was $1,600 a year, or about $38,000 in today's dollars.

Contrary to its gentle image of later years, created largely through one of the great public relations machines in corporate history, Ma Bell in its first few decades was close to a public menace—a ruthless, rapacious, grasping "Bell Octopus," as its enemies would describe it to the press. "The Bell Company has had a monopoly more profitable and more controlling—and more generally hated—than any ever given by any patent," one phone company lawyer admitted.[8] Jewett came into the business nearly thirty years after Alexander Graham Bell patented the telephone; by that point approximately two million subscribers around the country, mostly in the Northeast, were using AT&T's phones and services. And yet the company was struggling. Bell's patents on the telephone had expired in the 1890s, and in the years after the expiration a host of independent phone companies had entered the business and begun signing up subscribers in numbers rivaling AT&T. By then the company's competitive practices—its unrelenting aggression, its flagrant disregard for ethical boundaries—had already won it a host of enemies. Almost from the day the Bell System was created, when Alexander Graham Bell became engaged in a multiyear litigation with an inventor named Elisha Gray over who actually deserved the patent to the telephone, the Bell company was known to be ferociously litigious.[9] In its later battles with independent phone companies, however, it would often move beyond battles in the courtroom and resort to sabotaging competitors' phone lines and stealthily taking over their equipment suppliers.

All the while, the company maintained a policy of "noncompliance" with other service providers. This meant that AT&T often refused to carry phone calls from the competition over its intercity long-distance lines. In some metro areas, the practice led to absurd redundancies: Residents or businesses sometimes needed two or even three telephones so they could speak with acquaintances who used different service providers.[10] In the meantime, AT&T did little to inspire loyalty in its customers. Their phone service was riddled with interruptions, poor sound quality, unreliable connections, and the frequent distractions of "crosstalk," the term engineers used to describe the intrusion of one signal (or one conversation) into another. In rural areas, phone subscribers had to make do with "party lines" that connected a dozen, or several dozen, households to the local operator but could only allow one conversation at a time. Subscribers were not supposed to listen in on their neighbors' conversations. Often they did anyway.

AT&T's savior was Theodore Vail, who became its president in 1907, just a few years after Millikan's friend Frank Jewett joined the company.[11] In appearance, Vail seemed almost a caricature of a Gilded Age executive: Rotund and jowly, with a white walrus mustache, round spectacles, and a sweep of silver hair, he carried forth a magisterial confidence. But he had in fact begun his career as a lowly telegraph operator. Thoughtfulness was his primary asset; he could see almost any side of an argument. Also, he could both disarm and outfox his detractors. As Vail began overseeing Bell operations, he saw that the costs of competition were making the phone business far less profitable than it had been—so much so, in fact, that Vail issued a frank corporate report in his first year admitting that the company had amassed an "abnormal indebtedness." If AT&T were to survive, it had to come up with a more effective strategy against its competition while bolstering its public image. One of Vail's first moves was to temper its aggression in the courts and reconsider its strategy in the field. He fired twelve thousand employees and consolidated the engineering departments (spread out in Chicago and Boston) in the New York office where Frank Jewett then worked.[12] Meanwhile, Vail saw the value of working with smaller phone companies rather than trying to crush them. He decided it was in the long-term interests of AT&T to buy independent phone companies whenever possible. And when it seemed likely a few years later that the government was concerned about this strategy, Vail agreed to stop buying up companies without government permission. He likewise agreed that AT&T would simply charge independent phone companies a fee for carrying long-distance calls.

Vail didn't do any of this out of altruism. He saw that a possible route to monopoly—or at least a near monopoly, which was what AT&T had always been striving for—could be achieved not through a show of muscle but through an acquiescence to political supervision. Yet his primary argument was an idea. He argued that telephone service had become "necessary to existence."[13] Moreover, he insisted that the public would be best served by a technologically unified and compatible system—and that it made sense for a single company to be in charge of it. Vail understood that government, or at least many politicians, would argue that phone subscribers must have protections against a monopoly; his company's expenditures, prices, and profits would thus have to be set by federal and local authorities.[14] As a former political official who years before had modernized the U.S. Post Office to great acclaim, Vail was not hostile toward government. Still, he believed that in return for regulation Ma Bell deserved to find the path cleared for reasonable profits and industry dominance.

In Vail's view, another key to AT&T's revival was defining it as a technological leader with legions of engineers working unceasingly to improve the system. As the business historian Louis Galambos would later point out, as Vail's strategy evolved, the company's executives began to imagine how their company might adapt its technology not only for the near term but for a future far, far away. "Eventually it came to be assumed within the Bell System that there would never be a time when technological innovation would no longer be needed." The Vail strategy, in short, would measure the company's progress "in decades instead of years."[15] Vail also saw it as necessary to merge the idea of technological leadership with a broad civic vision. His publicity department had come up with a slogan that was meant to rally its public image, but Vail himself soon adopted it as the company's core philosophical principle as well.[16] It was simple enough: "One policy, one system, universal service." That this was a kind of wishful thinking seemed not to matter. For one thing, there were many systems: The regional phone companies, especially in rural areas, provided service for millions of Americans. For another, the closest a customer could get to telephoning long distance was a call between New York and Chicago. AT&T did not have a universal reach. It didn't even have a national reach.

AT&T'S ENGINEERS HAD BEEN VEXED by distance from the very beginning. The telephone essentially converted the human voice into an electrical signal; in turn-of-the-century phones this was done by allowing sound waves produced by a voice to vibrate a taut diaphragm—usually a disc made of thin aluminum—that was backed by another thin metal disc. A mild electric current ran between the two discs, which were separated by a chamber filled with the tiny carbon granules Edison had invented. As sound waves from a voice vibrated the top diaphragm, waves of varying pressure moved through the granules below it. The varying pressure would in turn vary the resistance to the electric current running between the metal discs. In the process sound waves would be converted to electric waves. On a simple journey, the electrified voice signal would then travel through a wire, to a switchboard, to another cable, to another switchboard, and finally to a receiver and a distant eardrum. But a telephone voice signal was weak—much weaker and more delicate than a telegraph's simple dot-dash signal. Even worse, the delicate signal would grow thinner—or "attenuate," to use the phone company's preferred term—after even a few miles.

In the telephone's first few decades, AT&T's engineers had found that different methods could move a phone call farther and farther. Copper wire worked better than iron wire, and stiff, "hard-drawn" copper wire seemed to work even better. Best of all was extremely thick-gauge hard-drawn copper wire. The engineers likewise discovered that an invention known as "loading coils" inserted on the wires could extend the signal tremendously. Finally, there were "repeaters." These were mechanical amplifiers that took the sound of a weakening voice and made it louder so the call could travel many miles farther. But you could only install a few repeaters on a line before the advantages of boosting a call's volume were undone by distortion and the attenuation of the signals. And that left the engineers at a final disconnect. The tricks of their trade might allow them to conquer a distance of about 1,700 miles, roughly from New York to Denver. A great impasse lay beyond.

In 1909, Frank Jewett, now one of the phone company's senior managers, traveled to San Francisco with his boss, John J. Carty, AT&T's chief engineer. They found parts of the city still in ruins. As Jewett recalled, "The wreckage of the [1906] earthquake and fire was still only partially cleared away and but the beginnings made on the vast rebuilding operations."[17] The men were there to determine how to repair the local phone system, but they also began discussing the possibility of providing transcontinental phone service—New York to San Francisco—in time for the Panama-Pacific International Exposition of 1914. Theodore Vail, who met Jewett and Carty there, was in favor of making a commitment, since it represented a clear step toward universal service. Carty and Jewett were more circumspect. Together they spent long days and nights debating the problem, usually continuing their discussions far past midnight. The men could see there were enormous, but surmountable, engineering challenges; they would, for instance, need a cable that could be effectively strung across the mountains and desert and survive the weather and stress. But there were also profound challenges of science. "The crux of the problem," Jewett wrote in describing his conversations with Carty, "was a satisfactory telephone repeater or amplifier":

Did we know how to develop such a repeater? No. Why not? Science hadn't yet shown us the way. Did we have any reason to think that she would? Yes. In time? Possibly. What must we do to make "possibly" into "probably" in two years?

And so on night after night without end almost.

Carty and Jewett eventually told Vail they would do it—and the task soon came to be Jewett's personal responsibility. That was risky on a number of counts. Jewett's talents were as a manager and social sophisticate; he was quick to apprehend technical problems but not necessarily equipped to solve them. On the other hand, he knew someone who could help.

Jewett returned to the University of Chicago in the fall of 1910 to visit his old friend Millikan, and he started the conversation without small talk. Jewett began, "Mr. John J. Carty, my chief, and the other higher-ups in the Bell System, have decided that by 1914, when the San Francisco Fair is to be held, we must be in position, if possible, to telephone from New York to San Francisco." To get through to San Francisco by the present methods was out of the question, he explained, but he wondered if perhaps Millikan's work—he pointed to some complex research on electrons—suggested that a different method might be possible. Then Jewett asked his friend for help. "Let us have one or two, or even three, of the best of the young men who are taking their doctorates with you and are intimately familiar with your field. Let us take them into our laboratory in New York and assign to them the sole task of developing a telephone repeater."[18]

Here was a new approach to solving an industrial problem, an approach that looked not to engineers but to scientists. The first person offered this opportunity was Millikan's lab assistant from the oil-drop experiment, Harvey Fletcher, who declined. Fletcher wanted to return home to Salt Lake City to teach at Brigham Young University. The next person was Harold Arnold, a savvy experimentalist who said yes, and who quickly joined the New York engineering group under Jewett.

Within two years Arnold came up with several possible solutions to the repeater problem, but he mainly went to work on improving an amplifier known as the audion that had been brought to AT&T in 1912 by an independent, Yale-trained inventor named Lee De Forest. The early audion was vaguely magical. It resembled a small incandescent light bulb, yet instead of a hot wire filament strung between two supporting wires it had three elements—a metal filament that would get hot and emit electrons (called a cathode); a metal plate that would stay cool and attract electrons (called an anode); and between them a wire mesh, or "grid." A small electrical current, or signal, that was applied to the audion's grid could be greatly amplified by another electrical current that was traveling from the hot cathode to the cool anode. Arnold found, through trial and error, the best materials, as well as a superior way to evacuate the air inside the audion tube. (He suspected correctly that a high vacuum would greatly improve the audion's efficiency.) Once Arnold had refined the audion, he, Jewett, and Millikan convened in Philadelphia to test it against other potential repeater ideas. The men listened in on phone conversations that were passed through the various repeaters, and they found the audion clearly superior. Soon to be known as the vacuum tube, it and its descendants would revolutionize twentieth-century communications.

The transcontinental line, complete with several new vacuum tube repeaters placed strategically in stations along the route, was finished in time for the Pacific exposition, which had been pushed back to 1915. Harold Arnold had improved the design so that the repeaters looked like spherical bulbs, with the three crucial elements inside, sitting upon a base from which three wires emerged. The continental link itself consisted of four copper wires (two for directing calls in each direction) that were strung coast-to-coast by AT&T linemen over 130,000 wooden poles. As a public relations stunt, Alexander Graham Bell, the inventor who had long since stopped having any day-to-day responsibilities at the company he founded, was stationed in New York to speak with his old assistant, Thomas Watson, in San Francisco.

"Mr. Watson, come here, I want you," the old man quipped, para-phrasing what he had said to Watson on the day the two discovered the working telephone in Boston nearly forty years before.

"It would take me a week to get there now," Watson replied.

It was a wry bit of stagecraft. For AT&T, it was also an encouraging sign that Vail's notion of universal service might indeed be possible—at least for customers who could afford to pay about $21 (about $440 in today's dollars) for a three-minute call to California.[19]

For Frank Jewett, meanwhile, the cross-country link proved that his cadre of young scientists could be trusted to achieve things that might at first seem technologically impossible. That led him to redouble his efforts to hire more men like Harold Arnold. Jewett kept writing to Harvey Fletcher, Millikan's former graduate student who was now in Salt Lake City, sending him every spring for five consecutive years a polite and persuasive invitation to join AT&T. In 1916, Fletcher finally agreed to leave Brigham Young and come work for Jewett. Millikan, meanwhile, didn't stop serving as the link between his Chicago graduates and his old friend. In late 1917, responding to an offer from Jewett for $2,100 a year, Mervin Kelly, now done counting oil drops, decided that he would come to New York City, too.

Two

WEST TO EAST

Fletcher and Kelly were joining a company whose size and structure seemed positively bewildering. AT&T was not only a phone company on its own; it contained within it a multitude of other large companies as well. Each region of the country, for instance, had its own local phone company—New England Telephone, for example, or Pacific Bell. These organizations were owned in large part by AT&T and provided service for local phone customers. But these so-called local operating companies didn't manufacture the equipment to actually make phone service work. For that there was Western Electric, another subsidiary of AT&T. On its own, Western Electric was larger than almost any other American manufacturing corporation. Its factories built the equipment that consumers could see (such as cables and phones), as well as equipment that was largely hidden from sight (such as switchboards). Finally, there was a third branch of AT&T. Neither the local phone companies nor Western Electric maintained the long-distance service that connected different regions and states together. For that, there was AT&T's Long Lines Department. Long Lines built and provided long-distance service to customers.

Both AT&T and Western Electric had large engineering departments.

There was a certain amount of duplication—and sometimes friction—between the two. Generally speaking, the standards and long-term goals of the Bell System were determined by engineers at AT&T. Western Electric's engineers, in turn, invented, designed, and developed all new equipment and devices.[1] In 1916, the year before Fletcher and Kelly arrived, Frank Jewett was appointed the chief of Western Electric's engineering division, which put him in charge of about a thousand engineers. Western's main building was located on West Street in New York City, on the western fringe of Greenwich Village, in an immense thirteen-story yellow-brick redoubt that looked out over the tugboat and ferry traffic of New York Harbor. The engineers on the waterfront comprised a twentieth-century insurgency in a receding nineteenth-century world. The fragrance of coffee beans drifted through the large sash windows of the plant from the roasting factories nearby. A rail line, serving the busy harbor docks, stretched north and south along West Street in front of the building. "The trains ran along West Street carrying freight to the boats," an employee there in the 1920s recalls. And oftentimes, "at dusk, a man with a lantern on horseback led the trains."[2]

Under Jewett, Western engineers worked mainly in expansive open rooms floored with maple planks and interrupted every few dozen feet by square stone pillars that supported the building's massive bulk. The elevators were hand-operated. All told, the rambling West Street plant comprised over 400,000 square feet—a figure that did not include the building's rooftop, which was also used by chemists for testing how various lacquers and paints and metals withstood the elements. In their first days at the Western Electric shop, Kelly and Fletcher encountered a small city of men, along with a number of female assistants. Vast rooms of the building were dedicated to diagramming new devices for production—men in crisp white shirts, sleeves rolled above their elbows, bent over rows and rows of drafting tables. Before a device was ready for the drafting room, though, it would have to pass through a lengthy and rigorous development process. West Street was a warren of testing labs for phones, cables, switches, cords, coils, and a nearly uncountable assortment of other essential parts. There were chemical laboratories for examining the properties of new materials, such as alloys for wire and sheathing for cables; there were numerous shops, meanwhile, cluttered with wires and dials and batteries, where legions of employees spent their days testing the effects of electrical currents and switching combinations or investigating new patterns of circuitry. Large sections of the labs were also devoted to the perfection of radio transmission, for it was believed (by Jewett's boss, John J. Carty, especially) that wireless transmission would be a thing of the future, a way "to reach inaccessible places where wires cannot be strung," or a way to someday create a commercial business linking New York to, say, London.

There was no real distinction at West Street between an engineer and a scientist. If anything, everyone was considered an engineer and was charged with the task of making the thousands of necessary small improvements to augment the phone service that was interconnecting the country. Yet the company now had a small division of men working in the department of research with Harold Arnold. This department was established just after Arnold began his work on a cross-country phone repeater; it had grown slowly and steadily in the time since. Frank Jewett and John J. Carty viewed the research team as an essential part of the phone company's business strategy.[3] These young scientists, many of whom came through Millikan, were encouraged to implement Theodore Vail's long-term vision for the phone company—to look beyond the day-to-day concerns that shaped the work of their fellow engineers (to think five or ten years ahead was admirable) and focus on how fundamental questions of physics or chemistry might someday affect communications. Scientific research was a leap into the unknown, in other words. "Of its output," Arnold would later say of his group, "inventions are a valuable part, but invention is not to be scheduled nor coerced." The point of this kind of experimentation was to provide a free environment for "the operation of genius." His point was that genius would undoubtedly improve the company's operations just as ordinary engineering could. But genius was not predictable. You had to give it room to assert itself.

. . .

JOINING WESTERN ELECTRIC, even as a PhD in physics, entailed indoctrination in the phone company's ways. In Harvey Fletcher's first year he was taught to climb telephone poles, install telephones, and operate switchboards. Kelly's experiences must have been similar, but his arrival also coincided with the company's deepening involvement in building equipment for the military during the final years of World War I. He and his wife, Katherine, lived in a small apartment on Edgecombe Avenue in upper Manhattan, where she would look out the window each day to follow the construction of the Cathedral of St. John the Divine, located on a hill a few dozen blocks south. Kelly, meanwhile, began work in Harold Arnold's division by sharing a lab office with a physicist named Clinton J. Davisson, whose friends called him Davy. Davisson was an almost spectral presence at the Labs. Taciturn and shy, he was physically slight. "His weight never exceeded 115 pounds," Kelly recalled, "and for many years it hovered around 100." Kelly believed Davy was quiet for a reason. He needed to minimize superfluous activity or argument so he could husband his "low level" of energy. Only by doing so, Kelly believed, could Davy direct it, vigorously, toward experimentation.

The two men were a peculiar contrast: the antic and robust Kelly paired with the wraithlike and slow-moving Davisson. Yet it didn't take long for Kelly to discover he was impressed. Davisson was a midwesterner, too—he was born in Bloomington, Illinois—and like Kelly he owed a debt to Millikan at Chicago, who had championed his career and had helped him win academic appointments at Purdue and Princeton before he came to Western Electric. Also, Davisson was a gifted experimentalist who had an almost unwavering commitment to what Kelly would later define as basic research—that is, research that generally had no immediate application to a product or company effort but (as in Davy's case) sought fundamental knowledge regarding the deeper nature of things, such as the behavior of electrons.

At Western Electric, Davisson's passion, not to mention his manner, made him something of an oddity. Industrial labs were less interested in basic research—that was better left to the academics—than in *applied* research, which was defined as the kind of investigation done with a specific product or goal in mind. The line wasn't always distinct (sometimes applied research could yield basic scientific insights, too), but generally speaking it was believed that basic research preceded applied research, and applied research preceded *development*. In turn, development preceded *manufacture*.

In Kelly and Davisson's first years of 1917 and 1918, the military demanded workable technology in Europe—radio sets, cable lines, and phones produced in mass quantities and built to a higher standard than the ones used in the home market so as to withstand the stresses of battle. Kelly and Davisson were assigned to work on resilient vacuum tubes, which were still so new to communications that they hadn't yet entered mass production. "The relatively few that were required for extending and maintaining [phone] service," Kelly would remember, "were made in the laboratories of the [Western Electric] Engineering Department." Thus on West Street the tubes needed to be designed and built, with the help of a team of expert glassblowers, and then tested for defects, one at a time. It was a development shop, in other words, with an eye on rapid deployment for urgent military needs. Until the end of the war, there wouldn't be time for applied research, let alone basic research.

Kelly and Davisson worked together "in an atmosphere of urgency," as Kelly recalled.[4] "The rapid tempo of the work, with the necessity of accepting partial answers and following one's nose in an empirical fashion, were foreign to [Davisson's] way of doing things." Still, Davisson seemed to accept the cut-and-try approach, along with the switch from research to development, without complaint. In a way, he and Kelly had largely regressed to the old inventive traditions of Edison. But in the process Kelly was learning some things about Davisson. If the Western Electric engineers in the tube shop confronted a baffling question, they would approach Davy, who would give a deep and thoughtful and ultimately convincing response—though it sometimes took him days to do so. Increasingly, Kelly recalled, he and the rest of the staff went to Davy as a matter of last resort. Western's physicists, like Kelly, could easily understand whether a new tube, or a new tube design, worked or failed, in other words. But they couldn't always easily understand why.

Davisson decided to stay at West Street when the war ended. He was allowed to carve out a position as a scientist who rejected any kind of management role and instead worked as a lone researcher, or sometimes a researcher teamed with one or two other experimentalists, pursuing only projects that aroused his interest. He seemed to display little concern about how (or whether) such research would assist the phone company. And he planned his experiments with such rigor and unhurried meticulousness that his output was considered meager, though in truth Davisson's work was often interrupted by his colleagues' questions. Frank Jewett had no illusions that his Western Electric shop was in the business of increasing human knowledge; they were in the business of increasing phone company revenues. By allowing Davisson a position on staff, though, Jewett and his deputy Harold Arnold recognized that Davy had financial value. If he was helpful to the researchers working on real-world problems, he was worth keeping around.

"He was perhaps my closest friend," Kelly later wrote. The two men ended up living a mile apart, in Short Hills, New Jersey, and whenever Davisson was ill with some unspecified malady—a common occurrence—Kelly would visit. "Invariably I would find him in dressing gown, writing pad on his knee and pencil in hand, smoking his pipe and puzzling over his problem." Davisson used to tell people he was lazy, but Kelly believed otherwise: "He worked at a slow pace but persistently." Years later, Kelly noted that Davy might well be called "the father of basic research" at Bell Labs. It was another way of saying that early on, long before either man had gained power or fame, Kelly recognized in Davisson not only a friend and gifted scientist but a model for what might come later.

BY THE TIME KELLY ARRIVED at AT&T, the U.S. government had begun to concur with Theodore Vail's arguments for his company's expansion. A group of senators issued a report noting that the phone business, because of its sensitive technological nature—those fragile voice signals needed a unified and compatible infrastructure—was a "natural monopoly." A House of Representatives committee, clearly sympathetic to the prospect of simply dealing with a single corporate representative, complained that telephone competition was "an endless annoyance."[5] In the Willis-Graham Act of 1921, the U.S. Congress formally exempted the telephone business from federal antitrust laws.[6]

By then, the so-called natural monopoly had grown even larger. Indeed, the engineering department at West Street had become so big (two thousand on its technical staff, and another sixteen hundred on its support staff) that AT&T executives agreed in a December 1924 board meeting to spin it off into a semiautonomous company. They chose the name Bell Telephone Laboratories, Inc. Some of their reasoning remains obscure. A short notice about the new labs in the *New York Times* noted that "the new company was said to [mean] a greater concentration upon the experimental phases of the telephone industry." The spin-off, in other words, was justified by the notion that scientific research at Bell Labs would play an increasingly greater role in phone company business.[7] Frank Jewett's private memos, meanwhile, suggest that the overlap between the AT&T and Western Electric engineering departments was creating needless duplications and accounting problems. By establishing one central lab to serve two masters, the phone company would simply be more efficient.[8]

On January 1, 1925, AT&T officially created Bell Telephone Laboratories as a stand-alone company, to be housed in its West Street offices, which would be expanded from 400,000 to 600,000 square feet. The new entity—owned half by AT&T and half by Western Electric—was somewhat perplexing, for you couldn't buy its stock on any of the exchanges. A new corporate board, led by AT&T's chief engineer, John J. Carty, and Bell Labs' new president, Frank Jewett, controlled the laboratory. The Labs would research and develop new equipment for Western Electric, and would conduct switching and transmission planning and invent communications-related devices for AT&T. These organizations would fund Bell Labs' work. At the start its budget was about $12 million, the equivalent of about $150 million today.[9]

As president of Bell Labs, Jewett now commanded an enormous shop. That an industrial laboratory would focus on research and development was not entirely novel; a few large German chemical and pharmaceutical companies had tried it successfully a half century before. But Bell Labs seemed to have embraced the idea on an entirely different scale. Of the two thousand technical experts, the vast majority worked in product development. About three hundred, including Clinton Davisson and Mervin Kelly, worked under Harold Arnold in basic and applied research. As Arnold explained, his department would include "the fields of physical and organic chemistry, of metallurgy, of magnetism, of electrical conduction, of radiation, of electronics, of acoustics, of phonetics, of optics, of mathematics, of mechanics, and even of physiology, of psychology, and of meteorology."[10]

From the start, Jewett and Arnold seemed to agree that at West Street there could be an indistinctness about goals. Who could know in advance exactly what practical applications Arnold's men would devise? Moreover, which of these ideas would ultimately move from the research department into the development department and then mass production at Western Electric? At the same time, they were clear about larger goals. The Bell Labs employees would be investigating anything remotely related to human communications, whether it be conducted through wires or radio or recorded sound or visual images. At the opening of the new U.S. patent office a few years after Bell Labs was set up, Frank Jewett, whose speeches were often long-winded and hyperbolic, found a way to explain the essential idea of his new organization. An industrial lab, he said, "is merely an organization of intelligent men, presumably of creative capacity, specially trained in a knowledge of the things and methods of science, and provided with the facilities and wherewithal to study and develop the particular industry with which they are associated." In short, he added, modern industrial research was meant to apply science to the "common affairs" of everyday life. "It is an instrument capable of avoiding many of the mistakes of a blind cut-and-try experimentation. It is likewise an instrument which can bring to bear an aggregate of creative force on any particular problem which is infinitely greater than any force which can be conceived of as residing in the intellectual capacity of an individual."

Buried within Jewett's long speech was a clear manifesto. The industrial lab showed that the group—especially the interdisciplinary group—was better than the lone scientist or small team. Also, the industrial lab was a challenge to the common assumption that its scientists were being paid to look high and low for good ideas. Men like Kelly and Davisson would soon repeat the notion that there were plenty of good ideas out there, almost too many.

Mainly, they were looking for good problems.

KELLY HEWED to his vacuum tube work in the years after World War I. As he ascended steadily into management, his job came to include responsibility for developing the vacuum tubes built by Bell Labs for Western Electric, which was to say he saw himself as being responsible for improving the most important invention of his lifetime up to that point. Tubes could do much more than amplify a weak phone signal or radio transmission: They could change alternating current into direct current, making them a crucial component in early radios and televisions, which received AC from the power grid but whose mechanisms required DC to operate. What's more, the tubes could function as simple and very fast switches that turned current on and off. Early in Kelly's tenure, his tube shop made fifteen different models. There were large water-cooled tubes the size of wine bottles that were used in high-power radio broadcast stations; small tubes for public-address systems; and the famed repeater tube that had been Harold Arnold's great contribution in bringing the transcontinental connection to bear.

Sometimes a vacuum tube was described as a cousin to the ordinary incandescent light bulb. In some respects this was true—for instance, both devices contained wires that were sealed inside a glass container. Yet the differences between them were far more pronounced. A factory could turn out tens of thousands of bulbs a day. But the daily output of, say, a repeater tube that allowed telephone conversations to be conveyed across the country was at best in the hundreds. What's more, the kind of tubes made in Kelly's shop had to be forged with a jeweler's precision. There was no room for error. If a light bulb failed, it would be easy to replace and not necessarily urgent; if a repeater failed, many conversations would, too. Money would be lost, maybe even lives.

Early in his career, Kelly wrote a long article explaining in meticulous detail how a vacuum tube was made. It reveals something of the nature of Bell Labs' work in general at the time, which aspired to be at the leading edge of what any company in the world could achieve, both in conceptual and manufacturing terms. The tube shop Kelly described was located in a building a dozen blocks south of the West Street labs in Manhattan, at 395 Hudson Street. There, in a gritty industrial neighborhood, his workers, men as well as women, labored behind lab benches in large rooms outfitted for assembly and production. To explain the process, Kelly used the example of the repeater tube known as the 101-D. Its production began with a glass pipe about the size of a man's pinkie. The pipe was heated and rotated on a machine so that its bottom opening could then be flared out. A different machine would take the top opening of the pipe, insert four long wires, and then heat the glass to seal the wires so they extended through the seal. The four wires resembled plant stalks poking through a hill of snow. This assembly was called the stem press.[11]

Next, a worker would attach, carefully and by hand, a solid glass rod atop the stem press, just behind the four wires that were already poking up. The glass rod was positioned vertically and was in turn superheated. A hand-operated machine was then used to insert in its hot, softened glass ten more wires. Two of these stuck out vertically, the other eight horizontally.

The assembly now looked more like a broken toy than an electronic device. It was a mass of wires, fourteen in all, poking out in all directions from a central glass core. But it wasn't done. First, the tube's glass core had to be heated and cooled and heated and cooled again in order to harden it. Afterward, a worker had to administer several chemical washes to remove grease and oil from the surface of the glass and wires. The faintest trace of impurities raised the risk of failure. Finally it was time for a worker to arrange the functional parts of the vacuum tube—the parts that would amplify phone signals—around the glass core and wires. These parts were the cathode, grid, and anode. It had taken the Bell scientists years to figure out, in Edisonian, trial-and-error fashion, which materials worked best. The anode was a tiny flattened, hollow box of sheet nickel; the grid was a mesh fashioned from nickel wire of several different diameters; the cathode was a ribbon of metal, M-shaped, made from a platinum-alloy core coated with other trace elements. All of these parts were heated in an oven to 1,000 degrees centigrade to burn off imperfections.

Afterward, the tube shop workers welded these parts to the ragged wires sticking out from the glass core. The contraption no longer looked disheveled. It looked like a device with a purpose. Every part was now tidily connected and wrapped tight. At this point an employee would insert what they had in front of them—the assembly of welded wires and metal plates anchored to the glass rod—into a round glass bulb roughly the size and shape of a conventional light bulb. Then they would heat the bottom of the bulb to create a closed seal.

A vacuum tube couldn't work without a vacuum inside. So the pumping began. It was a complex, four-stage process requiring several different pumps, all of the machines designed within Kelly's tube shop itself. The goal was to eliminate the air inside—to "approximately one-millionth of an atmosphere," as Kelly would explain—through a hole on top of the bulb. But afterward a few other steps still remained: The inside of the bulb, for instance, had to now be heated to about 800 degrees centigrade for further improvements in the vacuum. And the hole on top of the bulb had to be sealed. Finally, a worker would connect four wires dangling from the bottom of the vacuum tube to a small cylindrical base and fasten the base to the bottom of the bulb. At last, after this final step, one could admire the finished vacuum tube, the 101-D, and get the impression of looking at a large but fabulously complex light bulb with an intricate miniature architecture of metal plates, posts, and wires inside.

Kelly called the tubes "miracle devices" that would usher in a great age of electronic communications. But he knew better than anyone how difficult they were to make: labor-intensive, complex, expensive. He knew they soaked up vast amounts of electricity to operate and gave off tremendous amounts of heat. Most of all he knew they had to be perfect, and often they weren't. "They were awfully hard to make and they broke all the time," his wife would recall. "He was always hoping there would be something." Something else, in other words, that could do what only tubes could do.[12]

IN THE LATE 1920s, work at the tube shop, as well as at Bell Labs, boomed along with the rest of the American economy. In the months after the stock market crash of 1929, when the black depths of the Great Depression weren't yet apparent, Kelly and a few other colleagues belonged to a buoyant "three-hours-for-lunch" club, a group of Labs employees intent on trying the newest Manhattan speakeasies (Prohibition was still in force) before the police could shut them down. But the business climate grew ever more dire. The astonishing drop in manufacturing jobs and the unrelenting misery in the American farm belt drove down phone subscriptions—and with them AT&T's revenue. In the course of three years, between 1930 and 1933, more than 2.5 million households, most of them Bell subscribers, disconnected from the phone grid. In 1932 alone, the number of telephones with Bell service dropped by 1.65 million. Western Electric laid off 80 percent of its workforce. The Labs, which had typically hired a few hundred young employees every spring, sending out a team of recruiters to speak with professors at colleges around the country in search of graduate students who might be well suited for industrial research, stopped hiring. And then, with a straitened budget, Frank Jewett, still the Labs' president, instituted pay cuts and a four-day workweek.

And then Harold Arnold died.

Jewett's research deputy, forty-nine years old, suffered a heart attack at 3 a.m. on a July morning at his home in Summit, New Jersey. Jewett soon appointed a successor: a tall, thoughtful, experimental physicist named Oliver Buckley who had spent much of his career at the Labs trying to address the special problems that affected "submarine" cable—that is, cable that went under water, connecting islands to the mainland, and was susceptible to a range of stresses that didn't affect ordinary underground phone cables. Buckley's dream was to run a transatlantic cable from North America to Great Britain, a project that the Depression and various technological challenges had placed on an indefinite hold.

Not long after Buckley moved up, Mervin Kelly did, too. Through his work in the tube shop, as both a researcher and production chief, he had extended the life of the Western Electric telephone repeater tubes from 1,000 hours to 80,000 hours, an impressive and cost-saving feat. In 1936, Kelly was appointed director of research. The Bell Labs hierarchy was now established for the next decade: Frank Jewett on top, Buckley below him, then Kelly. Though Kelly was not technically in charge, that mattered little. As events would show, he would lead regardless of his rank or station.

KELLY'S PROMOTION, in the mid-1930s, coincided with a slight easing in the Great Depression. Phone subscriptions picked up, and so did telephone company revenues. At that point, Kelly successfully argued for extra funding to hire a group of scientists for his research department. He had his pick of almost anyone. For one thing, the Labs' reputation had been burnished over the past few years by the work of Kelly's old office mate, Davy Davisson. He had won fame in his profession—and in 1937, the Nobel Prize—for his experiments in what was called electron diffraction. (In an experiment, Davisson had bombarded a piece of crystalline nickel with electrons, and the results demonstrated a theory first put forward by the Austrian physicist Erwin Schrödinger that electrons moved in a wave pattern.) For Kelly's new hires, however, a good salary likely mattered more than Davisson's notoriety. Kelly had funding at a time when almost no one else did. The country's universities had drastically pared their budgets and teaching positions were almost impossible to come by. And even where research or teaching positions could be found, colleges were offering only a fraction—a half or, at best, two-thirds—of the Labs' starting salary of $3,000 a year. "I had already figured that $2,600 was practically putting me up in the state of a rajah," said one of the recruits, Dean Woolridge, a student under Millikan (who had now moved from the University of Chicago to Caltech). "$3,000 was just fantastic."[13]

It was curious, in a way, who they were, these men coming to Bell Labs in New York. Most had been trained at first-rate graduate schools like MIT and Chicago and Caltech; they had been flagged by physics or chemistry or engineering professors at these places and their names had been quietly passed along to Kelly or someone else at the Labs. But most had been raised in fly-speck towns, intersections of nowhere and nowhere, places with names like Chickasa (in Woolridge's case) or Quaker Neck or Petoskey, towns like the one Kelly had come from, rural and premodern like Gallatin, towns where their fathers had been fruit growers or merchants or small-time lawyers. Almost all of them had found a way out—a high school teacher, oftentimes, who noticed something about them, a startling knack for mathematics, for example, or an insatiable curiosity about electricity, and had tried to nurture this talent with extra assignments or after-school tutoring, all in the hope (never explained to the young men but realized by them all, gratefully, many years later) that the students could be pushed toward a local university and away from the desolation of a life behind a plow or a cash register.

The young Bell Labs recruits had other things in common. Almost all had grown up with a peculiar desire to know more about the stars or the telephone lines or (most often) the radio, and especially their makeshift home wireless sets. Almost all of them had put one together themselves, and in turn had discovered how sound could be pulled from the air.

Kelly had personally hired two young PhDs from MIT, William Shockley and Jim Fisk, both of whom would have vast impacts on the Labs' future. Others came from Caltech, such as Woolridge and an engineer named John Pierce. A chemist named William Baker was hired from Princeton. Pierce and Baker would also have tremendous influence over Bell Labs' destiny. Another Caltech graduate was a physicist named Charles Townes, who'd been raised on a farm near Greenville, South Carolina. To grow up that way, he would later explain, made you "pay attention to the natural world, to work with machinery, and to know how to solve practical problems and fix things innovatively, with what is on hand." In Townes's view, those "farms and small towns are good training grounds for experimental physics."[14]

This was not necessarily an isolated opinion; a young experimental physicist who had come to the Labs a few years before Townes felt the same way. Walter Brattain grew up in rural Washington State, in Walla Walla. He had spent an entire year before college herding cattle in the mountains near his home, sleeping alone for months on end in a tent with a rifle. (When he left Washington for graduate school in Minnesota, he hopped a freight train to get there.)[15] In regard to his skills as a physicist, Brattain would later say it was important "that my maternal grandfather was [a] flour miller by trade, that my paternal great grandfather, Andrew McCalley, was also a flour miller by trade, and [that] I spent a consider-able [part] of my youth—a lot of years of high school and while I was at college—in a flour mill run by my father." Brattain could take apart a car engine easily and put it back together with equal ease.[16]

A certain fearlessness about life characterized the recruits. Charles Townes had been given $100 by Bell Labs to make the trip from California by rail, a sum he figured could go much further if he improvised. He took a Greyhound bus from Los Angeles to Tucson, and once there he bought a ticket for a cheap train to Mexico City. Before leaving on his trip he'd bought an accordion from a German student, "a rather ardent Nazi follower who spent a fair amount of time telling us all what a vital job Hitler was doing." And so Townes sat on the Mexico City train in third class in the summer of 1939, "on slatted wood benches that were none too comfortable, and played a Nazi's accordion and sang songs with Mexican fruit pickers on their way home from the fields in the United States." He felt nervous about eating the local food at the stops mostly he was afraid of dysentery—and for two days he lived on bottled beer.

From Mexico City he traveled to the Guatemala border, but could go no farther when he discovered a bridge was closed. So instead he went up to Acapulco, not yet a tourist destination, and rented a hut on the beach for fifty cents a night, where he spent the days swimming in the warm tropical waters. Then another cheap train to Texas. Then a bus to see his family in South Carolina. And then finally another bus to get to New York City. "The $100 from Bell Labs," he recalled, "just about exactly covered the trip's total cost."[17]

During their first few days in New York, the new "members of the technical staff"—MTSs, as they were called—learned their way around West Street. They were summoned to listen to speeches by Labs vice president Buckley, delivered from detailed note cards, and research chief Kelly, delivered from memory with his eyes closed, as was his habit, welcoming them to Bell Labs. But mostly they met with their supervisors—in Townes's case, Harvey Fletcher; in Bill Shockley's case, Clinton Davisson—to try and hash out what kind of work they would be doing. At one point during the first few days the freshmen were asked to sell the rights to their future patents, whatever these might be; their research, wherever it took them, was to benefit Bell Labs and phone subscribers. None of the young men refused. And in exchange for their signatures, each was given a crisp one-dollar bill.

Three

SYSTEM

The physicists that Kelly hired toward the end of the Great Depression—Shockley, Fisk, Woolridge, Townes, and all the rest—already knew how easily ideas could move from one side of the earth to the other. Usually the ideas came inside an envelope, printed in a formidable journal—*Annalen der Physik* from Germany, for instance, or *Physical Review* from New York—transported by the mail trains to New England, the Midwest, or the West Coast, where the package would be eagerly received by young physicists at places like Harvard, Chicago, or Caltech. The ideas also came to willing readers, in clear and eloquent English, via a publication named the *Bell System Technical Journal*, where a physicist named Karl Darrow, another former student of Millikan's, had a gift for summarizing what he called "contemporary advances" in science, such as the newest model of the structure of the atom. Darrow had trembling hands. This left him unsuited to experimentation. Before Harold Arnold died, however, he had recognized in Darrow a useful skill for disseminating information. "I was thinking that I ought to look for a place in the academic world," Darrow recalled, when Arnold "told me I might remain and do what I pleased."[1] The catch was that Darrow shouldn't expect his Bell Labs salary to rise as high as that of the engineers working more directly on phone company business—a fair enough trade-off to Darrow. From then on his job involved traveling to Europe in the summers and effectively serving as an intermediary between scientific ideas there and in the United States. More often than not, his writings addressed the behavior of matter and energy at the tiny, molecular—that is, *quantum*—level. Quantum mechanics, as it was beginning to be called, was a science of deep surprises, where theory had largely outpaced the proof of experimentation. Some years later the physicist Richard Feynman would elegantly explain that "it was discovered that things on a small scale behave *nothing like* things on a large scale." In the quantum world, for instance, you could no longer say that a particle has a certain location or speed. Nor was it possible, Feynman would point out, "to predict *exactly* what will happen in any circumstance." To describe the actions of electrons or nuclei at the center of atoms, in other words, was not only exceedingly difficult. One also had to forsake the sturdy and established laws of Newtonian physics for an airy realm of imagination.[2]

Increasingly, during the late 1920s and early 1930s, ideas arrived in the flesh, too. Some years Karl Darrow would visit California to lecture; some years students in various locations would learn from a physics professor named John Van Vleck, who was permitted to ride the nation's passenger trains free of charge because he had helped work out the national rail schedules with exacting precision. It also was the case that a scholar from abroad (a 1931 world tour by the German physicist Arnold Sommerfeld, for instance) would bring the new ideas to the students at Caltech or the University of Michigan. Indeed, the Bell Labs experimentalist Walter Brattain, the physicist son of a flour miller, was taking a summer course at Michigan when he heard Sommerfeld talk about atomic structure. Brattain dutifully took notes and brought the ideas back to New York. At West Street, he gave an informal lecture series to his Bell Labs colleagues.

Every month, as it happened, seemed to bring a new study on physics, chemistry, or metallurgy that was worth spreading around—on the atomic structure of crystals, on ultra-high-frequency radio waves, on films that cover the surface of metals, and so forth. One place to learn about these ideas was the upper floor of the Bell Labs West Street offices, where a large auditorium served as a place for Bell Labs functions and a forum for new ideas. In the 1920s, a one-hour colloquium was set up at 5 p.m. on Mondays so that outside scholars like Robert Millikan and Enrico Fermi or inside scholars like Davisson, Darrow, and Shockley—though only twenty-seven years old at the time—could lecture members of the Bell Labs technical staff on recent scientific developments. (Albert Einstein came to West Street in 1935, but was evidently more interested in touring the microphone shop with Harvey Fletcher than giving a talk.)[3] Another place to learn about the new ideas was the local universities. The Great Depression, as it happened, was a boon for scientific knowledge. Bell Labs had been forced to reduce its employees' hours, but some of the young staffers, now with extra time on their hands, had signed up for academic courses at Columbia University in uptown Manhattan. Usually the recruits enrolled in a class taught on the Columbia campus by a professor named Isidor Isaac (I. I.) Rabi, who was destined for a Nobel Prize.

And there was, finally, another place on West Street where new ideas could now spread. Attendance was allowed by invitation only. Some of the Labs' newest arrivals after the Depression had decided to further educate themselves through study groups where they would make their way through scientific textbooks, one chapter a week, and take turns lecturing one another on the newest advances in theoretical and experimental physics. One study group in particular, informally led by William Shockley at the West Street labs, and often joined by Brattain, Fisk, Townes, and Woolridge, among others, met on Thursday afternoons. The men were interested in a particular branch of physics that would later take on the name "solid-state physics." It explored the properties of solids (their magnetism and conductivity, for instance) in terms of what happens on their surfaces as well as deep in their atomic structure. And the men were especially interested in the motions of electrons as they travel through the crystalline lattice of metals. "What had happened, I think, is that these young Ph.D.'s were introducing what is essentially an academic concept into this industrial laboratory," one member of the group,

Addison White, would tell the physics historian Lillian Hoddeson some years later. "The seminar, for example, was privileged in that we started at let's say a quarter of five, when quitting time was five." The men had tea and cookies served to them from the cafeteria—"all part of the university tradition," White remarked, "but unconventional in the industrial laboratory of that day." The material was a challenge for everyone in the group except Shockley, who could have done the work in his sleep, Woolridge would recall. Out of habit, the men addressed one another by their last names. According to Brattain, it was always *Shockley* and *Woolridge*—never Bill and Dean, and never Dr. Shockley and Dr. Woolridge.

As the study group wound down for the evening, the men would often make their way over to Brattain's Greenwich Village apartment for a drink. By then it was 8 or 9 p.m.—time for dinner at a restaurant in the Village and then bed. Shockley lived nearby in an apartment on West 17th Street. "I don't think we had the idea then that some of the sort of things that later have become so central in the technology—that they were around the corner," he would recall. "There's no telling how far off they were."[4] By outward appearances, the study group was merely comprised of telephone men who were intent on learning new ideas. They weren't yet famous enough to set their own hours. They were expected to be back at West Street the next morning at 8:45 sharp, each wearing a crisp white shirt, jacket, and tie.

IN LATER YEARS it would sometimes be construed, thanks in part to AT&T's vast publicity apparatus, that scientists came to the Labs in the 1930s and 1940s for the good of science. But that was an incidental dividend of their work. Mervin Kelly hired the best researchers he could find for the good of the system. The new recruits were no longer asked to climb telephone poles and operate switchboards. But all were given long seminars in their first few weeks on how the Bell System worked. Oliver Buckley, the Labs vice president, told his new employees, "Our job, essentially, is to devise and develop facilities which will enable two human beings anywhere in the world to talk to each other as clearly as if they were face to face and to do this economically as well as efficiently."[5] It was reminiscent of Theodore Vail's dictum of "one policy, one system, universal service." But it likewise suggested that the task at hand was immense. Already in the Bell System there were about 73 million phone calls made each day—and the numbers kept climbing.[6] In the earliest days of AT&T, company engineers realized the daunting implications of such growth: The larger the system became, the larger the challenges would be in managing its complexity and structural integrity. It was also likely that the larger the system became, the higher the cost might be to individual subscribers unless technologies became more efficient. To scientists like Jewett, Buckley, and Kelly, that the growth of the system produced an unceasing stream of operational problems meant it had an unceasing need for inventive solutions. But the engineers weren't merely trying to improve the system functionally; their agreements with state and federal governments obliged them to improve it economically, too. Every employee on West Street was therefore encouraged to take a similar perspective on the future: Phone service not only had to get better and bigger. It had to get cheaper.

Not everyone took Ma Bell's corporate adages at face value. By the late 1930s, in fact, AT&T was in the midst of a federal investigation that focused closely on whether it was overpaying for phone equipment from Western Electric, and thus overcharging phone users as a result. Some of the findings that came out of the multiyear inquiry—summarized in a scathing portrait of the company, written by a federal lawyer named N. R. Danielian, entitled *A.T.&T.: The Story of Industrial Conquest*—portrayed the Bell System as a monstrous entity focused less on public service than on maintaining its stock price and rate of expansion. Danielian painted an ugly picture of how Ma Bell executives had used propaganda—books, periodicals, short films—to enhance their corporate image during the 1920s. In his view, moreover, AT&T's size and dominating nature raised the question of whether it was actually an "industrial dictatorship" obscured by a scrim of civic-mindedness. "The [Bell] System," Danielian pointed out, "constitutes the largest aggregation of capital that has ever been controlled by a single private company at any time in the history of business. It is larger than the Pennsylvania Railroad Company and United States Steel Corporation put together. Its gross revenues of more than one billion dollars a year are surpassed by the incomes of few governments of the world. The System comprises over 200 vassal corporations. Through some 140 companies it controls between 80 and 90 percent of local telephone service and 98 percent of the long-distance telephone wires of the United States." The Bell System owned the wires involved in certain aspects of radio transmission, Danielian added, and had become involved in a host of other pursuits, such as equipment for motion pictures. Its needs for raw materials added up to "hundreds of millions of dollars" annually; its deposits in banks involved "almost a third of the active banks in the United States"; its investors numbered nearly a million. It was also, not incidentally, the largest employer in the United States.[7]

Kelly would maintain—sometimes under oath, in front of a state or federal utility commission—that Bell Labs' purpose was to give AT&T and its regional operating companies "the best and most complete telephone service at the lowest possible cost." He could talk for long stretches—easily for thirty minutes at a time, and with deep conviction— about the virtues of the Bell System and the scientific research it paid for at his laboratory. The difficulty was to reconcile his views with Danielian's. Perhaps the only way to do so was to accept that there was no reconciliation. The truths about the Bell System, and in turn Bell Labs, were not so much mutually exclusive as simultaneous. The overseers of the phone company, those top-hatted executives at AT&T, were mercenary and aggressive and as arrogant as any captains of industry. But the phone service offered to subscribers was reliable and of high quality and not terribly expensive. That was a point even Danielian conceded. AT&T's aggressive strategy to patent its inventions, meanwhile, made it difficult for individuals and smaller companies to compete; it was also a tool for generating profits. But Danielian likewise acknowledged that the discoveries at Bell Labs had been essential to the progress of society at large. "They have not only made things better, but have created new services and industries," he wrote of the scientists and engineers. "They have also made significant contributions to pure science. For these, no one would wish to deny just praise."

The larger point in all of this was that Bell Labs, for all its romantic forays into the mysteries of science, remained an integral part of the phone business. The Labs management made an effort to isolate its scientists from the gritty day-to-day political concerns of the business. But the managers themselves had to keep track of how the technology and politics and finances of their endeavor meshed together. Indeed, they could never forget it. As long as the business remained robust—and it was the primary job of people like Mervin Kelly to keep the business robust—so did the Labs.

IN THE FIRST DECADE of the twentieth century, the transcontinental phone line had been one example of how the challenges of expanding the phone system led to inventions like the repeater tube. But it was only one example. Following the rapid development of the telephone business in the early twentieth century, everything that eventually came to be associated with telephone use had been assembled from scratch. The scientists and engineers at Bell Labs inhabited what one researcher there would aptly describe, much later, as "a problem-rich environment."[8] There were no telephone ringers at the very start; callers would get the attention of those they were calling by yelling loudly (often, "ahoy!") into the receiver until someone on the other end noticed. There were no hang-up hooks, no pay phones, no phone booths, no operator headsets. The batteries that powered the phones worked poorly. Proper cables didn't exist, and neither did switchboards, dials, or buttons. Dial tones and busy signals had to be invented. Lines strung between poles often didn't work, or worked poorly; lines that were put underground, a necessity in urban centers, had even more perplexing transmission problems. Once telephone engineers realized they could also broadcast messages via radio waves, they encountered a host of other problems (such as atmospheric interference) they had never before contemplated. But slowly they solved these problems, and the result was something that soon came to be known, simply and plainly, as the system.

The system's problems and needs were so vast that it was hard to know where to begin explaining them. The system required that teams of chemists spend their entire lives trying to invent new, cheaper sheathing so that phone cables would not be permeated by rain and ice; the system required that other teams of chemists spend their lives working to improve the insulation that lay between the sheathing and the phone wires themselves. Engineers schooled in electronics, meanwhile, studied echoes, delays, distortion, feedback, and a host of other problems in the hope of inventing strategies, or new circuits, to somehow circumvent them. Measurement devices that could assess things like loudness, signal strength, and channel capacity didn't exist, so they, too, had to be created—for it was impossible to study and improve something unless it could be measured. Likewise, the system had to keep track of the millions of daily calls within it, which in turn required a vast, novel infrastructure for billing. The system had to account for it all.

"There is always a larger volume of work that is worth doing than can be done currently," Kelly said, which was a way of acknowledging that work on the system, by definition, could have no end. It simply kept expanding at too great a clip; its problems meanwhile kept proliferating. For one person to call another required the interrelated functioning of tens of thousands of mechanical and electronic elements, all of them designed and developed by Bell Labs, and all of them manufactured by Western Electric. What's more, almost every part of the system was designed and built to stay in service for forty years. That entailed a litany of durability tests at the West Street plant on even the most trifling of system components. Labs engineers invented a "dropping machine" to simulate "the violence of impact" of a receiver dropped into its cradle tens of thousands of times. They fashioned a "woodpecker machine," meant to resemble "that industrious bird in action," to test the scratch-resistant qualities of varnishes and finishes. They fabricated what they called an "artificial mouth," resembling a freestanding microphone, to test the aural sensitivity of handsets; they created a machine with a simulated finger to mimic the demands of button-pushing and dialing. And it wasn't enough to merely measure the durability of a telephone dial; other teams of engineers had to calibrate and measure, to a level approaching perfection, the precise *speed* at which the dial rotated.

Some men at West Street specialized in experimenting on springs for switchboard keys, others in improving the metal within the springs. AT&T linemen bet with their lives on the integrity of the leather harnesses that kept them tethered at great heights—so Labs technicians established strength and standards for the two-inch leather belts (limiting "the content of Epsom salts, glucose, free acid, ash and total water-soluble materials") and improving the metal rivets and parts. Millions of soldered joints held the system together—so Labs engineers had to spend years investigating which fluxes and compounds were best for reinforcing anything from seams on sheet metal to lead joints to copper wires to brass casings. AT&T lines carried transmissions from the Teletype, a machine that could send and translate written messages over long distances—so Labs engineers likewise found it necessary to invent a better teletypewriter oiler, a small square oil can, named the 512A tool. And the Labs engineers were not necessarily content with designing any oil can; this one had to be built with a complex inner mechanism for dispensing up to (but no more than) fifteen drops of lubricant. The 512A was an example of how, if good problems led to good inventions, then good inventions likewise would lead to other related inventions, and that nothing was too small or incidental to be excepted from improvement. Indeed, the system demanded so much improvement, so much in the way of new products, so much insurance of durability, that new methods had to be created to guarantee there was improvement and durability amid all the novelty. And to ensure that the products manufactured by Western Electric were of the proper specifications and quality, a Bell Labs mathematician named Walter Shewhart invented a statistical management technique for manufacturing that was soon known, more colloquially, as "quality control." His insights not only guided the manufacture of items within the Bell System for the next few decades, but in time were applied to improve industrial processes and products around the world.[9]

The system demanded that a small branch of the Labs was established in western New Jersey, in the country village of Chester. The men there were to study the outdoors deterioration of telephone equipment. Lodgepole pine trees from five western states had been determined by Labs engineers to be the most useful for poles, and so telephone men in Chester buried the pine phone poles ten feet deep and spent decades studying their degradation. At the same time, they mixed a witches' brew of stains and fungicides, applied them to the buried poles, and graded their effectiveness. They found it necessary, too, to investigate the behavior of gophers, squirrels, and termites, which gnawed through wood and cables and were fingered as the cause of hundreds of thousands of dollars in losses every year. (One strategy the men discovered could rebuff the pesky gopher: steel tape on cables.) The cables seemed to require a variety of other types of study, too. A Bell Labs engineer named Donald Quarles, who was in charge of the Chester plant, wrote a long treatise entitled "Motion of Telephone Wires in the Wind." His men made rigorous, multiyear tests on the proper spans (how far should the poles be spaced apart?), proper lashing (how tight should the wires be tied together?), proper vertical spacing between horizontal strings of wires (company practice suggested twelve inches, but engineers discovered eight inches could be enough to prevent abrasion). Many of the system's most important cables, meanwhile, were not strung through the air but ran underground. For burying wire, the men in Chester had to develop new processes involving special tractors they invented and splicing techniques. Other Labs engineers focused on undersea cables, which required not only special materials and techniques but special ships, outfitted with enormous spools of cable in their massive holds, that could lay the cable smoothly on the sea bottom.

The system demanded that the Labs men go anywhere necessary to test or acquire proper materials. A sturdy telephone cable that carried hundreds of calls at the same time would do so at different frequencies, much as daylight carries within it different colors of the spectrum. "In order to get the economies resulting from putting a bundle of dozens or hundreds of telephone conversations in one conductor," Kelly explained, "you have got to have very intricate and complex equipment at the ends of those circuits to combine all of these different telephone conversations in the one single bundle and then at the other end to unscramble them so that each conversation shall go where you want it to go."[10] The scrambling and unscrambling was done at either end with electronic filters, which separated channels, just as a prism can divide light. The essential components of these filters were quartz plates cut from quartz crystals. The men at the Labs had discovered that the best quartz for this purpose came from Brazil—"the only place in the world that has the quality of crystals of sufficient size to do this work," Kelly said. And so the Bell Labs managers set up an extraordinary supply chain so they could get the perfect quartz, so they could make the perfect quartz filters, so they could try to perfect the system that, by its very nature, could never be perfected.

WE USUALLY IMAGINE that invention occurs in a flash, with a eureka moment that leads a lone inventor toward a startling epiphany. In truth, large leaps forward in technology rarely have a precise point of origin. At the start, forces that precede an invention merely begin to align, often imperceptibly, as a group of people and ideas converge, until over the course of months or years (or decades) they gain clarity and momentum and the help of additional ideas and actors. Luck seems to matter, and so does timing, for it tends to be the case that the right answers, the right people, the right place—perhaps all three—require a serendipitous encounter with the right problem. And then—sometimes—a leap. Only in retrospect do such leaps look obvious. When Niels Bohr—along with Einstein, the world's greatest physicist—heard in 1938 that splitting a uranium atom could yield a tremendous burst of energy, he slapped his head and said, "Oh, what idiots we have all been!"[11]

A year earlier, Mervin Kelly assigned William Shockley to a training program for new employees that included time in the vacuum tube laboratory. One day, Kelly stopped by Shockley's West Street office, possibly to visit with Davisson, with whom Shockley shared the office, and began to talk. Shockley later recalled:

> I was given a lecture by then–research director Dr. Kelly, saying that he looked forward to the time when we would get all of the relays that make contacts in the telephone exchange out of the telephone exchange and replace them with something electronic so they'd have less trouble.[12]

In a system that required supreme durability and quality, there were, in other words, two crucial elements that had neither: switching relays and vacuum tubes. As we've seen, tubes were extremely delicate and difficult to make; they required a lot of electricity and gave off great heat. Switches—the mechanisms by which each customer's call was passed along the system's vast grid to the precise party he was calling—were prone to similar problems. They were delicate mechanical devices; they used relays that employed numerous metal contacts; they could easily stop working and would eventually wear out. They were also, because they clicked open and closed, far slower than an electronic switch, without moving parts, might be. Kelly had set an intriguing goal that lingered in Shockley's mind as he finished his indoctrination program and turned back to studying the physical properties of solid materials on his own and with his study group. Kelly's articulation of a solution—a product, in essence—was fairly straightforward, even if the methods for creating such a product remained obscure: Perhaps the Labs could fashion solid-state switches, or solid-state amplifiers, with no breakable parts that operated only by way of electric pulses, to replace the system's proliferating relays and tubes. For the rest of his life Shockley considered Kelly's lecture as the moment when a particular idea freed his ambition, and in many respects all modern technology, from its moorings.

. . .

WHEN HE WAS OLDER, when he had become famous for his scientific achievements and infamous for his unscientific views on race, Bill Shockley would recount his past and point to what he called "irregularities." There were, he would concede, certain irregularities about his childhood—that he was homeschooled until he was about eight, for instance, or that his parents moved so often and so arbitrarily that it was sometimes difficult to explain why he attended a particular school or lived in a particular place. Other irregularities he didn't readily concede. As a toddler, Shockley—"Billy" to his parents, May and William—would experience tantrums that put him beyond the reach of consolation. As his father dutifully related in his diaries, his son's emotional outbursts were often uncontrollable. Billy would slap at his parents, throw stones at other boys, bark like a dog. "Billy always gets angry because he is thwarted or denied something," his father noted in May 1912, when his son was just two years old, in a prescient journal entry entitled "Billy's rages." A week later, he observed that "when he is good, he is very good indeed; and when he is bad he is horrid."[13]

Shockley was an only child, and a solitary one. When he was three years old, May and William had settled in Palo Alto, California, to begin eking out a middle-class life in the same small city where May had been raised. Shockley's father was a mining engineer with a thoughtful demeanor and substantial assets from earlier in his career, but in Palo Alto he was often short on both employment and money. His fitful work schedule left him home often enough to tutor his son in math and encourage his early curiosity about science. But Shockley would say that his largest influence was a neighbor named Pearley Ross, a professor at Stanford who worked with X-rays and whose young daughters were Shockley's main companions. Ross taught Shockley the fundamentals of physics.

In his teens, Shockley's family moved from Palo Alto to Los Angeles, where he attended high school before enrolling at Caltech. Slight in frame (at five foot eight) but in taut physical condition (he was devoted to calisthenics and swimming), he cut a memorable figure as an undergraduate. His childhood rages had subsided, replaced by a geniality that hid a relentless competitive edge and an occasional and savage asperity. The science prodigy seemed to have a compulsive need to charm, to entertain, to challenge the dull conventionality of academia, often in a way that subtly merged humor and aggression. Shockley schooled himself in parlor tricks and amateur magic. Sometimes he would use it to entertain a crowd at parties; other times he would use it to interrupt a sober affair or gently humiliate a lecturer. Bouncing balls materialized from nowhere, flowerpots exploded, bouquets popped suddenly from his sleeve in place of a handshake—incidents that created a distraction from the seriousness of institutional life while turning attention back on Shockley. *How did he do that?* It was no wonder he loved to construct intricate practical jokes as well. In one Caltech class, Shockley, with the help of some fellow students and a few faculty members, concocted a successful scheme to enroll an entirely fictitious student named, in one Caltech student's recollection, Helvar Skaade. The target was the class professor, Fritz Zwicky, who was known for his casual attitude on matters of class attendance. "All these tests were open book exams typically," Dean Woolridge, one of Kelly's young recruits who also attended Caltech, would later recall. "You could use any books that you wanted to. The procedure was for the professor to come in and write down the questions on the board, and Zwicky always had five problems, and then he would leave the room and come back at the end of the hour." Shockley arranged for one copy of the exam to be taken out of the classroom, solved expertly by himself and a team of graduates who had already taken the class, signed by Helvar Skaade, and then returned in time to be handed in. Skaade, the mysterious young genius, answered all the questions brilliantly except for the last one, to which he responded, "Hell, I'm too damn drunk to write anymore." Skaade got an A-minus, the highest grade in the class.[14]

Shockley came east with an adventurous flourish that burnished his personal mythology. In the summer of 1932, he and an acquaintance, Fred Seitz, drove from California in Shockley's 1929 DeSoto Roadster. Shockley was on his way to MIT, where he had decided to pursue a PhD.

Seitz was going to get his physics PhD at Princeton—the men agreed Shockley would drop him off there. For the trip, Shockley brought a loaded pistol that he kept in the glove compartment. They selected a southern route that took them through Arizona, New Mexico, Texas, and Arkansas, and they barely survived the trip. They encountered nights of torrential rains that obliterated the desert highways; in Kentucky, they narrowly averted a deadly head-on collision when they encountered two trucks racing toward them around a mountain pass, taking up both lanes of a two-lane road. "By the grace of the Lord, I had just enough shoulder to squeeze by the oncoming truck with perhaps an inch to spare," Seitz recalled. "To the best of my knowledge I have never been closer to instant death than in those few seconds."[15] A few days later, on a moonlit night, Shockley dropped his new friend off in New Jersey. Princeton's campus struck him as extremely attractive. When he arrived at MIT the next day, a stiff wind was blowing factory fumes into his face and the campus buildings appeared more industrial than academic. He wondered if he'd made a mistake. But MIT nevertheless turned out to be a good experience for Shockley. It gave him a strong background in quantum mechanics and introduced him to two friends who would prove crucial to his later career: Jim Fisk, a classmate, and Philip Morse, a professor.

To know Shockley, even in his twenties and thirties, was to be confused by him. Was he likable? "In a way," says one of his former colleagues, Phil Anderson. An infectious energy and a boundless enthusiasm for physics had a tendency to pull colleagues into his orbit and allow them to overlook, at least for a while, his marauding ego. He could be fantastically good company—warm, witty, entertaining. He loved sharing a drink or two and frequently invited friends for rock-climbing trips or vacations in upstate New York with his wife and young daughter. He had an extraordinary talent for instruction and could be surprisingly generous with his time. "I would go visit him in the evenings in his apartment in Manhattan," Chuck Elmendorf, a Caltech graduate who joined Bell Labs in 1936, recalls. "And I would just sit there on the edge of his couch and he would just teach me physics every night. He was decent, wonderful, pleasant." In a more formal work environment, moreover, being around

Shockley meant being dazzled. "He was the quickest mind I've ever known," adds Anderson, a theoretical physicist who went on to win a Nobel Prize. Even at the Labs, a place where everyone was fast on their feet, Shockley was faster. "His intellectual power was such that when Shockley said something," recalled his colleague Addison White, "I recognized it was right."[16]

There was something in particular about the way he solved difficult problems, looking them over and coming up with a method—often an *irregular* method, solving them backward or from the inside out or by finding a trapdoor that was hidden to everyone else—to arrive at an answer in what seemed a few heartbeats. Mervin Kelly had sensed this gift right away when he had visited MIT and met with Shockley in 1936. Shockley would later say, "I can recall talking to Kelly and being impressed that he called up, used the telephone to call all the way down to New York City, to find out if he'd be authorized to make me an offer, because I had to decide right then and there."[17]

In Kelly's research department on West Street, Shockley found he could go mostly where his curiosity led him, which was often to solid-state physics. He likewise found that at the Labs the experimentalists and theoreticians were encouraged to work together, and that chemists and metallurgists were welcome to join in, too. The interactions could be casual, but the work was a serious matter. Every new member of the technical staff was given a stock of hardcover lab notebooks that were bound in cloth and leather and filled with two hundred lined pages. In most offices, recalls Walter Brown, an experimental physicist who worked under Shockley, there was a notebook table, "maybe twelve by eighteen inches, standing on a three-legged stand on the floor, painted black. It was intended to hold a notebook for recording details of experiments and their results [as well as] ideas and plans for the future. Results or ideas that one thought were potentially valuable were witnessed and signed by another engineer for documentation of the timing of the idea." The scientists were not permitted to rip out pages. Nor were they encouraged to attach loose sheets of paper into the notebook. "No erasures," says Brown. "Lines through mistakes—initialed by who drew the lines."[18]

Also, the notebooks were issued with registered numbers that were matched to each scientist and were tracked by supervisors and Labs attorneys. There was to be no confusion about who did what. The notebooks were proof for gaining a patent.

At some point in late 1939, Shockley had settled on an idea for how to make an electronic amplifier—much like the old repeater tube that Harold Arnold had improved—but this time out of solid materials. The production of vacuum tubes had improved since the days when Kelly ran the tube shop, but the essential problems remained: They were still fragile, they still consumed much electricity and produced much heat. The first attempts at making a solid-state amplifier, as Shockley was trying to do, involved simply copying the architecture of vacuum tubes. Shockley recalled later that his "first notebook entry on what might have been a working [solid-state amplifier] was as I recall late 1939."[19] It was actually December 29, 1939. Shockley had concluded by then that a certain class of materials known as semiconductors—so named because they are neither good conductors of electricity (like copper) nor good insulators of electricity (like glass), but somewhere in between—might be an ideal solid replacement for tubes. Under certain circumstances semiconductors are also known to be good "rectifiers"; that is, they allow an electric current passing through them to move in only one direction. This property made them potentially useful in certain kinds of electronic circuits. Shockley believed there could be a way to get them to amplify a current as well. He intuited that one common semiconductor—copper oxide—was a good place to start.

As a physicist, Shockley was far better as a theoretician than an experimentalist. On the other hand, Walter Brattain, his colleague at West Street, was about as good an experimentalist as could be found at Bell Labs. With good reason, Brattain prided himself on being able to build anything. "He came to me one day and said that he thought that if we made a copper-oxide rectifier in just the right way, that maybe we could make an amplifier," Brattain recalled. "And I listened to him. I had a good esprit de corps with him, and so after he explained, I laughed at him." Brattain, it turned out, had already tried a variation on the idea with another colleague. But when he saw how intent Shockley was on trying out his idea, Brattain went along, pledging that he would make a prototype to Shockley's precise specifications. In the early winter months of 1940, Brattain built a couple of units to Shockley's specifications. "It was tested and the result was nil," he recalled. "I mean, there was no evidence of anything."[20]

But Shockley wasn't convinced his idea was wrong. He would speculate later about what might have occurred had he continued to develop that particular amplifier experiment without interruption. But as it happened he couldn't. In fact, few people at the Labs could carry on their customary work anymore. The news from Europe—beginning with Germany's invasion of Poland in 1939, and its invasion of Belgium, France, and the Netherlands in the spring of 1940—put an end to business as usual.

Four

WAR

By the middle of 1940, the research department at Bell Labs stopped doing research as nearly all of the Labs' work—about 75 percent of it—was redirected toward developing electronic devices for wartime, first to help the Allies in Europe, and soon after to assist the U.S. Army and Navy. Frank Jewett, Bell Labs' president, began spending almost all his time in Washington advising political leaders on how the country's scientists might contribute to the war effort. Oliver Buckley, the Labs' vice president, mostly focused on their obligation to maintain phone company operations. "Buckley in essence handed over Bell Laboratories to [Mervin] Kelly during World War II," one Bell Labs researcher recalled.[1] Indeed, Buckley let Kelly—whose research department was essentially dismantled for the duration of the war and replaced by a multitude of development projects—run the day-to-day business of the laboratories from that point on.

One of the Labs' first assignments as the war began in Europe resulted from Jewett's political connections:[2] finding out at the government's behest whether it was actually possible, in light of several recent theoretical papers on nuclear reactions, to create a weapon out of ordinary uranium.[3] At Kelly's request, Shockley and Jim Fisk, his friend from

MIT who had just joined the physics department at the Labs, were asked to take time off to prepare a report. Shockley did most of the calculations. And while the two men quickly concluded that uranium in its natural state could not make a devastating weapon, they theorized that by placing "piles" of a specially enriched uranium preparation close together one might be able to create a sustained, low-level reaction. Put simply, they'd figured out how to make a nuclear reactor.[4] The men tried to patent the idea, but met with resistance from the government and the patent courts. As Fisk would recall, the reason was that the physicists Enrico Fermi and Leo Szilard "had essentially the same idea and probably at about the same time. We may have been earlier, they may have been earlier, I don't know. I don't think that anybody will ever know. But they were working hard on this and we were doing this simply as an exercise to answer the question."[5]

Like Kelly, Shockley rarely lingered over any one project. That he had figured out the essential concepts for nuclear power on his own (actually, the idea came to him while he was taking a shower) merely seemed an intriguing interlude in a frenetic schedule. Indeed, his schedule was too busy for him to do anything else but keep moving. Beginning in 1940, Kelly assigned Shockley to a secret effort to help develop applications for a new technology known as radar. Other members of the technical staff were asked to put aside their research work, too, in the same way that Kelly and Davisson had done two decades earlier during World War I.

There was no use challenging the Labs executives on the matter. For one thing, it was Kelly giving the orders. He still carried forth the brusque manner that had defined him as a young man, but he was now a boss who wielded power forcefully, and sometimes fearsomely. For another thing, researchers in Kelly's department rarely had any objections to pitching in on the war effort. One day not long after he began his war work, Charles Townes was walking through Times Square when a complete stranger came up to him: "You're not in uniform. Shame! A man your age ought to be in uniform, and helping out."[6] Townes was working fifty- and sixty-hour weeks to do precisely that. To the general public, however—largely unaware that this war would depend as much on technology as strategy—their role remained obscure. For a while, at least, so did the devices they were working on.

ON THE HOME FRONT, the war overturned the established norms of science, engineering, and business. Mervin Kelly had long regarded scientific research as a pursuit of the unknown that inherently defied corporate and political regimes. Science had no true owners, only participants and contributors. By the early 1940s, the great discoveries of his lifetime, including the work of his friends Robert Millikan at Chicago and Davy Davisson at Bell Labs, were such that they transcended borders and nationalities. The results of their work were widely shared, discussed, and augmented, as Kelly thought they should be, through international gatherings and cooperative, investigative efforts. Engineering, however, was different. Kelly defined it as the application of science to a problem affecting society. Engineers dipped into the "common reservoir" of science on behalf of their own industries and countries. In peacetime, that meant they focused on making profitable commodities like automobiles and telephone equipment; in wartime, that meant they focused on building military communications equipment as well as ships, planes, and munitions. At the same time, wartime engineers had an additional responsibility. They were charged not only with building everything better but building everything faster, which meant striving to improve their processes as well as their products. In the summer of 1943, Kelly wrote an article for an engineering publication called *The Bridge* that directed attention to the contributions of the American engineers working on the Allied war effort. In particular he remarked on the speed with which U.S. industry had caught up with the military economies of the Axis powers. "Progress has been made in some fields of technology in a four-year interval," he pointed out, "that, under the normal conditions of peace, would have required from ten to twenty years." Much of what Bell Labs was now building hadn't existed four years before. "In this astonishing short period of time," he added, "we have developed, designed, placed in manufacture and expanded to unprecedented high rates the production of a substantial portion of the tools of warfare that our Army and Navy now are so successfully employing." If the peacetime needs of the rapidly expanding phone system were an unceasing force for new inventions, in other words, war was turning out to be an even greater force.

One obvious reason was that war created an air of urgency. An example Kelly liked to cite was the supply of quartz crystals for telephone filters—the crucial equipment used to scramble and unscramble transmissions at either end of a cable.[7] In the late 1930s, the supply of quartz from South America slowed noticeably. The countries of the world had already begun competing for valuable resources. "We had been working in the period before the war in growing crystals in the laboratory," Kelly explained, "growing out of solutions crystals which would have the same electrical and mechanical properties as quartz." War accelerated these developments. Technicians in fact discovered that by placing a small "seed" in a vat of chemicals—the scientists tried a hundred different formulas before settling on one—artificial crystals could be grown to the length of six inches, resembling in their glass tanks huge clusters of rock candy. At that point the fully grown crystals were removed from the tanks, the *New York Times* explained, "sawed into thin wafers, ground carefully to precise thicknesses, mounted in special holders and installed in electrical circuits."[8] It was a good example, in Kelly's view, of how America's scientists and engineers had responded to the war effort. The Labs ultimately produced hundreds of thousands of synthetic crystals.

The war also unleashed a great gush of money toward new technology. In an effort to quicken the development of equipment as much as possible, beginning around 1941 the federal government began directing hundreds of millions of dollars toward engineering organizations like Western Electric and Bell Labs. In the first few years after Pearl Harbor, in fact, Bell Labs took on nearly a thousand different projects for the military—everything from tank radio sets to communications systems for pilots wearing oxygen masks to enciphering machines for scrambling secret messages—leading Kelly to expand his staff by several thousand.

The Labs actually doubled its size from about forty-six hundred before the war to nine thousand during it. At the West Street offices, Oliver Buckley wrote, "there is hardly room to turn around." Elevators were so jammed with employees that it was difficult to squeeze on. A six-day workweek became the norm.

Not all the new employees fit the profile of the old-time telephone engineer. Some seven hundred Bell Labs staff members had gone into active military service. ("There are so many Laboratories people on Guam at one time or another," the *Bell Laboratories Record* noted, "that they might have opened a Western Pacific branch.") This led Labs executives to hire hundreds of women to replace the men. What's more, for the first time, the Labs began to hire Jews, bucking a strain of anti-Semitism that ran deep within the AT&T establishment, though not, apparently, within Kelly. The precise reasons for the shift remained ambiguous, though some Labs members later reflected that by 1940 the specter of war had trumped the Bell System's ugly traditions of bigotry. A slightly different explanation was that a meritocratic organization such as the Labs could perceive a competitive disadvantage of passing over the best scientists on religious grounds, an error they might have already made with the young Richard Feynman, a former colleague of Shockley's at MIT who would eventually be drafted into the Manhattan Project.[9] Whatever the explanation, some of the older and most hidebound scientists at the Labs found discomfort in this aspect of the Labs' evolution, as well as in Kelly's war mobilization effort. Lloyd Espenschied, for example, an inventor of the thick coaxial cables that carried phone conversations between major cities and a senior advisor to Oliver Buckley, was an avowed isolationist. In the midst of the war, an investigator for the War Department named G. E. Schwartz visited Espenschied at West Street. "We were led into this mess largely by British propaganda, Jewish propaganda, Roosevelt imperialism," Espenschied told him in response to his questions.

> He then said "I suppose you are anti-Semitic." I told him he could take that supposition if he wished, that frankly I do not like Jews.

He then asked me why I did not like them and I told him that it was because of their racial characteristics. To his question of why was this, I responded that he would have to ask the Jews themselves that question, looking him straight in the face as the Jew that he appeared to be. I told him that to my mind all this was part of the problem as a result of different peoples being thrown together rather rapidly in this age of increased mobility and shrinking distances.[10]

Frank Jewett and Oliver Buckley were appalled by Espenschied's "incredible stupidity"—though it was not clear whether their displeasure related to the content of Espenschied's opinions or his willingness to share them. "His error lies in his insistence on expressing his views intemperately," Jewett noted in a private memo about the exchange.[11] Espenschied was forbidden from having any contact with work on the war, or with any Labs employees involved with such work. He was not dismissed, however—a report on the incident was filed away by Buckley and stamped as "Confidential F.B.I." Mostly it seemed that Jewett and Buckley were primarily concerned that one of their most accomplished engineers might be a blemish on the Labs' reputation for patriotism and national service.[12]

Probably they worried needlessly. Thousands of men and women under Kelly were working intently on military applications, and the public was increasingly aware of their contributions. Press coverage of various aspects of what was known as "the physicists' war" was glowing. Still, for many members of the technical staff, the wartime work required a difficult philosophical transformation. The ideas of scientists thrive on publication and broad dissemination; but the ideas of engineers, especially during wartime, thrive only if secrecy is maintained. For the first few years of World War II, the word "radar"—its name stood for "radio detection and ranging"—was almost unknown by the general public, earning it the description of a *New York Times* reporter as being "the war's most fabulously and zealously guarded secret."[13] This was not quite true—the Manhattan Project was kept under tighter wraps—but it was generally the case that Kelly's men at Bell Labs, many of whom worked on radar systems, were forbidden to speak with anyone, even their wives, about their work, and that virtually all details about radar and its sister technology for underwater detection, sonar, were closely held for the duration of the war. Harvey Fletcher, for instance, who contributed his knowledge of acoustics to the Bell Labs sonar work during the early 1940s, actually refrained from speaking of his involvement for fifty years after victory was declared. Kelly never offered any details about his war work and likely either destroyed or purposely discarded his personal files on military matters. Secrecy began to cloak his responsibilities, as well as his power. When a three-hundred-page internal history of Bell Labs' World War II work was later compiled, his name was never mentioned.

THE SCIENTIFIC PRINCIPLES OF RADAR were fairly straightforward, even if the details of the technology weren't. A memo to Bell Labs employees explained that radar could be defined as a powerful electronic "eye" that used high-frequency radio echoes to determine the presence and location of unseen objects in space: "Specifically, a radar system does the following: (1) it generates high-power electrical waves; (2) it projects these waves from an antenna, usually in a narrow beam; (3) it picks up the waves which reflect back from objects in its range; and (4) converts these into a pattern on a fluorescent screen." The memo might have added that the waves traveled and bounced back at the speed of light, about 186,000 miles per second. Thus, in evaluating the time it took for radar waves to leave and then echo back to a radar antenna, a set could also calculate the distance of an unseen object based on the knowledge that distance equals rate multiplied by time. (Radar equipment was designed to divide in half the distance a wave traveled, since operators didn't need the distance to and from the object, only the distance in one direction.) It all happened instantaneously. An object one thousand yards from the radar set would give its echo six-millionths of a second later.[14]

Scientists who worked on radar often quipped that radar won the war, whereas the atomic bomb merely ended it. This was not a minority view. The complexity of the military's radar project ultimately rivaled that of the Manhattan Project, but with several exceptions. Notably, radar was a far larger investment on the part of the U.S. government, probably amounting to $3 billion as contrasted with $2 billion for the atomic bomb. In addition, radar wasn't a single kind of device but multiple devices—there were dozens of different models—employing a similar technology that could be used on the ground, on water, or in the air. Perhaps most important, radar was both an offensive and a defensive weapon. It could be used to spot enemy aircraft, guide gunfire and bombs toward a target, identify enemy submarines, and land a plane at night or in thick fog. Its origins for domestic military uses dated back to the 1930s, when several scientists at the Naval Research Laboratory discovered that when they directed a radio pulse from a transmitter toward planes passing overhead some of the waves were reflected back. In 1937, the Navy approached Bell Labs to help it refine the technology. Those first applications were primitive and riddled with problems, but by the start of the war radar sets were being used in Great Britain, where an intricate network of coastal radar stations helped the British defend against the onslaught of German bombers. The U.S. military also used radar stations in the Pacific—indeed, the Japanese squadrons flying toward Pearl Harbor were picked up well before they arrived. The officers monitoring the stations disregarded their readings, thinking the blips to be friendly aircraft.[15]

In time difficulties like this were overcome—a method was devised to distinguish between friendly and enemy aircraft—but the rapid evolution of radar between 1937 and 1943, the same speedy development that had awed Kelly in assessing the innovative forces of war, was rife with frustrations. In early 1940, if one could by chance eavesdrop on a group of Bell Labs radio engineers discussing the ideal radar set, one would hear described a technology that with the help of vacuum tubes could transmit very brief electromagnetic pulses (perhaps a thousand per second) in a very focused beam of waves that measured perhaps ten or fifteen centimeters in length. This was a fraction of the length of regular radio waves, the ones that brought music and news, which were sometimes a hundred meters long. It was also an ideal and not a reality. At the start of the war in Europe, the vacuum tubes that powered the early radar sets mostly sent out longer waves measuring a meter or more in length; such waves were too diffuse to help "pilots home in on their quarry."[16] And when American scientists attempted to create sets that could emit shorter waves of thirty or forty centimeters, they discovered that their vacuum tubes lacked the power to send out a strong enough signal. "The big problem in radar is to generate enough power to get a detectable echo from a distant point," *Time* magazine explained. "Of the total energy sent out in a radar beam scanning the skies, only a tiny fraction hits the target (e.g., a plane), and a much tinier echo gets back to the receiver. Engineers estimate that if the outgoing energy were represented by the sands of a beach, the returning echo would be just one grain of sand."[17]

How to create a device to send out shorter waves, and with more power? There was a solution to this problem—it came by ship, in a veritable black box, by way of a secret mission to the United States by British scientists in the late summer of 1940. In the box was something called a cavity magnetron, a metal device invented by two physicists at the University of Birmingham that resembled a small fishing reel. "Unlike conventional vacuum tubes," a Bell scientist explained, "with their components exposed in a glass envelope, the new tube was an inscrutable copper cylinder with cathode leads and a coaxial line emerging from it." The magnetron whirled electrons inside its six or eight circular internal cavities to produce short waves of ten centimeters and transmissions of great power. It was brought to Kelly at the West Street labs not long after the British mission came to the United States—the idea was that Western Electric would be the logical company to mass-produce the device if Bell Labs could refine the technology and design. On October 6, Kelly watched as the magnetron was demonstrated for the first time in the United States at a small Bell Labs branch office in Whippany, New Jersey.[18] The engineers in the room were dumbstruck by its power output. It wasn't just an incremental improvement in radar engineering. Luis Alvarez, a physicist who wasn't present that day but would later work directly on the magnetron designs, pointed out that the invention improved upon current technologies by a factor of three thousand. "If automobiles had been similarly improved," he noted, "modern cars would cost about a dollar and go a thousand miles on a gallon of gas."[19]

THE DEVELOPMENT WORK on magnetrons—the work that preceded their manufacture—was done in concert at MIT and Bell Labs. At MIT, an ad hoc group of scientists and engineers, eventually numbering several thousand, worked in a secure campus building with blacked-out windows known as the Rad Lab. At Bell Labs, many of the scientists on the radar project worked under Jim Fisk, who had already won favor for his work with Bill Shockley on uranium. Kelly directed him to set up a workshop in a building near the West Street offices that had once been a biscuit factory. The horses for the New York Police Department were housed nearby; Fisk would later recall that his senses were often sharpened by the "man-eating flies" that shuttled between his offices and the stables. At first Bell Labs' technicians took X-ray photographs of the British magnetron so they could create blueprints; in those photographs the device looks like a small, circular metal plate that contains a linked chain of eight smaller circles around its inner edge. It resembled an old-style film projector reel. But it still wasn't immediately clear how to manufacture devices for production. Engineers at the Labs knew that the gulf between an invention and a mass-produced product could in some cases be extraordinary, even insurmountable. "Could the new magnetron be reproduced quickly and in quantity?" Fisk wondered. "Was its operating life satisfactory? Could its efficiency and output power be substantially increased? Could one construct similar magnetrons at [wavelengths of] forty centimeters, at three centimeters, even at one centimeter?"[20]

In some respects, the work ahead of him was, by necessity, at odds with Kelly's philosophical preferences. Instead of using scientific research for development so that he could then make a device, Fisk was reverse-engineering—analyzing an existing device so he could work out a research plan he would then use for development. He quickly saw that he couldn't just make a magnetron bigger or smaller to get different outputs in microwave power and wavelength; engineering the device for ships and planes and other uses involved a careful consideration of everything about the magnetron—the size and number of its internal cavities, the shape of those cavities, its input voltage, and so on. Fisk carried around with him a notebook he filled with sketches and ideas. Often he consulted with Clinton Davisson, the thin, quiet researcher who stayed far away from development but was always willing to ponder a difficult problem, especially if it had to do with electrons.

Having just turned thirty years old, Fisk was now charged with perhaps the most important scientific project in the United States. Only a couple of years before, when he was finishing graduate school and beginning to look for a job, he had visited the research department at West Street to see his old MIT friend Bill Shockley. After he left, Shockley turned to a colleague and said, "Remember I told you. If that man gets hired, we'll all be working for him in ten years." Fisk himself had no managerial aspirations when he began at Bell Labs, but the war began to shape the course of his career in unexpected ways. "When we got all through," he said of the various researchers doing wartime magnetron work, "some of us emerged as something else."[21] Kelly had nonetheless seen executive potential in Fisk from the start, when he'd offered him a job the first time they met over lunch at a midtown Manhattan hotel. "Jim was Mervin's protégé," Kelly's son-in-law recalls, noting that Fisk was soon a frequent guest at Kelly's house in Short Hills, New Jersey, joining other regulars in the Kelly home like Davisson, Buckley, and, during his occasional visits to the United States, the physicist Niels Bohr.[22] In some ways the two had much in common—agile minds and a natural talent for decision-making. Both loved golf and gardening, though Kelly grew flowers and Fisk grew vegetables. Fisk—slender, patrician, polite, unflappable—was a more polished version of his mentor. Always with a cigar in hand, he had the soft touch and twinkle that Kelly lacked. On long train rides with colleagues (frequently they would go from New York to Western Electric factories in Chicago), Fisk would sometimes produce a bottle of Southern Comfort and pass it around.[23] He was fond of putting his colleagues on mailing lists of doctors peddling dubious tonics. When a friend of his went away for a week to a New Hampshire resort, Fisk sent to him a series of telegrams that resembled the instructions of an underworld gangster to one of his cronies.[24] At the magnetron workshop in the old biscuit factory, Fisk sometimes wore a striped train engineer's cap and, on occasion, striped overalls to meetings. "After all, we're engineers," he would say. His friend Larry Walker recalled that at the biscuit factory work was done in a large open first-floor room divided by waist-high partitions. Sitting in an office, Walker remembered, "one saw only the upper half of the people as they passed." It was Fisk's occasional habit, "if he caught someone's eye as he was leaving, to continue walking but gradually [bend] his knees farther and farther. Few who saw it can have forgotten the sight of Mr. F. apparently disappearing in a hole in the ground, eyes firmly ahead, chin up."

Fisk was liked the same way that Kelly was feared. Soon enough, he became known for "Fiskian" aphorisms: "When you don't know what to do," he would say, "do *something*." Or: "We have now successfully passed all our deadlines without meeting any of them." The magnetron project was starting to look like that, as the men worked out the fiendish difficulties, often during all-night sessions, of not only putting the original model in production but modifying the device for new applications. The low point for Fisk's team was the day after Pearl Harbor, as the men sat among stacks of nonworking magnetrons (they had poor vacuum seals, apparently) and listened to the grim news in the Pacific.[25] At one point during the work, Fisk—perhaps seeking a diversion from the grinding six-days-a-week schedule—came up with the notion that the propulsion system of sharks was worth studying for applications to naval warfare. He decided to try to requisition $50,000 for a swimming pool, to be constructed in the basement of the biscuit building, from Bell Labs management. It was a joke with an appearance of plausibility (in fact, many years later at Bell Labs, biological systems would become an acceptable research pursuit). Fisk's request rose through the executive ranks, receiving several green lights, until it got to Kelly, who instantly rejected it.[26]

FISK AND HIS STAFF eventually worked out fifteen different and successful magnetron designs for new radar instruments. His team passed its finished work on to Western Electric, which ultimately manufactured more than half of the radar sets used in World War II. Bill Shockley had been involved with the radar work at the start, but for the star Bell scientist, the war brought a profusion of other opportunities. In early 1942, Shockley received an offer from his friend Philip Morse, an MIT professor who was now in Washington organizing something called the Anti-Submarine Warfare Operations Research Group, commonly referred to as ASWORG. Morse asked Shockley to become the group's research director, and Kelly agreed to give his prize physicist a leave of absence from Bell Labs. Essentially a small think tank, the ASWORG group was staffed by statisticians and physicists and even a chess grandmaster. Together, they used sophisticated probability calculations to solve military problems. At the start, they were charged with figuring how to better detect and destroy the elusive German U-boats that were making passage through the Atlantic deadly. The work—conceptual rather than experimental—suited Shockley well: He and his colleagues looked at statistical information to figure out more effective methods of combat. An early problem: Why did bombs dropped from Allied airplanes have no effect on German submarines that had come to the surface? Shockley realized that the bombs were set to explode at a depth of seventy-five feet—too deep to sink a sub that was on the surface. He suggested changing the depth charge to thirty feet. Morse noted that within two months the change increased by a factor of five the number of subs sunk by Allied bombs.

Shockley and his staff at ASWORG rarely went out into the field. But the military men grudgingly admitted the effectiveness of the insights coming from these desk-bound civilian scientists. With the help of an IBM data processing system, the team assumed record-keeping duties for the entire U.S. antisubmarine effort. They brought in a computer expert and several insurance actuaries to analyze data on "hits" and "misses."[27] Shockley became immersed in these efforts. As his Washington schedule became increasingly demanding, in fact, Shockley moved to the capital, eventually taking a room at the University Club and coming home by train on occasional weekends to see his family in Madison, New Jersey. His marriage was not going well. He barely saw his daughter or newborn son. A tendency to push himself to exhaustion, an old habit, returned. His new habit was to try and organize a life that was so absurdly busy it couldn't possibly be organized. From the war onward until the end of his life, Shockley began to keep several different sets of calendars and diaries, some for his military work, some for his scientific work, and some for his home life. The war diaries suggest that his days were a blur of ideas and appointments and phone calls, all tucked between exercise regimens, doctor's appointments, train trips, and lunch and dinner meetings.[28] It hardly helped that his travel schedule was frenetic, too. As a testament to his high security clearance, he was authorized by the secretary of war to ride any commercial airplane in the country at any time. Often his schedule took him around the country as well as to Europe.[29]

The work, the travel, the wartime pressure—all of it bore down on him. In addition, Shockley was struggling to manage a few dozen men, some of whom in his opinion were not as bright or committed as he regarded himself. And here, perhaps for the first time, a clear sign of Shockley's limitations emerged. Whatever his friend Jim Fisk found easy and natural—relaxing a roomful of scientists with some inspired slapstick, for instance, or giving men freedom to do their work as they chose—Shockley found difficult. He simply could not get the hang of managing people. Some fifty years later, Shockley's biographer, Joel Shurkin, found among his private papers a sealed envelope from this period containing a note informing his wife that he had just attempted suicide. He had played Russian roulette with a revolver. "There was just one chance in six that the loaded chamber would be under the firing pin," he wrote, before adding, with characteristic precision, that "there was some chance of a misfire even then." He apologized to her that he hadn't found a better way of solving their domestic problems.[30] Shockley never gave her the note and returned to his work on the war.

BELL LABS VICE PRESIDENT Oliver Buckley wrote a form letter in January 1944 to the Bell Labs staff members on leave with the military services. The war hadn't yet turned decisively in the Allies' favor—D-Day and the invasion of Europe were still six months away—but the news was increasingly encouraging. "We hear a lot these days about 'post war planning,' and hear of some others doing research and development for the post-war period," Buckley wrote. "As far as we are concerned, that is pretty well out. We want to see this war won before we start working on what is to follow."[31]

It was not quite true. Nearly a year before, for instance, Mervin Kelly began writing a twenty-nine-page memorandum—"A First Record of Thoughts Concerning an Important Post-War Problem of the Bell Telephone Laboratories and Western Electric Company"—that was circulated to Buckley and a number of other executives at the Labs in May 1943. Kelly's point was not to outline precisely what new products the company should make at the close of war. Rather, he felt compelled to outline his vision of how Bell Labs, once it regrouped, would fare within an electronics industry that was sure to grow exponentially after the war's conclusion. "It is reasonable to expect that in the decade following the war there may well be changes of even greater total significance than those of the past thirty years," Kelly wrote. Radar, he added, had already opened up vast new opportunities in the business of radio wave and microwave devices. Kelly also thought it likely that the telecommunications industry was destined to resemble, in its nature as well as its products, sister industries like radio and television. Before the war this had not been the case: Bell Labs researched and designed equipment for the highly specialized nature and problems of telephone service. But to Kelly, the era at hand would require different approaches. Deep within the long memo, he noted, "We have been a conservative and non-competitive organization. We engineer for high quality service, with long life, low maintenance costs, [and a] high factor of reliability as basic elements in our philosophy of design and manufacture. But our basic technology is becoming increasingly similar to that of a high volume, annual model, highly competitive, young, vigorous and growing industry."[32] In other words, there would soon be a revolution in electronics. And as he saw it, Bell Labs would need to lead it rather than join it.

Kelly wanted his old team back—the team he had handpicked in the late 1930s. First, he pursued Bill Shockley. In January 1944 Shockley had left his position with the Navy for a job at the War Department. Now working with a group considering radar problems on B-29 bombers, he had maintained his frenetic schedule, beginning with a world tour of B-29 bases overseas, "going by way of England and Italy, and visiting American and British bombing establishments en route," the *Bell Laboratories Record* noted. "After six weeks at the B-29 base in India, he proceeded via Australia to 21st Bomber Command headquarters in the Marianas, then across the Pacific to the United States, where he arrived in January 1945." Not long after, Kelly invited Shockley to Bell Labs for a series of meetings. The two men discussed Shockley's possible return to industrial research, and Shockley thereafter agreed to leave Washington and return to Kelly's shop. At first he began working part-time, visiting the Labs frequently in the late winter and spring of 1945. The war was finally winding down—Germany surrendered to the Allied armies that May. It had been several years since Shockley had done any work in physics, and the meetings and discussions at Bell Labs reawakened in him an old interest in some of the technologies he had been pursuing—the solid-state amplifier, for instance—just before the war.[33] He returned to full-time work at the Labs immediately after the atomic bombs were dropped on Hiroshima and Nagasaki. Kelly had told him he had a new project in mind.

Five

SOLID STATE

I n technology and business—and in men's lives, too—the war cleaved the past from the future. Shockley, for instance, would not be returning to the Labs' West Street offices in Manhattan. Kelly's postwar plans for him, as well as for Bell Labs, depended on a move across the Hudson River. For the past five years, in between all of his wartime work, Kelly had devoted large blocks of time to orchestrating the construction of an immense complex of buildings in the wooded hills of New Jersey. Situated in a neighborhood called Murray Hill that was set within the quiet suburban towns of New Providence and Berkeley Heights, the new office was to be located about twenty-five miles from New York. "It was decided in the mid-thirties to initiate a gradual exodus from the city," Kelly recalled some years later, explaining that the aging labs at West Street had been deemed "non-functional" by that point. A 225-acre tract in Murray Hill had already been acquired. In July 1930 Frank Jewett, the Labs president, had called a public meeting near the future location to announce the land acquisition and describe a plan to build a research facility.[1] Those plans had been put on hold until the worst years of the Depression had passed, and in 1938 the idea of a grand suburban laboratory was revived by Jewett, then–vice president Oliver Buckley, and research chief Kelly.

Some of their reasons were purely scientific. For years Bell Labs had been operating small satellite facilities at far-flung locations around New Jersey—near the shore in the towns of Holmdel and Deal, for instance, and in the forested hills near the North Jersey town of Whippany. Long-wave and shortwave radio researchers at those outposts needed distance from the interference of New York City (and from one another) to do proper research and measurements. Murray Hill was put in a similar context: A move to the suburbs would allow the physics, chemistry, and acoustics staff to conduct research in a location unaffected by the dirt, noise, vibrations, and general disturbances of New York City.

It was also true, however, that Bell Labs was growing at such a clip it had simply overrun its facilities. In a private memo to Jewett, Buckley noted that "these buildings at Murray Hill [would] secure relief from an unsatisfactory and hazardous housing situation in New York City." A new laboratory, in other words, would allow for the ultimate consolidation of the organization's research staff, which was now spread out in a number of rented buildings and even in the basement of 463 West Street. In his memo, Buckley could have added another reason for the move, but it would have been unnecessary: The new site was close to his home and Jewett's. Buckley lived a few minutes away in the town of Maplewood; Jewett was in Short Hills, as were Kelly and Davisson. Shockley lived in nearby Madison as well. The executives were building a new laboratory in their own backyard.

With a price tag of about $4.1 million, the new building was not conceived as an ordinary laboratory.[2] By the late 1930s, Murray Hill had become a research and development project in itself, designed with meticulous care by Buckley and Kelly and a number of other engineers so that it would not only relieve Bell Labs' congestion problems but would organize its scientists in an unusual configuration that could be expanded on a far grander scale in future years. Kelly, Buckley, and Jewett were of the mind that Bell Labs would soon become—or was already—the largest and most advanced research organization in the world. As they toured industrial labs in the United States and Europe in the mid-1930s, seeking ideas for their own project, their opinions were reinforced. They wanted the new building to reflect the Labs' lofty status and academic standing—"surroundings more suggestive of a university than a factory," in Buckley's words, but with a slight but significant difference. "No attempt has been made to achieve the character of a university campus with its separate buildings," Buckley told Jewett. "On the contrary, all buildings have been connected so as to avoid fixed geographical delineation between departments and to encourage free interchange and close contact among them."[3] The physicists and chemists and mathematicians were not meant to avoid one another, in other words, and the research people were not meant to evade the development people.

By intention, everyone would be in one another's way. Members of the technical staff would often have both laboratories and small offices—but these might be in different corridors, therefore making it necessary to walk between the two, and all but assuring a chance encounter or two with a colleague during the commute. By the same token, the long corridor for the wing that would house many of the physics researchers was intentionally made to be seven hundred feet in length. It was so long that to look down it from one end was to see the other end disappear at a vanishing point. Traveling its length without encountering a number of acquaintances, problems, diversions, and ideas would be almost impossible. Then again, that was the point. Walking down that impossibly long tiled corridor, a scientist on his way to lunch in the Murray Hill cafeteria was like a magnet rolling past iron filings.

MURRAY HILL'S FIRST BUILDING—Building 1, as it was eventually known—officially opened in 1942.[4] Inside it was a model of sleek and flexible utility. Every office and every lab was divided into six-foot increments so that spaces could be expanded or shrunk depending on needs, thanks to a system of soundproofed steel partition walls that could be moved on short notice. Thus a research team with an eighteen-foot lab might, if space allowed, quickly expand their work into a twenty-four-foot lab.

Each six-foot space, in addition, was outfitted with pipes providing all the basic needs of an experimentalist: compressed air, distilled water, steam, gas, vacuum, hydrogen, oxygen, and nitrogen. And there was both DC and AC power. From the outside, the Murray Hill complex appeared vaguely H-shaped. Most of the actual laboratories were located in two long wings, each four stories high, which were built in parallel and were connected by another wing. Set back a thousand feet from the street, Building 1—finished in limestone and buff-colored brick and roofed with copper sheeting that in a few years' time would acquire a green patina— was fronted by a vast apron of lawn. Buckley and Kelly wanted quiet, and quiet is what they got. The roads on all four sides of the complex were lightly traveled, and behind the building were hundreds of acres of protected forest set in a county reservation of rolling hills. In the mornings, before the nine hundred or so scientists and technical assistants arrived, the massive building set amid the greenery had a grand and rarefied hush. The Bell Labs executives had not only built a new lab; they had built a citadel.

As American industry made the transition to postwar work, scientists from around the country made a pilgrimage. "Its use in our military research activities during the war years demonstrated its great worth," Kelly said with pride in the mid-1940s, as if it were another Bell Labs invention, which in fact he thought it was. "It has attracted so much attention that representatives of more than eighty industrial laboratories have visited it and have obtained detailed information from us about its special functional features," he added. Many laboratories patterned after Murray Hill were now under construction in the United States; European research managers and architects had come to study the building, too. Meanwhile the location, and not just the building, had drawn the interest of other companies, leading a number of industrial labs to locate nearby. One newspaper dubbed the New Jersey suburbs "research row."

Bell Labs was Kelly's shop now—officially, that is. In 1944 Frank Jewett had elevated himself to Bell Labs chairman, a mostly honorary position, as the war and Washington politics consumed his attention. Buckley had meanwhile assumed the presidency, a rank that put him on a busy circuit of speaking engagements around the country while also forcing him to tend to AT&T corporate obligations in New York City. Kelly became executive vice president, giving him complete control over day-to-day operations. One of his first acts was reorganizing the Murray Hill staff one day in July 1945. On that morning, several supervisors were demoted, while men younger and better versed in solid-state physics, such as Bill Shockley, were promoted. Dean Woolridge told the science historian Lillian Hoddeson that "there were rumors that Kelly was reorganizing for peacetime and everybody knew something was going to happen when he announced a big meeting." Many, if not all, of the supervisors in the research department were called together in a large room with Kelly. "He sat up there at the head of the room and he read off 'From now on thou shalt do this and thou shalt have this particular group and you're going to move over here and do this kind of work and you're going to do this.'" Kelly had worked out the new organizational structure in minute detail. The meeting took most of the day.

Essentially Kelly was creating interdisciplinary groups—combining chemists, physicists, metallurgists, and engineers; combining theoreticians with experimentalists—to work on new electronic technologies. But putting young men like Shockley in a management position devastated some of the older Labs scientists. Addison White, a younger member of the technical staff who before the war had taken part in Shockley's weekly study group, told Hoddeson he nevertheless considered it "a stroke of enormously good management on Kelly's part." He even thought it an act of managerial bravery to strip the titles from men Kelly had worked with for decades. "One of these men wept in my office after this happened," White said. "I'm sure it was an essential part of what by this time had become a revolution."[5]

No documentation, and no private papers, have ever surfaced to fully explain Kelly's rationale.[6] Was he at this point looking for an actual invention that he could now clearly articulate, or was he just acting on a hunch that something useful was waiting to be found, and that if it weren't something big, then surely combining men of such talent in a state-of-the-art building would produce a breakthrough? There exists only one true founding document of the revolution: an authorization form that Kelly signed to fund the new groups he had organized. Under Bell System protocol, work at Bell Labs had to be billed to either AT&T, Western Electric, or the local operating companies like Pacific Telephone. Different types of research were classified as a "case," and each case was in turn approved (or, rarely, challenged) by corporate management. Vacuum tube development, for instance, had its own case number, as did basic physics research. This was one reason why Kelly had quashed Fisk's swimming pool requisition: It could hardly be justified under an existing case number, and even if it could, a new case for building a swimming pool was bound to be met at AT&T headquarters with a skepticism rivaling Kelly's own.

On June 21, 1945, Kelly had signed off on Case 38139. "A unified approach to all of our solid state problems offers great promise," he wrote. "Hence all of the research activity in the area of solids is now being consolidated in order to achieve the unified approach to the theoretical and experimental work of the solid state area." The point of the new effort, in case it was hard to see through the jargon, was "the obtaining of new knowledge that can be used in the development of completely new and improved components" of communications systems. At the end of the six-page document, Kelly noted that he did not anticipate that the solid-state work would bring immediate results to the telephone business. On the other hand, he reasoned, the research was "so basic and may well be of such far-reaching importance" to the business that it was imperative that the phone company supply the funding. The initial cost of the program was put at $417,000, most of which comprised salaries for the group members. The authorization was signed by Harvey Fletcher, Jim Fisk, and Kelly himself. It was billed to AT&T.[7]

One by one, meanwhile, the men who had left West Street for some military reason or another—for the battlefields, for work in Washington, for a stint in Whippany or at the biscuit building in Manhattan to design radar sets—were filtering back. Walter Brattain had spent the first half of the war working in Washington on submarine detection and the second half at Bell Labs working on other matters of military engineering.

In Kelly's new order, he would be joining a solid-state research group led mainly by Bill Shockley. The notebook Brattain had used to chart his semiconductor experiments before the war—notebook number 18194— had its last entry in West Street on November 7, 1941. Four years later, in the new Murray Hill building, Brattain picked it up again and opened to page 40. "The war is over," he wrote.[8]

IN LATER YEARS, it would become a kind of received wisdom that many of the revolutionary technologies that arose at Bell Labs in the 1940s and 1950s owed their existence to dashing physicists such as Bill Shockley, and to the iconoclastic ideas of quantum mechanics. These men could effectively see into the deepest recesses of the atom, and could theorize inventions no one had previously deemed possible. More fundamentally, however, the coming age of technologies owed its existence to a quiet revolution in materials. Indeed, without new materials—that is, materials that were created through new chemistry techniques, or rare and common metals that could now be brought to a novel state of ultrapurity by resourceful metallurgists—the actual physical inventions of the period might have been impossible. Shockley would have spent his career trapped in a prison of elegant theory.

A few of the scientists at Bell Labs grasped the fact earlier than others. Just as the men in Thomas Edison's old laboratory would tinker with animal hoofs and horsehair in their inventions, for the first three or four decades of the phone system its engineers worked mostly with everyday substances—wooden blocks hewn from the finest bird's-eye maple to mount switching equipment, or a natural latex known as gutta-percha for waterproofing cables. Increasingly, however, new ideas, and new projects, ran up against the limitation of nature's design. "We had such specific requirements that ordinary raw materials had an agonizing time meeting them," William Baker, who joined the Labs as a chemist just before World War II, explained.[9] The solution, as Baker described it, was to literally create new types of matter. When the Labs' chemists needed to help design impermeable underwater cables, for example, one possibility was gutta-percha. But gutta-percha had drawbacks, including its extreme expense. For an undersea cable to Catalina Island, off the coast of California, the Labs' chemists began looking for an alternative. Natural rubber was considered too soft in its pure state, and when it was treated with sulfur to toughen it—that is, to "vulcanize" it—the men had to address the problem that sulfur corrodes copper and would undoubtedly degrade the vital wires within the undersea cable. Only after the chemists determined that they could purify the rubber in a complex manner and then create fine silica flour as an insulator could the cable go into production.

Those working with metals wrestled with challenges similar to those working with rubber and plastics. During the Labs' early days, while Mervin Kelly was running the tube shop in lower Manhattan, metallurgists began to focus on whether special coatings on metal filaments inside the vacuum tubes could vastly improve their performance or durability. The work required them to delve into the uses of obscure elements and alloys, and to conceive—successfully—of arcane processes to heat and cool their mixtures. A similar example arose when Labs scientists tried to improve the thin diaphragm in a phone receiver—the metal disc that vibrates in response to a speaker's voice. They ultimately created something called Permendur, an alloy of cobalt and iron that was spiced with around 2 percent of the element vanadium. But the metallurgists soon realized that Permendur was only part of the solution to finding a better diaphragm. If the ingredients in the alloy weren't pure—if they happened to contain minute traces of carbon, oxygen, or nitrogen, for instance—Permendur would be imperfect.[10] "There was a time not so long ago when a thousandth of a percent or a hundredth of a percent of a foreign body in a chemical mixture was looked upon merely as an incidental inclusion which could have no appreciable effect on the characteristics of the substance," Frank Jewett, the first president of the Labs, explained. "We have learned in recent years that this is an absolutely erroneous idea."[11]

It was understandable, then, for an engineer or scientist to regard the era of postwar electronics much like many people at Bell Labs did: New devices were waiting to be discovered through the combination of basic research and the intellectual force of physicists and engineers working together. The world of electronics could move forward through new ideas and the meticulous work of development engineers. "I think that there are vistas ahead that [are] as large or larger than the past," Mervin Kelly declared in May 1947.[12] A more cautious view maintained that the potential new age depended on finding a solution to a cosmic puzzle. Progress, in both technology and business, depended on new materials, and new materials were scattered about the earth in confusion. There were substances that might be useful on their own or combined into compounds—that alone comprised an infinite number. And each of these potential substances could, in turn, be rendered in a laboratory into an almost unimaginable range of purities. One might conceivably find the right material, with the right level of purity, for a new device. Or one might not.

THE PURSUIT OF PURITY at Bell Labs went further than Frank Jewett or Mervin Kelly ever imagined. In 1939, the year Shockley and Brattain had tried and failed to make a solid-state amplifier out of copper oxide, the chemists and metallurgists at the Labs had already begun looking closely at the class of materials called semiconductors. In particular they focused their inquiries on silicon. Already physicists had concluded that many semiconductors exhibited a unique set of behaviors related to their atomic structure. Atoms have a nucleus packed full of protons and neutrons that are surrounded by bands of vibrating electrons. In a good conductor—copper, for instance—the band farthest from the nucleus has only one or two electrons, which means this band is mostly empty. The outermost electrons are often free to bounce around and move to neighboring copper atoms. In a good insulator—glass, for instance—the opposite holds true; the band farthest from the nucleus has seven or eight electrons, which means it is mostly full. The electrons are therefore held in fixed positions. These contrasts translate into differences in how the materials conduct electric current, which might be thought of as a "flow" of electrons through a solid material. In a conductor, electrons move freely. In an insulator, they don't. As the Bell Labs researchers would describe it, the electrons in an insulator essentially "act as a rigid cement" to bind together the atoms in the solid.

Semiconductors—as their name implies, neither conductors nor insulators—are a curious case. In their outer band, atoms that comprise these substances have somewhere between three and five electrons, and they seem to exhibit qualities that are different from those of either a conductor or an insulator. Early in the twentieth century, physicists noted that these materials became better electrical conductors as their temperature increased—the opposite of what happened with metals (and good conductors) like copper. In addition, they could in some circumstances produce an electric current when placed under a light—what was known as a photovoltaic effect. Perhaps most compelling, the materials could rectify, meaning they allowed electric signals to pass in one direction only (they could, in other words, convert alternating electric current, AC, to direct current, DC).[13] This was a useful property familiar to almost any Bell engineer. The early crystal wireless radios that so many Labs scientists grew up with depended on semiconductor crystals like silicon. Silicon crystals would process the incoming radio signal, transforming a weak AC signal into DC, so it could be heard through a headphone.

Three Bell Labs researchers in particular—Jack Scaff, Henry Theurer, and Russell Ohl—had been working with silicon in the late 1930s, mostly because of its potential for the Labs' work in radio transmission. Scaff and Theurer would order raw silicon powder from Europe, or (later) from American companies like DuPont, and melt it at extraordinary temperatures in quartz crucibles. When the material cooled they would be left with small ingots that they could test and examine. They soon realized that some of their ingots—they looked like coal-black chunks, with cracks from where the material had cooled too quickly—rectified current in one direction, and some samples rectified current in another direction. At one point, Russell Ohl came across a sample that seemed to do both: The top part of the sample went in one direction and the bottom in the other. That particular piece was intriguing in another respect. Ohl discovered that when he shone a bright light on it he could generate a surprisingly large electric voltage. Indeed the effect was so striking, and so unexpected, that Ohl was asked to demonstrate it in Mervin Kelly's office one afternoon. Kelly immediately called in Walter Brattain to take a look, but none of the men had a definitive explanation.[14] "In discussing these mysteries Ohl and I decided we needed to characterize them in some way," the metallurgist Scaff later explained.[15] During a phone call, the two men decided to call one type of silicon p-type (for positive conduction) and the other n-type (for negative).

It wasn't necessarily clear at the start why this was so—or whether it was even important. By the early 1940s, however, Scaff and Ohl were increasingly sure that the two differing types of silicon were the product of almost infinitesimal amounts of different impurities. Atoms within semiconductors bond easily with a number of other elements. Scaff and his colleagues knew that when they cut n-type silicon (atomic number 14) into smaller pieces on a power saw, for instance, they could smell something they were sure was phosphorus (atomic number 15). None of the measurement equipment could pick up the taint, but their noses could.[16] Later, the men also determined that p-type silicon often had faint traces of the elements aluminum (13) or boron (5).

This was the beginning of a larger insight. Ultimately the metallurgists Scaff and Ohl agreed that certain elements added to the silicon (such as phosphorus) would add excess electrons to its outer band of electrons; those extra electrons could, in turn, move around and help the silicon conduct current, just as they might in a conductor such as copper. This was n-type silicon. On the other hand, certain other elements added to the silicon (such as boron) created additional empty spaces for electrons in the outer band—these became known as holes. These so-called holes, much like electrons, could also move about and conduct current, like a stream of bubbles moving air through a liquid. This was p-type silicon. For Scaff and Theurer—and, in time, the rest of the solid-state team at Bell Labs—one way to think of these effects was that purity in a semiconductor was necessary. But so was a controlled impurity. Indeed, an almost vanishingly small impurity mixed into silicon, having a net effect of perhaps one rogue atom of boron or phosphorus inserted among five or ten million atoms of a pure semiconductor like silicon, was what could determine whether, and how well, the semiconductor could conduct a current. One way to think of it—a term that was sometimes used at the Labs—was as a *functional* impurity.

When the war arrived, Ohl, Scaff, and some of the other metallurgists continued to hone their techniques with silicon. So, too, did scientists at a number of other universities and industrial labs. Silicon crystals, in particular, were found to be a vital component in the airborne radar sets that the Labs was designing and that Western Electric was making. Just as in crystal wireless sets, the silicon diodes, as they were known, could perform a vital function in radar receivers for their ability to process (that is, rectify) incoming radar signals. The Labs' metallurgists were now able to buy highly purified silicon from DuPont and reliably integrate traces of impurities by adding them to the "melt" before a substance cooled.

Indeed, this was the situation Shockley confronted upon his return to Bell Labs, as he began working with his solid-state physics group. His team could have continued working with a semiconductor such as copper oxide, which was what Shockley and Brattain had tinkered with before the war. But early on, the solid-state group decided they were interested in understanding better what seemed to them the most fundamental and representative materials in the class. "Although silicon and germanium were the simplest semi-conductors, and most of their properties could be well understood in terms of existing theory, there were still a number of matters not completely investigated," an internal history of the solid-state research project later explained. Germanium had been discovered in Europe in the late 1800s, and as with silicon, wafers made from the element had proven useful during the war for radar detectors. The small amounts the men worked with came from tiny military stockpiles in Missouri, Oklahoma, and Kansas.

Silvery and lustrous in appearance, germanium was so rare that one Bell Labs scientist remarked that before 1940 only a handful of people in the world had actually ever seen it.[17] Moreover, it was until that time considered largely useless. A Bell Labs metallurgist named Gordon Teal had wondered if therein resided its virtue. "A research man," he later remarked, "is endlessly searching to find a use for something that has no use."[18]

MEN ARE A KIND OF MATERIAL, TOO.[19]

"It was probably one of the greatest research teams ever pulled together on a problem," Walter Brattain would later say. When he first reviewed the list of who would be working with silicon and germanium in the new solid-state group with Shockley at Murray Hill—roughly every month, the Labs' staff received typed organizational charts of their department's personnel—Brattain read it over twice. *There isn't an S.O.B. in the group*, he thought to himself, pleased with the prospect of joining in. Then after a minute he had a second thought: *Maybe I'm the S.O.B. in the group.*[20]

It was entirely possible. Brattain, rarely at a loss for words, was known to be reflexively argumentative. He no longer carried a rifle, but a career in the most rarefied precincts of science had done nothing to smooth the cantankerous attitude he'd brought with him from the frontier of Washington State.[21] Wiry in stature, with slender mustache and thinning gray hair combed straight back from his forehead, Brattain nevertheless had an endearing side. He was good company and—unlike Shockley—often self-deprecating; he was also collegial, garrulous, and unusually game for an experiment or a round of golf. He had deft hands as well as deft ideas on how to carry out lab work. So did another experimentalist in the group, Gerald Pearson. The two men shared a large laboratory on the fourth floor of Building 1 in Murray Hill. Brattain was intent on investigating the surfaces of semiconductor materials—what happened on those surfaces, for instance, when small slabs or cylinders of silicon or germanium were exposed to heat or connected to various kinds of circuits. Pearson was intent on exploring the "bulk" properties of the materials. That is, he wanted to investigate their interiors. The distinctions could seem irrelevant to a layman, but in the subatomic world of solids the behavior of surfaces and interiors was not necessarily the same.

Other scientists from the Labs joined the solid-state group at varying levels of involvement—chemists, circuitry experts, metallurgists, technical assistants. Shockley naturally assumed the role of the leading theorist of the group; at the start it was his ideas that drove experimentation. But Shockley believed the group was incomplete. "Bill Shockley is one of the ablest theoretical people around anywhere and always had been," Jim Fisk, who directed the wartime magnetron work at the Labs, later told Lillian Hoddeson. "But he recognized that we needed more and we decided jointly that there were probably only three people in the country that qualified here."[22] Of the three, Shockley and Fisk set their sights on one man in particular—John Bardeen, who was just finishing up a stint at the Naval Ordnance Laboratory in Maryland, where he'd been stationed for wartime work on mines and submarines. The field of solid-state physics was new enough, and small enough, that Bardeen already had a number of connections with Bell Labs scientists. He'd met Shockley in the mid-1930s, while he was a postgraduate student at Harvard and Shockley was a grad student at MIT; around the same time, he'd become friends with Fisk, who also held a postgraduate fellowship at Harvard. Bardeen likewise knew Walter Brattain through Brattain's younger brother, with whom he had gone to graduate school.

Fisk and Shockley convinced Mervin Kelly to hire Bardeen, and Kelly in turn offered Bardeen a salary significantly higher than any university could pay him. Bardeen joined the Labs in October 1945.[23] The large volume of the war work had created crowded conditions at Building 1 in Murray Hill; a new building—Building 2—was planned but not yet begun. When Bardeen arrived there was no office ready for him. So he did the next best thing: He set up his theoretical shop in the lab shared by Brattain and Pearson.

One could readily see that the men on the solid-state team were distinct in their talents as well as personalities. There was Shockley, a lightning-quick aggressor; Brattain, a skeptical and talkative experimentalist; and Pearson, an easygoing and steady presence. Bardeen—with thinning dark hair, a medium build, and a soft physicality that belied his athleticism—wasn't just quiet. He barely spoke. And when he did speak it was often in something best described as a mumble-whisper. Bardeen and Shockley differed not only in temperament but in their scientific approaches. Shockley enjoyed finding a hanging thread so he could unravel a problem with a swift, magical pull and then move on to something else. Bardeen was content to yank away steadfastly, tirelessly, pulling on various corners of a problem until the whole thing ripped open. Sometimes he did loosen up in conversation—a few beers would often do the trick—but mainly he watched and listened and worked various angles. Brattain realized that when Bardeen did choose to interpret data or ask a question, a profundity was likely to tumble forth. The group recognized this not long after. When Bardeen talked, everyone else immediately stopped to listen.

Bardeen and Brattain—brain and hands, introvert and extrovert—discovered they worked well together. Indeed, they sometimes seemed hardly able to function independently. "I was the experimentalist and Bardeen was the theorist," Brattain would later recall, "and in fact there were occasions where I had to go to another department and Bardeen was left in the lab and he was anxious to get the experiment done and I said, 'Well, there it is, John, I'll be back in about an hour.' And I'd come back in about an hour and John would be gone and I'd ask the other people in the lab what happened and they'd say, 'Oh, he worked for about five minutes and said, "Oh, damn!" and left.'"[24]

Their work together was further buoyed by the exchange of ideas within the larger solid-state group, which would gather sometimes once a day—and at least once a week—in meetings often led by Shockley, to exchange thoughts and review experiments. "I cannot overemphasize the rapport of this group," Brattain said. "We would meet together to discuss important steps almost on the spur of the moment of an afternoon. We would discuss things freely. I think many of us had ideas in these discussion groups, one person's remarks suggesting an idea to another."[25] The group would carry their discussions into lunch in the cafeteria as well. Or they would get in their cars and drive a few miles south along Diamond Hill Road, a narrow, sinuous county highway, to visit a small hamburger joint called Snuffy's. The Bell Labs cafeteria didn't serve beer. Snuffy's did.[26]

The formal purpose of the new solid-state group was not so much to build something as to *understand* it. Officially, Shockley's men were after a basic knowledge of their new materials; only in the back of their minds did a few believe they would soon find something useful for the Bell System. In many respects their work was orders of magnitude more difficult than what usually went on at the Labs. Some years later, the semiconductor historians Ernest Braun and Stuart Macdonald would note that "everything in a solid happens on such a minute scale that not even a microscope, even an electron microscope, can resolve the elementary processes. The scientist must operate at a level of abstraction of which the untrained mind is not capable in order to visualize processes which cannot be seen." So the theorists at Bell Labs worked on blackboards, attempting to "see," at a subatomic level, the surfaces and interiors of semiconductor crystals; the experimentalists, in turn, tested the theorists' blackboard predictions at their lab benches with carefully calibrated instruments that recorded what happened within tiny pieces of silicon or germanium set on a breadboard—a wooden plank with holes—and wired into battery-powered circuits. Then the theorists would in turn try to interpret the data that emerged from the experimentalists' attempts to investigate the theorists' original ideas.

The circle was virtuous but not always illuminating. Early on, for instance, Shockley had formulated a theory on something he called the "field effect." The team agreed it seemed like a sound idea, suggesting that an electrical current applied to the surface of a suitably prepared semiconductor slice should change (that is, increase) the conductivity of the semiconductor slice and result in an amplifier. But it didn't. "My calculations showed that very substantial modulation of the resistance should occur," Shockley later noted. "None was observed. On 23 June 1945, I wrote that the effects were at least 1,500 times smaller than what I predicted should have been observable." He was vexed. And for close to a year, any attempts to make the field effect work failed. Shockley would eventually call this period "the natural blundering process of finding one's way."[27]

Whatever it was, by January 1946 the solid-state group, venturing far beyond the traditional methods of trial-and-error invention, had found neither enlightenment nor promise. As Brattain's lab mate Gerald Pearson would later note, they were groping in the dark.

Six

HOUSE OF MAGIC

In the early fall of 1947, Bardeen and Shockley made their short drives to Murray Hill every morning—Shockley from the town of Madison, Bardeen from the town of Summit. Brattain would come to Murray Hill by carpool from nearby Morristown. For more than a year their solid-state group had been scientifically vigorous and socially cohesive—Shockley and Bardeen had even spent several weeks together in the summer of 1947 visiting various European laboratories. By late autumn, with the newly planted trees on the Murray Hill campus nearly bereft of leaves, the solid-state team began an experimental regimen on silicon and germanium slices that offered a steady progression of insights.

The failures of the previous year had been instructive. In the spring of 1946, in an effort to explain the failure of Shockley's field effect, John Bardeen had spoken up, and something profound had tumbled out. He had offered a theory relating to what he called the "surface states" on semiconducting materials. In simplest terms, he postulated that when a charge was applied to a semiconductor, the electrons on its surface were not free to move the same way the electrons in the interior might. Instead (in Shockley's words) these surface electrons "were trapped in surface states, so that they were immobile." The result was that the surface state created a frozen barrier between any outside voltage and the material's interior.[1] This insight had changed the direction of their research. If one were to make an amplifier out of solid materials—that is, if one wanted to pass a current through a piece of silicon or germanium and boost that current by a third input current, much like what transpires within a vacuum tube—there was now clear agreement: One first had to break through the wall of the surface state. Bardeen's notion, as Brattain later recalled it, liberated the team.

To wander into the stately entrance of Bell Labs' Murray Hill building in mid-November 1947—into the new building's airy art deco lobby, up a short flight of wide stairs, past the guard station, up eight half flights of granite stairs bounded by a smooth, continuously curving wooden handrail, turning off the fourth-floor landing to walk down the impossibly long, brightly lit, polished red terrazzo hallway, and then on toward Walter Brattain's laboratory, room 1E455—was to wander into a realm of opaque chatter and experimentation. Brattain's lab was furnished with workbenches weighted with bulky calibration equipment, all of it adorned with myriad dials and meters. The circulating air was often infused with a sharp burn of smoke from cigars, cigarettes, and soldering irons. The men there spoke of their concerns in another language: "dipoles" and "space-charge layers" and "hysteresis" on the surface of the small semiconductor crystals they bent over to work on. But in large part their quest had come down to how to clean or prepare the surface of their materials—they were at that point using many varieties of silicon—so that either the electrons, or the holes within, or both, would move freely when a current passed through the silicon.

On November 17, Brattain and an electrochemist in the solid-state group, Robert Gibney, explored whether applying an electrolyte—a solution that conducts electricity—in a particular manner would help cut through the surface states barrier. It did. Shockley would later identify this development as a breakthrough and the beginning of what he called "the magic month." In time, the events of the following weeks would indeed be viewed by some of the men in terms resembling enchantment—the team's slow, methodical success effecting the appearance of preordained destiny. For men of science, it was an odd conclusion to draw. Yet Walter Brattain would in time admit he had "a mystical feeling" that what he ultimately discovered had been waiting for him.[2]

Within a week of the experiment with electrolytes, Brattain recalled, "Bardeen walked into my office one morning" and suggested a "geometry" for building a solid-state amplifier. "I said, 'Let's go to the laboratory and do it.'" It involved a drop of electrolyte fluid, and a "point-contact" wire piercing the drop and touching the surface of the semiconductor slice. With Bardeen sometimes looking over his shoulder, Brattain and a lab assistant began building a rough prototype—a slab of silicon with a metal point pushed down into it. After some early tests of running a current through the point and into the slab setup, the contraption indicated a slight gain in power. This seemed promising. That night—a Friday, November 21—Brattain told the members of his carpool that he had taken part in the most important experiment in his life. Over the weekend, though, he had second thoughts, worrying that if he and Bardeen were indeed onto something it was of such importance that it must not become gossip. The next day he rode with the group—Monday—he asked them to forget what he had told them about his semiconductor work. "I had to swear them to secrecy," he said.[3]

Bardeen, too, seemed to think they were close; he spent the weekend making notations in his notebook outlining what might be happening within the silicon crystal. The two men let Shockley know about the recent developments. Then Brattain came down with the flu.

THE DEVICE THAT BRATTAIN was fiddling with in his lab by mid-December was a small and inelegant thing: a tiny slice of semiconducting material, about one-fourth the size of a penny, lying flat on a metal base. At Bardeen's suggestion they had switched from using silicon to n-type germanium. A wire was connected to the base. What looked like a tiny arrowhead—it was actually a small, triangular piece of plastic—pushed down into the top face of the germanium slice. But in fact the plastic arrowhead wasn't touching the germanium. Brattain had wrapped the edge of the arrowhead in a thin gold foil—"some of the prewar gold that I still had around," he said later—and by doing so had created a gold, V-shaped wire. With a razor he had then cut a slice in the golden V, precisely at the sharp end of the V, so that there was an almost imperceptible gap. Rather than a V-shaped wire, in other words, he had thus created two separate wires—"points," as they were called at the time— that were pushing into the germanium. Bardeen had told Brattain that a narrow space between the two gold points was essential. He estimated that they could create an amplifier if the points were spaced only a few thousandths of an inch apart.

Brattain then connected the top ends of each of the points, located at the top corners of the arrowhead, to separate wires, each of which led to separate batteries. He was forming a simple circuit. The men had been experimenting for the past two weeks with various setups—they had tried different metals (n-types and p-types of germanium and silicon), different types of points (with different distances between them), sundry chemical preparations, and a variety of electrolyte solutions. Each configuration had yielded new insights and small degrees of amplification. And by the morning of December 16 it had come to this. Brattain connected it up. "I found if I wiggled it just right," he explained later, so that both points of the gold had contact with the germanium slab, he could get a remarkable effect. What was happening within the device was exceedingly complex—a subatomic movement of electrons and holes brought about by the impurities in the metal and the current introduced through the gold points. At the time, even Brattain and Bardeen didn't understand it. But the net amplification was the important fact. Years later, William Shockley would comb through Brattain's old notebooks in an effort to tell the story of how this amplifier had been born. It was on December 16, Shockley concluded, that Brattain's entries demonstrated the real achievement: a very significant power gain.

Over the next few days, Brattain and Bardeen refined their device in preparation for a demonstration to the Bell Labs management. It was scheduled for the afternoon of December 23, 1947. Mervin Kelly was not invited. A believer in granting a degree of autonomy to researchers, he had not asked about, and had not been kept apprised of, Bardeen and Brattain's work. What's more, there was a tendency at Bell Labs to confine important developments to middle management for a purgatorial period, lest word of a breakthrough reach upper management too soon. The concern was that research that appeared to be important could turn out, upon closer inspection, to be nothing of the sort. Thus the practice was for a supervisor to move any big news up a step—a week or two at a time, in Brattain's recollection—only after he was convinced of its importance. The worst scenario would be telling Kelly without being sure.[4]

On Christmas Eve 1947, Brattain wrote an account of the previous day's events in his notebook. Amid the lined pages, he made a quick drawing of the amplifier he had built. He drew another diagram of the device after he and some colleagues had arranged for a voice signal to pass through the circuit that ran through this ugly little amplifier. Brattain wrote, "This circuit was actually spoken over and by switching the device in and out a distinct gain in speech level could be heard." There was no noticeable change in quality. The power gain was eighteen times or more. "Various people witnessed this test and listened," Brattain added. In the room that day were Bardeen, Brattain, and Shockley; Brattain's lab partner Pearson; R. B. Gibney, the chemist who had helped; and H. R. Moore, a circuit expert who had also pitched in. Harvey Fletcher, the head of physical research—who along with Mervin Kelly had come from Millikan's labs so many years before and was finishing his career now at the Labs—was also present. As with all important entries in the scientists' notebooks, Brattain's entry ended with a signature and verification by third parties: *Read & understood by G. L. Pearson Dec. 24, 1947 and H. R. Moore Dec. 24, 1947.*

Another man in the room that afternoon was Ralph Bown, the vice president of research. Bown was an affable and well-regarded manager who had been promoted to research chief when Kelly moved up to become the Labs' executive vice president. Bown was more open and more approachable than Kelly. At Murray Hill a large arcade connected the two main buildings. Bown would often hold court in that arcade, and people would come by to chat with him.[5] At Bardeen and Brattain's demonstration, Bown was the most important and skeptical of the attendees. He reacted to what he saw with a challenge. "The acid test of an amplifier," Bown declared, "is whether it can be made to oscillate."[6] In other words, if it is believed that a device can truly produce more power than it takes in—the very definition of an amplifier—then there is a way to check its authenticity. It is done by changing the circuit so that the wires are arranged in a certain manner (the output is "fed back" into the input). What then gets produced is a consistent wavering—that is, oscillating—signal, like a sine wave. In communications systems, oscillating elements are fundamental: They form the basis for everything from a telephone's dial tone to the broadcasting of radio waves. At Bown's request, Brattain and his colleagues soon made their amplifier do exactly that. And at that point the research boss was satisfied. In Bown's personal account of the episode, he admitted that he was immediately convinced the device was not only real but important. It was, he privately suspected, more than an amplifier, more than a switch, more than a replacement for the vacuum tube. But on Christmas Eve, with the year almost over and with a powerful snowstorm bearing down on the East Coast, knowing precisely what it portended, or indeed how it all fit together, would require some serious consideration, as well as some time.

Bardeen and Brattain's device, Bown simply noted, was "a basically new thing in the world."

THE NEW THING NEEDED a new patent. In early 1948, probably a month after the invention, Mervin Kelly was informed of the breakthrough.[7] At that point, Bardeen and Brattain began working with the Bell Labs lawyers on assembling an application. The two men—Brattain especially—worried that other scientists in the United States or abroad might patent a similar device first. One possibility was the research team at Purdue University, which had been experimenting with semiconductors all through the war. Any Bell scientist knew about the spooky and coincidental nature of important inventions. The origins of their entire company—Alexander Bell's race to the patent office to beat Elisha Gray and become the recognized inventor of the telephone—was the text-book case.

The new thing needed a new name, too. A notice was circulated to thirty-one people on the Bell Labs staff, executives as well as members of the solid-state team. "On the subject of a generic name to be applied to this class of devices," the memo explained, "the committee is unable to make [a] unanimous recommendation." So a ballot was attached with some possible names. "Triode" seemed to have a natural appeal, since the new device had three main elements (the two points and the base), much like a vacuum tube had three main elements (the cathode, anode, and grid). Vacuum tubes, in fact, were often called triodes. The recipients were asked to number, in order of preference, the possibilities:

___ Semiconductor Triode

___ Surface States Triode

___ Crystal Triode

___ Solid Triode

___ Iotatron

___ Transistor

_____ (Other Suggestion)

Bell Labs engineers had become fond of the suffix "-istor": Small devices known as varistors and thermistors had already become essential components in the phone system's circuitry. "Transistor," the memo noted, was "an abbreviated combination of the words 'transconductance' or 'transfer,' and 'varistor.'"[8] To the company brass, the other names had some winning aspects, too. Iotatron, for instance, "satisfactorily conveys the sense of a minute element." Semiconductor triode, the memo noted, was a "fairly good name," if a bit unwieldy. But when the ballot results came in, transistor was the clear winner.

This so-called transistor needed, finally, to be better understood. "The most gifted electronic engineers from all parts of our laboratory," Bown, the head of research, recalled, "were brought into a concerted plan to study the device from all angles and to use it to do the various things an amplifier should be able to do." It was true, as well as obvious, that those minute impurities within the germanium had helped to amplify a voice signal, just as a vacuum tube might. But how the holes, and electrons, moved within the semiconductor slab when a voltage was applied needed far more study.

In early May 1948 the transistor was officially designated a classified Bell Telephone Laboratories technology—"BTL Confidential." To those with knowledge of the device, Ralph Bown sent around a lengthy protocol to instill an envelope of security.[9] The transistor work, and anything relating to it, was given a code name: Surface States Phenomena. Yet there were doubts about how long the Labs could maintain secrecy—or even how long it should. Almost from the start, Labs executives agreed that they should show the device to the military before any public debut but should try to resist any orders to contain the device as a military secret. On the other hand, the executives—Kelly, Bown, et al.—doubted they could keep the transistor rights to themselves once the device became public knowledge. For one thing, AT&T maintained its monopoly at the government's pleasure, and with the understanding that its scientific work was in the public's interest. An audacious move to capitalize on the transistor, should it turn out to be hugely valuable, could well invite government regulators to reexamine the company's civic-mindedness and antitrust status. What's more, sharing the technology with competitors in the electronics field might be a positive development. AT&T could earn licensing fees from the patent. And if Bell Labs could gain a head start of a few months, it could take a lead over the competition and reap further rewards as a host of outside engineers and scientists worked to improve its functionality. There was in this strategy "a modicum of self interest," according to Bown. "Who is in a better position than the originator to recognize and profit from further advances?"

At 9 a.m. on May 26, 1948, Bardeen, Brattain, Shockley, and a number of Bell staffers held a meeting on the transistor in Ralph Bown's spacious Murray Hill office. A noticeable degree of paranoia was setting in. The men talked about the possibility that others, perhaps at Purdue, were "skirting close to this subject in their work" and would announce it first.[10] There was agreement that they didn't want to maintain secrecy any longer than it would take to secure their patents, which were almost ready for application. The men decided to continue apace with the patent application and that Bardeen and Brattain would quickly write a letter to the journal *Physical Review* describing their device—in effect staking out their ground in the scientific community. It was also decided that the Labs' public relations department would begin writing a press release and discuss ideas for a public presentation.[11]

By mid-June 1948, the patent application on the transistor had been filed and a press conference on the new device had been scheduled for the end of the month at the large auditorium in Bell Labs' West Street offices in Manhattan. Crafting the news release had been something of a nightmare, with at least a half dozen cooks (Shockley, Bardeen, Bown, and Brattain included) adding to the broth. "It has been rewritten at least N times, where $N > 6$," Shockley noted.[12] The transistor was not much larger than the tip of a shoelace, the news release said, and more than a hundred of them could fit easily in an outstretched hand. Yet it was quite complicated. The explanation of the device took up seven typed pages.[13] Bardeen and Brattain's letter to the *Physical Review* that announced their breakthrough, meanwhile, was impenetrable to all but an accomplished solid-state physicist. If anyone really wanted to know what the scientists had accomplished over the past few years, they would need a world-class understanding of metallurgy, quantum physics, and electrical engineering.[14]

WHEN HE WAS A TEENAGER in the 1920s, William Shockley took a high school mathematics exam that haunted him for an entire weekend. "It reflects a personality aspect which may be common to a number of persons who work hard at technical endeavors," he later explained. It wasn't that Shockley felt he had failed his math test. "I was especially concerned about my relative grade to that of another student—one who was not at all remarkable," he said. "All weekend I worried over whether I had done better than he."[15]

As Bardeen and Brattain closed in on the transistor in the autumn of 1947, Shockley, their supervisor, had become increasingly interested in their work, sometimes offering suggestions and periodically receiving updates from the men. When the breakthrough came in December, Shockley would admit to a complex set of reactions—"I must confess to a little disappointment that I hadn't been more personally involved in it," he later admitted.[16] On another occasion he conceded, "My elation with the group's success was balanced by not being one of the inventors."[17] Perhaps far more than Bardeen and Brattain, Shockley understood the implications of the new device; he had been dreaming of its possible existence since Kelly had visited his office, so many years before, to talk about his idea of an electronic switch.

Walter Brattain would later recall that shortly after the December 23 transistor demonstration for the Bell Labs executives, Shockley called both Bardeen and Brattain into his office, separately, to say, "Sometimes the people who do the work don't get the credit for it."[18] In Brattain's telling, Shockley seemed confident that he himself could write a patent that covered his field effect idea and that would overshadow subsequent work done by his two colleagues. This would prove to be impossible for a number of reasons. First, many people at the Labs already knew that Bardeen and Brattain had built the first transistor together, and their lab notebooks could verify it. But in a patent search on Shockley's field effect, it also appeared that an inventor named Julius Lillienfield had come upon a similar idea two decades before. There was little evidence that Lillienfield had made a working model of his device, or that the device outlined in his patent would work. What's more, there was no likelihood that he had any theoretical understanding of semiconductors, let alone a knowledge of the movements of holes and electrons at the subatomic level. Still, legally speaking, Lillienfield had been there first. And—though they used a different approach—so had Bardeen and Brattain.

Shockley's Christmas was a holiday of torment. He left New Jersey at the end of the week for a conference in the Midwest. "On New Year's Eve I was alone in Chicago between two meetings that came so close together that a return to New Jersey seemed impractical," he later explained. In fact his mind was afire. Alone in his hotel room, he went to work. "In 2 days I wrote enough to fill a bit more than 19 notebook pages. My notebook was at the Laboratories and I used a pad of paper and mailed the disclosures back to my co-supervisor, S.O. Morgan, who witnessed them and asked Bardeen to do the same. Later these pages were rubber-cemented into my notebook."[19] For the next three weeks, Shockley kept up a furious pace. By late January he had come up with a theory, and a design, for a transistor that both looked and functioned differently than Bardeen and Brattain's. Theirs had been described as the point-contact transistor; Shockley's was to be known as the junction transistor. Rather than two metal points jammed into a sliver of semiconducting material, it was a solid block made from two pieces of n-type germanium and a nearly microscopic slice of p-type germanium in between. The metaphor of a sandwich wasn't far off. Except the sandwich was about the size of a kernel of corn.[20]

In some respects, Shockley's idea was illicit. At Bell Labs, there were boundaries, much like surface states, that members of the technical staff were not meant to pass through. Eccentricity—not wearing socks, say, or using company time to build gadgets that had perhaps not even a glancing relationship to the phone business—could be forgiven. Other behaviors could not. MTSs were never to seduce the secretaries. They were not to work with their doors closed. They were not to refuse help to a colleague, regardless of his rank or department, when it might be necessary. And perhaps most important, the supervisor was authorized to guide, not interfere with, the people he (or she) managed. "The management style was, and remained for many years, to use the lightest touch and absolutely never to compete with underlings," recalls Phil Anderson, a physicist who joined Bell Labs soon after the transistor was developed. "This was the taboo that Shockley transgressed, and was never forgiven."[21] To Addison White, another manager who years before had been a privileged member of Shockley's solid-state study group, the forces that drove

Shockley to compete were clear to those who knew him. White said, "I've never encountered a more brilliant man, I think. And he just wasn't going to sacrifice that in the interests of the members of his group."[22]

Shockley kept the design a secret for another month. At a conference with the solid-state group in mid-February, however, a colleague of Shockley's named John Shive stood up to inform the group about his recent findings that related closely to some of Shockley's new ideas for the junction transistor. Knowing the alertness of the group—Bardeen and Brattain were in the audience—Shockley sensed that within a few minutes someone would make the leap to propose something akin to the theoretical construct, known as "minority carrier injection," that he had developed on his own the month before. "From that point on," he noted, "the concept of using p-n junctions rather than metal point contacts would have been but a small step and the junction transistor would have been invented."[23]

So Shockley made the leap, literally. He jumped from his seat and proceeded to give a presentation to the group on his newest theories and design. "I felt I did not want to be left behind on this one," he recalled. Many of the men were dumbstruck. The solid-state group that Shockley led had been built upon the principles of an open exchange of ideas, and Shockley had apparently ignored those principles. At the same time, it was hard not to be awed—the men were witnessing another breakthrough on the level of Bardeen and Brattain's earlier work. Did it matter whether it was the product of Shockley's brilliance and effort, or his cunning and bruised ego?

By Shockley's calculations, the junction transistor was almost certainly superior to the point-contact transistor as a practical device. But there was a problem. Unlike the point-contact transistor, the junction transistor could not actually be built by anyone at the labs. It was still theoretical. And it looked to be exceedingly challenging for the metallurgists to create the materials—that tiny sandwich of n-p-n germanium. An irony, at least for that moment, was that Shockley's phantom invention (the junction transistor) had improved upon another invention (the point-contact transistor) that wasn't useful in any meaningful sense of the word. Lest anyone forget, the point-contact transistor was a device that had never been manufactured, had never been sold, and was still so secret that perhaps only a few dozen people in the world knew it existed.

THE UNVEILING of the two most important technologies of the twentieth century—the atomic bomb and the transistor—occurred almost exactly three years apart. The nuclear test blast at the Trinity site in the New Mexico desert took place at 5:29 a.m. on July 16, 1945. It was in many respects a demonstration of the power, and the terror, of new materials; a baseball-sized chunk of purified metal—about eleven pounds of newly discovered plutonium—could level a midsized city.[24] The transistor, too, was a demonstration of the power of new materials—less than a gram of germanium containing a slight impurity—but its significance was far less obvious. Its unveiling, a modest affair in Manhattan's Greenwich Village on June 30, 1948, offered only the most obvious suggestions—the new device was a replacement for the vacuum tube. It was smaller, more rugged, and used less power. The analogy most often used was that the transistor was like a faucet. Rather than water, it could either switch electricity on or off or make it pour out in a torrent. A small turn on the handle, so to speak, could produce big effects.

Ralph Bown, the head of research at Bell Labs who had demanded that Bardeen and Brattain make their transistor circuit oscillate before he would concede its legitimacy, led the conference. Shockley—gifted with a deep, sonorous voice that exuded confidence and calm amusement—handled most of the questions.

Each member of the audience was given a pair of headphones. Bown, tall and elegantly dressed, stood alongside a huge, human-sized replica of a point-contact transistor. The demonstration had three highlights: First, the attendees experienced the amplification properties of the transistor as Bown's voice was switched (and boosted) through its circuitry. Next, the audience heard a radio broadcast from a set constructed with transistors rather than vacuum tubes. Finally, a transistor was used to generate a frequency tone, thus showing it could oscillate. Bown and his colleagues had spent the past six months considering the potential applications of the transistor. They had no intention of soft-pedaling their device. As the transistor historians Michael Riordan and Lillian Hoddeson would later recount it, Bown told the audience that this "little bitty" thing could do "just about everything a vacuum tube can do, and some unique things which a vacuum tube cannot do."[25]

Most newspapers couldn't discern the value of the tiny device. The *New York Times*, in a famous lapse of editorial judgment, relegated a report on the West Street demonstration to a four-paragraph mention on page 46, in a column called "The News of Radio." Oliver Buckley, the Bell Labs president, chose to keep a copy of the *Times* story, either out of amusement or chagrin, in his personal files until his death. Yet there is little evidence that the Bell scientists were daunted by a perceived lack of excitement. Any apathy on the public's part was balanced by enthusiasm within the electronics industry, whose executives were given a special presentation soon after the public announcement. Earnest, chummy, beseeching letters—to Bown; to Kelly; to Buckley; to Shockley; to Bardeen; to Brattain; to anyone, in fact, with any connection to the solid-state work—began arriving at Bell Labs from executives in all corners of the electronics business, begging for samples of the new device. RCA wanted one, and so did Motorola, and so did Westinghouse and a host of other radio and television manufacturers. Moreover, the announcement had piqued the interest of the academy, leading professors at Harvard, Purdue, Stanford, Cornell, and a half dozen other schools to request a sample of the device for their own laboratories. "It appears that Transistors might have important uses in electronic computer circuits," Jay Forrester, the associate director of MIT's electrical engineering department, wrote to Bown in July 1948. "In view of this fact, we would like to obtain some sample transistors when they become available in order to investigate their possible applications to high-speed digital computing apparatus."[26] Whether Bown, Shockley, or Kelly had considered how the transistor might be used as a logic circuit in a computer—vacuum tubes were already being used, with mixed success, due to their tremendous energy requirements and fragility—the Forrester letter surely validated such applications. "We are interested that you think the transistor may be useful in connection with computing apparatus," Bown quickly responded. The men at the Labs, he added, would be happy to hear from the MIT scientists about how the device could be used or improved for computing purposes.[27]

It was still far too early for the press or the public to visualize how the invention of the transistor might pay off in practical terms. In time, however, the Bell scientists were confident that the public would figure out what the academics, electronics executives, and the Bell scientists already knew. On the day after the unveiling, Buckley had taken out a sheet of stationery to scribble a note to Bell Labs chairman Frank Jewett, now in declining health, who was vacationing at his summer home on Martha's Vineyard. He attached the long news release on the transistor. "The attached press release explains my recent hint of things to come," Buckley wrote his friend and former boss, who had apparently not been informed about the developments. "This looks very important to us."

THE LANGUAGE that affixes to new technologies is almost always confusing and inexact. If an idea is the most elemental unit of human progress, what comes after that? For instance, had Brattain and Bardeen made a discovery, or an invention? The distinctions could be real enough. A discovery often describes a scientific observation of the natural world— the first observation of Jupiter's moons, for example, or the isolation of a bacteria that causes a deadly plague. Also, a discovery could represent a huge scientific achievement but an economic dead end. In the early 1930s, for instance, at the Bell Labs radio facility in Holmdel, New Jersey, a young engineer named Karl Jansky created a movable antenna to research atmospheric noise. With this antenna, he observed a steady hiss emanating from the Milky Way. In this moment, Jansky had essentially started the field of radio astronomy—a discovery that paid a lasting dividend to his and Bell Labs' renown. On the other hand, it never led to any kind of profitable telecommunications invention or device.[28]

John Bardeen, the most careful of men, referred to his transistor work as a "discovery" of "transistor action"; he and Brattain had effectively observed in their experiment how a current applied to a slightly impure slice of germanium could hasten the movement of microscopic holes inside and thus amplify a signal. An invention, by contrast, usually refers to a work of engineering that may use a new scientific discovery—or, as is sometimes the case, long-existing ones—in novel ways. Shockley considered the transistor device, in its various forms (both point-contact and junction, for instance), to be an invention. If there was any doubt, he asserted, the legal protections ultimately awarded to these devices verified their status. The U.S. patent office wasn't in the business of licensing discoveries, only inventions.

To those with an open mind, of course, the transistor could be considered a breakthrough of both science and engineering—in effect both a discovery and an invention. What seemed fair to say, though, was that the transistor was not yet an innovation.

The term "innovation" dated back to sixteenth-century England. Originally it described the introduction into society of a novelty or new idea, usually relating to philosophy or religion. By the middle of the twentieth century, the words "innovate" and "innovation" were just beginning to be applied to technology and industry.[29] And they began to fill a descriptive gap. If an idea begat a discovery, and if a discovery begat an invention, then an innovation defined the lengthy and wholesale transformation of an idea into a technological product (or process) meant for widespread practical use. Almost by definition, a single person, or even a single group, could not alone create an innovation. The task was too variegated and involved.

The Labs executives were familiar with the difficulties ahead. Funding and resources necessary for the transistor's innovation would not be a problem—being attached to the world's biggest monopoly took care of that. Still, a product like the transistor could ultimately fail for technical reasons (if it proved unreliable) or for manufacturing reasons (if it proved difficult to reproduce consistently or cheaply). Also, it might be the case that there was no market for a new device: Why not continue to keep using vacuum tubes if they remained cheaper and more dependable than point-contact transistors?

In the late 1940s, finding a market for the new device may have been the least of the Labs' concerns. "If they could be made, they could be sold," the technology historians Ernest Braun and Stuart Macdonald noted about the new transistor. "If nobody else bought them, certainly the vast Bell empire itself would form an adequate market."[30] Thus it was the technical and production hurdles that seemed most formidable. And for all its publicity, the new point-contact transistor was useless as a practical device. A wave of the hand or a spell of humidity could alter its performance; it was so delicate and unpredictable that it would sometimes cease working if someone slammed a door nearby. "Making a few laboratory point-contact transistors to prove feasibility was not difficult," Ralph Bown explained. "But learning how to make them by the hundreds or thousands, and of sufficient uniformity to be interchangeable and reliable, was another problem."[31] By Kelly's orders, the transistor's innovation involved a handoff from Shockley's group in the research department to a team in the Labs' much larger development department. The development expert who was chosen for this responsibility—a brilliant, bullying, hard-drinking engineer named Jack Morton—not coincidentally had the admiration of both Kelly and Shockley. Morton, then thirty-five years old, had arrived at Bell Labs, like so many others, from a small midwestern school in the mid-1930s. Morton remembered a meeting with Kelly, who called him into his office in the midsummer of 1948 to say, "Morton, I think you know something about transistors. Don't you?" Morton summoned his courage and replied that he knew they were pretty important.

"No time to waste," Kelly told him. "I'm going to Europe for the next month, Morton, and when I get back I'd like to see your recommendations as to how we should go about developing this thing. Goodbye."[32]

Morton would eventually think more deeply about the innovative process than any Bell Labs scientist, with the possible exception of Kelly. In his view, innovation was not a simple action but "a total process" of interrelated parts. "It is not just the discovery of new phenomena, nor the development of a new product or manufacturing technique, nor the creation of a new market," he later wrote. "Rather, the process is all these things acting together in an integrated way toward a common industrial goal."[33] One of Morton's disciples, a Bell Labs development scientist named Eugene Gordon, points out that there were two corollaries to Morton's view of innovation: The first is that if you haven't manufactured the new thing in substantial quantities, you have not innovated; the second is that if you haven't found a market to sell the product, you have not innovated.[34] But these realizations would come together later. After hearing Kelly's orders to produce a road map for transistor production, Morton spent the next twenty-nine days in a state of terror. On the thirtieth he settled on a development plan.

BY THE SUMMER OF 1949, Morton's team, in conjunction with the Labs' metallurgists, had fabricated five thousand working germanium transistors. Many were given to the military or as complimentary samples to academics. Nearly a thousand were used at Bell Labs to study the properties of germanium. To Morton, the essential challenges in manufacturing the devices were "reliability," "reproducibility," and "designability." He planned to set up a production line at a Western Electric plant in Pennsylvania, but before that could happen he had to improve the consistency of the germanium. One problem with the early devices was that the germanium was cut from a polycrystalline ingot. In this ingot, the multiple crystals created imperfections within the structure that compromised transistor performance. The ideal material to slice up for transistors would be a perfect single crystal, with all the atoms in the germanium arranged in symmetrical and uninterrupted order, like apple trees stretching hither and yon in an infinite orchard. The problem was that nature didn't provide perfect single crystals.

In late 1949, the Bell Labs metallurgist Gordon Teal had an idea of how to make large single crystals of germanium in a device he designed that resembled a drill press. By dipping a tiny "seed" of pure germanium into a "melt" of the element, and then slowly, gently "pulling" it from the melt, Teal believed he could fabricate a large and perfect crystal that could in turn be cut into pieces for better point-contact transistors. Teal's bosses were skeptical, so he worked in secret on his process at Murray Hill, on borrowed equipment in a borrowed laboratory, from 5 p.m. in the afternoon until 3 a.m.[35] Eventually his methods worked so well that Jack Morton gave Teal his full support. Perhaps more important, the advances in crystal pulling soon allowed Teal and his colleague Morgan Sparks to grow junction transistors for Shockley—the device Shockley had theoretically predicted several years before, beginning with his late-night scribbling in the Chicago hotel on New Year's Eve. Shockley later acknowledged that the two materials scientists had provided "the essential missing ingredient" that made his idea possible.

By the summer of 1951, Jack Morton's team had thus readied Bardeen and Brattain's point-contact transistor for large-scale production. The manufacture of the device roughly coincided with Shockley's demonstration of the first junction transistors at a public unveiling at the West Street auditorium. The newest invention, hailed as clearly superior to the point-contact transistor in terms of its efficiency and performance (it used only one-millionth of the power of a typical vacuum tube), was "a radically new type of transistor which has astonishing properties never before achieved in any amplifying device."

Shockley, much to his delight, was now the public face of Bell Laboratories' research as well as the personification of the three-year-old transistor age. His fabled research group that had given rise to the device—*one of the greatest research teams ever pulled together on a problem*, as Brattain had put it—had largely collapsed, however. John Bardeen, frustrated by Shockley's muscular efforts to monopolize the Labs' semiconductor research, had decided to leave for a professorship at the University of Illinois at Urbana. Walter Brattain, too, had made his displeasure with Shockley known.

One afternoon, Mervin Kelly invited Brattain over to his home in Short Hills to discuss the matter. They likely met in Kelly's study, where he saw all his visitors—a large and stately room, clad in dark wood paneling, with a large fireplace and big windows that looked out over Kelly's backyard tulip gardens.[36] A servant could be summoned through the push of a button on the floor. Brattain listed his frustrations with the Labs and with Shockley, whom Kelly depended on as a conduit for information. Some of Brattain's complaints came as a surprise to Kelly, yet the boss swung back anyway with a powerful backhand. "He's a tough customer," Brattain would later recall with some respect. "I stated my case, and [he] pretty thoroughly knocked me down on every question I raised." But when Brattain mentioned to Kelly that he knew precisely when Shockley invented the junction transistor, the tone changed. In veiled terms Brattain was suggesting to Kelly that he and Bardeen could somehow complicate the transistor patents based on the fact that their invention came well before Shockley's, and "if we ever went on the stand in a patent fight" they could not lie about what they knew.[37]

Thereafter, Kelly made sure that Brattain was blessed with a nearly unfettered freedom at Bell Labs. Brattain was no longer involved in the transistor work, but he no longer had to report to Shockley, whom he now considered intolerable as a manager. Apparently, there was an S.O.B. in the solid-state group after all.

NOT LONG AFTER the transistor's unveiling at the West Street auditorium, the Labs began to spread its new invention around. In later years, corporations would give calculated thought and effort to this process, which would become known as the *diffusion* of new technology. The executives at Bell Labs, however, were making things up as they went along. The top managers had already agreed that they were compelled to share and license the transistor device. The political logic—the appeasement of government regulators—was overwhelming. And an open-door policy had other advantages, too. Bell Labs' breakthrough burnished its reputation as a national resource. For those doubting that the monopoly granted to AT&T, its parent company, would result in any large-scale scientific and public benefits, here was contravening proof.[38]

In the late 1940s, the Labs executives were simply content to pass out samples of the new device without explaining how they were made. In-

deed, as Kelly traipsed through northern Europe in the summer of 1948—
the trip he took after telling Jack Morton he wanted to hear a production
plan when he got back—he handed out transistors like a beneficent grandfather passing out gifts of hard candy.[39] By September of that year, Oliver
Buckley, the Labs' president, was writing to Kelly, then staying at the
Savoy Hotel in London, to say that the Labs would soon send out samples
to academic and industrial scientists. "The plan," Buckley wrote to his
deputy, "is to make a gift of two Transistors put up in a nice little box
marked as a gift from the Laboratories."[40] The invention hadn't made a
whit of difference yet in the machinery of the world, of course. From a
public relations perspective, though, the little transistor was a godsend.

By the time Jack Morton had ironed out some of the production problems with the point-contact transistor in the early 1950s, the Labs was
ready to move ahead in licensing the technology. Shockley, seemingly
immune to normal human fatigue and now without question the most
eminent solid-state physicist in the world, had already written and published a five-hundred-page book—*Electrons and Holes in Semiconductors*—
that would serve for decades as a definitive guide to scientists and engineers
working with the new materials. And in 1951 and 1952, the Labs began
sponsoring multiday conventions at Murray Hill, attended by hundreds of
scientists and engineers from around the world, who were interested in
licensing transistor rights. At the conventions, Jack Morton gave the guests
a brief overview of the transistor and Gerald Pearson followed with a brief
tutorial on transistor theory. The next two days were given to in-depth
presentations on different types of transistors and their applications. The
cost for licensing the transistor technology was $25,000. A free exception
was made for companies that wanted to use the devices for hearing aids.
This was in deference to AT&T founder Alexander Graham Bell, who had
spent much of his career working with the deaf.

"IT IS THE BEGINNING of a new era in telecommunications and no one can
have quite the vision to see how big it is," Mervin Kelly told an audience
of telephone company executives in 1951. Speaking of the transistor, he

added that "no one can predict the rate of its impact." Kelly admitted that he wouldn't see its full effect before he retired from the Labs, but that "in the time I may live, certainly in 20 years," it would transform the electronics industry and everyday life in a manner much more dramatic than the vacuum tube. The telecommunications systems of the future would be "more like the biological systems of man's brain and nervous system." The tiny transistor had reduced dimensions and power consumption "so far that we are going to get into a new economic area, particularly in switching and local transmission, and other places that we can't even envision now."[41] It seemed to be some kind of extended human network he had in mind, hazy and fantastical and technologically sophisticated, one where communications whipped about the globe effortlessly and where everyone was potentially in contact with everyone else.

There had been whispers in the electronics industry about whether Bell Labs' enthusiasm over the transistor was overblown; the reported difficulty in manufacturing the devices only added to the skepticism. Whether it was a shortcoming or an advantage, Kelly's confidence was almost certainly rooted in his early experiences. He remembered the endless days and nights constructing vacuum tubes in lower Manhattan, the countless problems in the beginning and then the stream of incremental developments that improved the tubes' performance and durability to once-unimaginable levels. He could remember, too, that as the tubes became increasingly common—in the phone system, radios, televisions, automobiles, and the like—they had come down to price levels that once seemed impossible. He had long understood that innovation was a matter of economic imperatives. As Jack Morton had said, if you hadn't sold anything you hadn't innovated, and without an affordable price you could never sell anything. So Kelly looked at the transistor and saw the past, and the past was tubes. He thereby intuited the future.

At the same time, there was one cautionary point he wanted to relate to fellow phone company executives. The transistor and other research projects at the Labs were hard work. Goals were set carefully, and then achieved by the process of experiment and calculation. "Bell Labs is no 'house of magic,'" Kelly warned, echoing the headline of a recent magazine story about the Labs that he had found repellent.[42] "There is nothing magical about science. Our research people are following a straight plan as a part of a system and there is no magic about it." People rarely disagreed with Kelly to his face. But to visitors, and sometimes to scientists, too, Bell Labs nevertheless was taking on a slightly magical air. And it was hard to deny that wholly unscientific factors—serendipity and chance, for example—played a part in the Labs' innovations. Hadn't Bardeen been out of luck in finding an office, for instance, and then just happened to camp out in Brattain's laboratory? Another example: Around the time Kelly was giving his speech to the phone company executives, a metallurgist named Bill Pfann was mulling over how to raise the purity of germanium to improve it further for transistor production. Pfann had returned to his office after lunch—"I put my feet on my desk and tilted my chair back to the window sill for a short nap, a habit then well established," he recalled. He had scarcely dozed off when he suddenly awoke with a solution. "I brought the chair down with a clack I still remember," he said.[43] Pfann envisioned passing a molten zone—a coil of metal, in effect, creating a superheated ring—along the length of a rod of germanium; as the ring moved, it would strafe the impurities out of the germanium.

Kelly would eventually tell people that Pfann's idea—it was called "zone refining," and was an ingenious adaptation of a technique metallurgists had used on other materials—ranked as one of the most important inventions of the past twenty-five years. Kelly didn't tell people it resulted from a man sleeping on the job. The process allowed the Labs' metallurgists to fabricate the purest materials in the history of the world—germanium that had perhaps one atom of impurity among 100 million atoms.[44] If that was too hard to envision, the Labs executives had a handy analogy to make it even more clear. The purity of the materials produced at Bell Labs, beginning in the early 1950s, was akin to a pinch of salt sprinkled amid a thirty-eight-car freight train carrying in its boxcars nothing else but sugar.

Seven

THE INFORMATIONIST

I t might have been said in 1948 that you either grasped the immense importance of the transistor or you did not. Usually an understanding of the device took time, since there were no tangible products—no proof—to demonstrate how it might someday alter technology or culture. But a few people could see it right away. The tiny device, its three wires peeking out, was sitting on the desk of Bill Shockley one day when a guest stopped in the midst of their conversation and asked what it was. "It's a solid-state amplifier," Shockley told his visitor. It worked like a vacuum tube, he added.[1] The visitor, a rail-thin mathematician in his early thirties who was known at Bell Labs as something of a loner, listened closely. He had a gaunt face and clear gray eyes; often he gave others the impression, always unspoken, that he was amused by whatever was being said. Very few things impressed him. But he would later explain that he immediately saw the import of what Shockley was saying. It mattered little that he was looking at the transistor in its earliest incarnation, before it was manufactured, before it even had a name.

At that point in time, employees at Bell Labs were informed of the secret device on a need-to-know basis. They were taken aside and briefed on the matter only if they were working with the metallurgical team on the purification of germanium, for instance, or if they had been drafted to work on the transistor's development and mass production with Jack Morton. But Shockley, who was notorious for the speed with which he judged colleagues as his intellectual inferiors, believed that his guest that day, Claude Elwood Shannon, was exceptional, a scientist vital to the Labs' reputation as an intellectual vanguard. Shannon deserved to know what the solid-state team had done. Almost anyone who spent time with the quiet, courteous Shannon, going back at least a decade, seemed to walk away with a similar impression. He was known to be retiring and eccentric. But above all, he was known to be special. "A decidedly un-conventional type of youngster," Shannon's advisor at MIT, the engineer-ing dean Vannevar Bush, described his young student a decade earlier.[2] "He is shy, personally likable, and a man who should be handled with great care."[3] Bush's assessment might have raised some questions—namely, why handle Shannon with great care? His thinness (five foot ten and 135 pounds) notwithstanding, it wasn't a question of physical fragil-ity: Shannon was athletic and energetic, never more so than when he was setting up machinery or ripping apart old electronic equipment to salvage parts for some kind of contraption he was building. Rather, it seemed to Bush and a handful of mathematicians who encountered Shannon in the late 1930s that he mightn't be just another exceedingly bright graduate student. He was something else entirely. One professor at MIT, informed in the late 1930s that young Shannon was taking piloting lessons, con-sidered intervening so the scientific community wouldn't risk losing him prematurely in an air crash.[4] There was, in other words, a quiet accord among the professors at MIT: People like Shannon come along so rarely that they must be protected.

At the University of Michigan and at MIT he had studied both math-ematics and electrical engineering, and it wasn't easy to say precisely where his genius resided. Some things about him actually suggested little in the way of conventional brilliance. When confronted with ordinary number problems—18 × 27, for instance—Shannon would work them out not in his head but on a blackboard.[5] He wasn't much for details;

sometimes he would solve problems in a way that showed surprising intuition but a mathematical approach that some colleagues found unsatisfactory or lacking in rigor. Above all, he almost seemed more interested in doing work with his hands than with his mind. He'd originally come east from his home state of Michigan because he had found a job listing by chance at the University of Michigan, where he was finishing his undergraduate degree. "There was this little postcard on the wall," Shannon later recalled, "saying that M.I.T. was looking for somebody to run the differential analyzer, a machine which Vannevar Bush had built to solve differential equations."[6] He applied for the job and got it.

The analyzer was an early "analog" computer that took up an entire room and required a crew of several operators. Yet it was a great leap ahead of any previous calculating machine; it could solve complex mathematical problems with revolutionary speed. The machine had a circuit of electronic switches that controlled sets of rods, pulleys, gears, and spinning disks, which assistants like Shannon had to constantly fiddle with. In a sense Shannon was a computer programmer: He would adjust the machine's rods and gears to correspond with the values in a numerical problem. The analyzer, set in motion, would then spit out the answer to equations not through a screen or a printout but with a mechanical pen on a sheet of graph paper.

At MIT, Shannon fell in love with his machine. And as he began working with the analyzer, he became especially intrigued by the electromechanical relays in its control circuit. These were magnetic switches that clicked open or closed when a current was applied or cut off. The open or closed position of the relays could stand in for a yes or no answer to a question. Or a string of relays could branch out in one logical direction or another, whereby the positions (open or closed) each stood for "AND" or "OR." One could thereby answer a complicated problem or execute a complicated set of commands. Shannon began to perceive a new way to think about the design and function of such circuits. He saw that one could make sense of them through an obscure branch of mathematics based on 0s and 1s—what was known as Boolean algebra

In the summer of 1937, Shannon left Cambridge for a few months to work at the Bell Labs office on West Street, where he continued to think about how relays, switching, and circuits fit into this notion. His choice of summer jobs was fortuitous: At the time, there was no place in the world better suited for studying electrical relays, which formed the switching backbone of the entire Bell System. With Vannevar Bush's endorsement, Shannon wrote up his insights upon his return to MIT. "I believed it was a classic, a comment which I very seldom make," Bush said of Shannon's thesis. But such praise, while unusual for Bush, soon seemed modest in comparison to the wider reception of Shannon's work. His paper demonstrated that designing logic circuits for a computer could be an efficient mathematical endeavor rather than a painstaking art. In 1939, the work won him a distinguished prize from an engineering society. "I was so surprised and pleased to receive the letter announcing the award," he wrote to Bush, "that I nearly fainted!"[7]

Like so many of his future colleagues at Bell Labs, Shannon had grown up in the Midwest—in tiny Gaylord, Michigan, population 3,000, in the state's northern tip, where by Shannon's own account it was "small enough that if you walked a couple of blocks, you'd be in the countryside."[8] Some of the buildings in Gaylord's modest downtown were built and owned by Shannon's father, a businessman and probate court judge; his mother was a principal of the town's high school. Gaylord's closest big cities, Grand Rapids and Detroit, were more than 150 miles to the south. Shannon's small-town innocence—*I nearly fainted!*—was unquestionably authentic. But so, too, was his lack of professional direction. As word spread, Shannon's slender and highly mathematical paper, about twenty-five pages in all, would ultimately become known as the most influential master's thesis in history.[9] In time, it would influence the design of computers that were just coming into existence as well as those that wouldn't be built for at least another generation. But this was off in a far distant future. Still only twenty-three years old, and not at all certain what to do with himself, the young man wrote to Vannevar Bush to ask what he should work on next.

VANNEVAR BUSH WAS just then moving from Cambridge to Washington to assume the presidency of the Carnegie Institution—at the time the premier private endowment in the United States for funding scientific research. Soon Bush would also begin lobbying President Franklin D. Roosevelt, successfully, to take charge of the United States' immense research and development efforts for World War II, effectively making him the most powerful scientist in the country. His stature happened to match his own healthy self-regard. Bush delighted in connecting students and friends to one another within his large social and professional web. What's more, inquiries such as Shannon's were useful in satisfying Bush's broad scientific curiosity. If Bush was interested in genetics, for instance, as he told Shannon he was, then Shannon could be a proxy: The young man might consider delving into the subject of human genes, and find a way to apply his mathematical skills to an analysis. There happened to be a laboratory in Cold Spring Harbor, New York—an affiliate of the Carnegie Institution, no less—where Bush suggested Shannon might work on his PhD research. Shannon agreed. "I had a very enjoyable summer working on my genetic algebra under Dr. Burks at Cold Spring Harbor and want to thank you for making it possible," he wrote to Bush a few months after. "The work came along very well and has been accepted here at Tech as a Ph.D. Thesis."[10]

This burst of professional success coincided with a change in his personal life. He had made overtures to an undergraduate from Radcliffe named Norma Levor, whom he'd met at a party in the MIT dorms. "He stood on the doorstep of his room and the living room," she recalls. "He didn't come in. Kind of shy. Didn't want to get into that. I threw popcorn on him. And he said, do you want to hear some great music?" A devoted clarinet player, Shannon had a large collection in his room of jazz and Dixieland records. He and Norma quickly fell in together. At Shannon's suggestion, they made love one evening—"he wooed me," Norma says—in the differential analyzer room, to which Shannon had a key. She thought that Claude, so thin, so angular, looked "Christ-like." They married in January 1940.[11]

That summer, Shannon took a temporary job at Bell Labs and the two moved to an apartment on Bank Street in Greenwich Village. "I am not at all sure that that sort of work would appeal to me," he had worried to Bush, "for there is bound to be some restraint in an industrial organization as to [the] type of research pursued."[12] But he found himself pleased with the arrangement—from his apartment he could walk to the Labs' offices at 463 West Street in the morning and to the local jazz clubs every night. Working in the mathematical research department, moreover, turned out far more pleasant than he imagined. Mostly he was considering the design of relay circuits, which related directly to the work he had done at MIT.

Those who met Shannon at the time usually came away with an impression similar to Vannevar Bush's: The young man was diffident, amiable, whip-smart. To those who knew him better, he could also be whimsical and prankish, with an insatiable love for games and gadgets. To those such as Norma who knew him better still he could be cold and remote, vanishing with some frequency into depressive sulks or abstraction. He had been exuberant when they first met—taking her up in the Piper Cub he was piloting, "scaring me shitless with his flying," and enthusiastically dragging her out to jazz shows in the Village and watching with rapt attention while barely touching his drink. Now he would sometimes sit in a chair for hours, a pack of cigarettes for company, without writing or speaking. In other instances—when he got into a minor accident in the couple's new Buick, a gift from Norma's father—he would break down in tears.

Rarely would Shannon talk to anyone about the things he was working through in his mind. In early 1939, he had hinted to Vannevar Bush in a letter that he had begun thinking about communications and the methods by which "intelligence" moves from place to place. In a sense, he'd been thinking about these things for most of his young life. As a boy he'd had a job delivering newspapers, he'd also delivered telegrams for Western Union, and he'd even set up a private telegraph line to a friend's house using a wire strung along a fence. Some years later as a student at the University of Michigan, Shannon had read a paper by a Bell Laboratories engineer named Ralph Hartley entitled "Transmission of Information" that made an enormous impression on him. Hartley had proposed ways to measure and think about the rate and flow of information from sender to receiver. Perhaps there were deeper and more fundamental properties, Shannon now wondered, that were common to all the different kinds of media—telephony, radio, television, telegraphy included.[13] In his letter to Bush, he hadn't gone far beyond what Hartley had put forward years earlier, but he hinted that he might, under the right circumstance and given some time, be able to work out some kind of overarching theory about messages and communications.

Nothing seemed to come of this during his summer at Bell Labs, and nothing seemed to come of it just after, as Shannon took a fellowship at the Institute for Advanced Study in Princeton, New Jersey, where Albert Einstein was in residence. "I poured tea for him," Norma recalls of Einstein, "and he told me I was married to a brilliant, brilliant man."

Shannon then returned to New York—accepting an offer to officially join the mathematics department at Bell Labs. "I could smell the war coming along," he would later recall, "and it seemed to me I would be safer working full-time for the war effort, safer against the draft, which I didn't exactly fancy. . . . I was trying to play the game, to the best of my ability. But not only that, I thought I'd probably contribute a hell of a lot more."[14] Like the other mathematicians and engineers at Bell Labs, Shannon spent all of his daytime hours working on communications technologies for the war. He and Norma had separated just prior to his return to New York, an event he found shattering. She considered him too depressed and difficult. During their year at Princeton, Norma recounts, "I tried to get him to go to an analyst but he wouldn't. He just got mad. He didn't want to leave the apartment. He just got grimmer and grimmer. I began to feel that unless he could do something about it, I couldn't stay with him."

During the war years, Shannon's colleagues at the Labs heard little, if anything, from him about communications, intelligence, and messages.

Now a bachelor again and living alone in a spare one-bedroom apartment on West 11th Street in Greenwich Village, Shannon would come home from work and labor over a personal project, writing equations and explanations in pencil, on lined sheets of paper, in a neat, compact script.[15] There was little or nothing in the icebox.[16] When he wanted to take a break he would put down his pencil, pick up his clarinet, and play.

THE MATH DEPARTMENT at Bell Laboratories had grown up around a single man, Thornton Fry, the son of a poor Ohio carpenter who was working on a PhD at the University of Wisconsin when Harold Arnold came through town on a recruiting trip in 1916. It was still a few years before the creation of Bell Labs, and Arnold, Western Electric's research chief, was looking for a young mathematician who could assist the engineers with the complex theory that often accompanied their switching and transmission plans. In a job interview, Arnold asked Fry a number of questions to test his knowledge about the era's most influential communications engineers. Did the young man know the work of Heaviside, Campbell, or Molina?[17] Fry shook his head. He didn't know a single one. Arnold must have nonetheless seen something encouraging in Fry. He gave him a job offer—$36 a week—and Fry immediately accepted.

Fry moved from Western Electric to the newly incorporated Bell Telephone Labs in 1925. Around 1930, he later recalled, the math group was expanded into a formal department staffed by several mathematics scholars.[18] At least as it was originally conceived, Fry's department was not supposed to do research; it was meant to be a consulting organization to the engineers, physicists, and chemists at the Labs who needed help. "At that time engineers of all types were pathetically ignorant of mathematics," Fry maintained, "so that anybody who could compute or quote a theory—even if he quoted it wrong—was admired by them. A mathematician was something like a nun; he was automatically admirable. He was different than other people."[19]

As the math department evolved and expanded, it became more than a mere consultancy. Many of its researchers became deeply involved in the conceptualization of new circuits; others became interested in applying statistical tools to quality control issues—mostly at Western Electric, which churned out phones and equipment at its factories—as well as to congestion problems within the company's vast switching system. More eccentric projects were sometimes deemed acceptable, too. George Stibitz, a mathematician under Fry whose work paralleled some of the ideas in Claude Shannon's master's thesis, decided to use telephone relays—those movable metal switches—in 1937 to build what appeared to be the world's first digital computer. (Unlike Vannevar Bush's differential analyzer, it did simple mathematical calculations using Boolean algebra and no gears or rods; the answer to a problem was displayed through light bulbs.) The job of a Bell Labs mathematician under Fry could thus vary tremendously. He could pursue his own interests without managerial interference. Or he (or she, since there were several women in the department now, too) could be asked to consider a question arising in another area, anything from a theoretical physicist's dilemma to a problem related to a device that the Labs was readying for a Western Electric assembly line. As Henry Pollak, a former director of the math department, recalls, it "was full of people who had a major area where they worked, just like I did, too. But they couldn't turn a good problem down. If one came by you dropped what you were doing and had fun with it. Our job was to stick our nose into everybody's business."[20]

By the early 1940s, when Fry asked Shannon to join his department, many Bell mathematicians were focused on the wartime problem known as fire control. Their challenge was to work out the complex mathematics behind the automatic firing of large guns that were trying to protect against enemy attacks—essentially, to create primitive computers that would gather information, largely through radar scans, on the location, speed, and trajectory of an incoming German rocket or plane. Their computers would immediately calculate the future position of the rocket or plane so that it could be intercepted and exploded in midair by shells or bullets. It took years to make this system workable, but it ultimately changed modern warfare. A defining moment came in 1944 during the defense of Great Britain against Hitler's V-1 rockets, known as "buzz bombs." Along the

English Channel, one Bell System historian noted, these gun directors intercepted 90 percent of the V-1 rockets aimed at London.[21]

Shannon contributed some vital ideas to the fire control effort. Eventually, though, he drifted toward several committees that were working on secret methods of communication, which was far more to his liking. Since childhood, he had been as interested in games as in mathematics; in some respects he still saw little difference between the two. "I was a great fan of Edgar Allan Poe's 'The Gold Bug' and stories like that," he later recalled to an interviewer.[22] The Poe story, written in 1843, revolves around the solution to a short cryptogram, whereby numbers and symbols—substitution ciphers, as cryptographers call them—are translated to individual letters. These letters in turn revealed secret coordinates near a huge tulip tree ("forty-one degrees and thirteen minutes northeast and by north") and, ultimately, a pirate's chest, once belonging to Captain Kidd, full of gold, diamonds, and rubies. Shannon often worked on similar cryptograms growing up, and his fascination with puzzles showed no perceptible ebb as he got older. At one point at Bell Labs, when Shannon read about a $10,000 crossword puzzle contest in the newspaper, he disappeared for a couple of weeks to the Labs library.[23] Thus to recruit Shannon into the wartime work on codes and codebreaking, in other words, was to pay him to work on his favorite hobby.

Shannon summarized his war work on secret communications in a 114-page opus, "A Mathematical Theory of Cryptography," which he finished in 1945. The paper was immediately deemed classified and too sensitive for publication, but those who read it found a long treatise exploring the histories and methodologies of various secrecy systems. Moreover, he had offered a persuasive analysis of which methods might be unbreakable (what he called "ideal") and which cryptographic systems might be most practical if an unbreakable system were deemed too complex or unwieldy. His mathematical proofs presented the few people cleared to read it with a number of useful insights and an essential observation that language, especially the English language, was filled with redundancy and predictability. Indeed, he later calculated that English was about 75 to 80 percent redundant. This had ramifications for cryptography: The less redundancy you have in a message, the harder it is to crack its code. And this also, by extension, had implications for how you might send a message more efficiently.[24] Shannon would often demonstrate that the sender of a message could strafe its vowels and still make it intelligible. To illustrate Shannon's point, David Kahn, a historian of cryptography who wrote extensively on Shannon, used the following example:

FCTSSTRNGRTHNFCTN

To transmit the message *fact is stranger than fiction* one could send fewer letters. You could, in other words, *compress* it without subtracting any of its content. Shannon suggested, moreover, that it wasn't only individual letters or symbols that were sometimes redundant. Sometimes you could take entire words out of a sentence without altering its meaning.

The completion of the cryptography paper coincided with the end of the war. But Shannon's personal project—the one he had been laboring on at home in the evenings—was largely worked out a year or two before that. Its subject was the general nature of communications. "There is this close connection," Shannon later said of the link between sending an encoded message and an uncoded one. "They are very similar things, in one case trying to conceal information, and in the other case trying to transmit it." In the secrecy paper, he referred briefly to something he called "information theory." This was a bit of a coded message in itself, for he offered no indication of what this theory might say.

ALL WRITTEN AND SPOKEN EXCHANGES, to some degree, depend on code— the symbolic letters on the page, or the sounds of consonants and vowels that are transmitted (encoded) by our voices and received (decoded) by our ears and minds. With each passing decade, modern technology has tended to push everyday written and spoken exchanges ever deeper into the realm of ciphers, symbols, and electronically enhanced puzzles of representation. Spoken language has yielded to written language, printed on a press; written language, in time, has yielded to transmitted language, sent over the air by radio waves or through a metal cable strung on poles. First came telegraph messages—which contained dots and dashes (or what might have just as well been the 1s and 0s of Boolean algebra) that were translated back into English upon reception. Then came phone calls, which were transformed during transmission—changing voices into electrical waves that represented sound pressure and then interleaving those waves in a cable or microwave transmission. At the receiving end, the interleaved messages were pulled apart—decoded, in a sense—by quartz filters and then relayed to the proper recipients.

In the mid-1940s Bell Labs began thinking about how to implement a new and more efficient method for carrying phone calls. PCM—short for pulse code modulation—was a theory that was not invented at the Labs but was perfected there, in part with Shannon's help and that of his good friend Barney Oliver, an extraordinarily able Bell Labs engineer who would later go on to run the research labs at Hewlett-Packard. Oliver would eventually become one of the driving forces behind the invention of the personal calculator.[25] Shannon and Oliver had become familiar with PCM during World War II, when Labs engineers helped create secret communication channels between the United States and Britain by using the technology. Phone signals moved via electrical waves. But PCM took these waves (or "waveforms," as Bell engineers called them) and "sampled" them at various points as they moved up and down. The samples—8,000 per second—could then be translated into on/off pulses, or the equivalent of 1s and 0s. With PCM, instead of sending waves along phone channels, one could send information that described the numerical coordinates of the waves. In effect what was being sent was a code. Sophisticated machines at a receiving station could then translate these pulses describing the numerical coordinates back into electrical waves, which would in turn (at a telephone) become voices again without any significant loss of fidelity. The reasons for PCM, if not its methods, were straightforward. It was believed that transmission quality could be better preserved, especially over long distances that required sending signals through many repeater stations, by using a digital code rather than an analog wave.[26] Indeed, PCM suggested that telephone engineers could create a potentially indestructible format that could be periodically (and perfectly) regenerated as it moved over vast distances.

Shannon wasn't interested in helping with the complex implementation of PCM—that was a job for the development engineers at Bell Labs, and would end up taking them more than a decade. "I am very seldom interested in applications," he later said. "I am more interested in the elegance of a problem. Is it a good problem, an interesting problem?"[27] For him, PCM was a catalyst for a more general theory about how messages move—or in the future *could* move—from one place to another. What he'd been working on at home during the early 1940s had become a long, elegant manuscript by 1947, and one day soon after the press conference in lower Manhattan unveiling the invention of the transistor, in July 1948, the first part of Shannon's manuscript was published as a paper in the *Bell System Technical Journal*; a second installment appeared in the *Journal* that October.[28] "A Mathematical Theory of Communication"— "the magna carta of the information age," as *Scientific American* later called it—wasn't about one particular thing, but rather about general rules and unifying ideas. "He was always searching for deep and fundamental relations," Shannon's colleague Brock McMillan explains. And here he had found them. One of his paper's underlying tenets, Shannon would later say, "is that information can be treated very much like a physical quantity, such as mass or energy."[29] To consider it on a more practical level, however, one might say that Shannon had laid out the essential answers to a question that had bedeviled Bell engineers from the beginning: How rapidly, and how accurately, can you send messages from one place to another?

"The fundamental problem of communication," Shannon's paper explained, "is that of reproducing at one point either exactly or approximately a message selected at another point." Perhaps that seemed obvious, but Shannon went on to show why it was profound. If "universal connectivity" remained the goal at Bell Labs—if indeed the telecommunications systems of the future, as Kelly saw it, would be "more like the biological systems of man's brain and nervous system"—then the realization of those dreams didn't only depend on the hardware of new technologies, such as the transistor. A mathematical guide for the system's engineers, a blueprint for how to move data around with optimal efficiency, which was what Shannon offered, would be crucial, too. Shannon maintained that all communications systems could be thought of in the same way, regardless of whether they involved a lunchroom conversation, a postmarked letter, a phone call, or a radio or telephone transmission. Messages all followed the same fairly simple pattern:

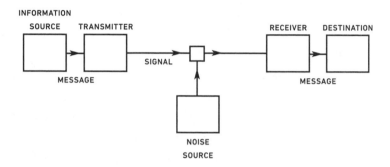

All messages, as they traveled from the information source to the destination, faced the problem of noise. This could be the background clatter of a cafeteria, or it could be static (on the radio) or snow (on television). Noise interfered with the accurate delivery of the message. And every channel that carried a message was, to some extent, a *noisy channel.*

To a non-engineer, Shannon's drawing seemed sensible but didn't necessarily explain anything. His larger point, however, as he proved in his mathematical proofs, was that there were ways to make sure messages got where they were supposed to, clearly and reliably so. The first place to start, Shannon suggested, was to think about the *information* within a message. The semantic aspects of communication were irrelevant to the engineering problem, he wrote. Or to say it another way: One shouldn't necessarily think of information in terms of *meaning.* Rather, one might think of it in terms of its ability to resolve uncertainty. Information provided a recipient with something that was not previously known, was not predictable, was not redundant. "We take the essence of information as the irreducible, fundamental underlying uncertainty that is removed by its receipt," a Bell Labs executive named Bob Lucky explained some years later.[30] If you send a message, you are merely choosing from a range of possible messages. The less the recipient knows about what part of the message comes next, the more information you are sending. Some language choices, Shannon's research suggested, occur with certain probabilities, and some messages are more frequent than others; this fact could therefore lead to precise calculations of how much information these words or messages contained.[31] (Shannon's favorite example was to explain that one might need to know that the word "quality" begins with q, for instance, but not that a u follows after. The u gives a recipient no information if they already have the q, since u always follows q; it can be filled in by the recipient.)

Shannon suggested it was most useful to calculate a message's information content and rate in a term that he suggested engineers call "bits"— a word that had never before appeared in print with this meaning. Shannon had borrowed it from his Bell Labs math colleague John Tukey as an abbreviation of "binary digits." The bit, Shannon explained, "corresponds to the information produced when a choice is made from two equally likely possibilities. If I toss a coin and tell you that it came down heads, I am giving you one bit of information about this event." All of this could be summed up in a few points that might seem unsurprising to those living in the twenty-first century but were in fact startling—"a bolt from the blue," as one of Shannon's colleagues put it[32]—to those just getting over the Second World War: (1) All communications could be thought of in terms of information; (2) all information could be measured in bits; (3) all the measurable bits of information could be thought of, and indeed should be thought of, *digitally*. This could mean dots or dashes, heads or tails, or the on/off pulses that comprised PCM. Or it could simply be a string of, say, five or six 1s and 0s, each grouping of numerical bits representing a letter or punctuation mark in the alphabet. For instance, in the American Standard Code for Information Interchange (ASCII), which was worked out several years after Shannon's theory, the binary representation for FACT IS STRANGER THAN FIC-TION would be as follows:

0100011001000001010000110101010000100000001001001010
1001100100000010100110101010001010010010000010100111
0010001110100010101010010001000000010101000100100001
0000010100111000100000010001100100100101000011010010
100010010010100111101001110

Thus Shannon was suggesting that all information, at least from the view of someone trying to move it from place to place, was the same, whether it was a phone call or a microwave transmission or a television signal.

This was a philosophical argument, in many respects, and one that would only infiltrate the thinking of the country's technologists slowly over the next few decades. To the engineers at the Labs, the practical mathematical arguments Shannon was also laying out made a more immediate impression. His calculations showed that the information content of a message could not exceed the capacity of the channel through which you were sending it. Much in the same way a pipe could only carry so many gallons of water per second and no more, a transmission channel could only carry so many bits of information at a certain rate and no more. Anything beyond that would reduce the quality of your transmission. The upshot was that by measuring the information capacity of your channel and by measuring the information content of your message you could know how fast, and how well, you could send your message. Engineers could now try to align the two—capacity and information content. For anyone who actually designed communications systems with wires or cables or microwave transmitters, Shannon had handed them not only an idea, but a new kind of yardstick.

SHANNON'S PAPER contained a claim so surprising that it seemed impossible to many at the time, and yet it would soon be proven true. He showed that any digital message could be sent with virtual perfection, even along the noisiest wire, as long as you included error-correcting codes—essentially extra bits of information, formulated as additional 1s and 0s—with the original message. In his earlier paper on cryptography, Shannon had already shown that by reducing redundancy you could compress a message to transmit its content more efficiently. Now he was also demonstrating something like the opposite: that in some situations you could increase the redundancy of a message to transmit it more accurately. One might think of the simplest case: Sending the message FCTSSTRNGRTHNFCTN FCTSSTRNGRTHNFCTN FCTSSTRNGRTHNFCTN—an intentional tripling—to improve the chances that the message was clearly received amid a noisy channel. But Shannon showed that there were myriad and far more elegant ways to fashion these kinds of self-correcting codes. Some mathematicians at the Labs would spend their careers working out the details, which Shannon was largely uninterested in doing.[33]

All the error-correcting codes were meant to ensure that the information on the receiving end was exactly or very close to the message that left the transmission end. Even fifty years later, this idea would leave many engineers slack-jawed. "To make the chance of error as small as you wish?" Robert Fano, a friend and colleague of Shannon's, later pointed out. "How he got that insight, how he even came to believe such a thing, I don't know." All modern communications engineering, from cell phone transmissions to compact discs and deep space communications, is based upon this insight.[34]

For the time being, knowledge of Shannon's work largely remained confined to the mathematics and engineering community. But in 1949, a year after the communications paper was published, it was published in book form for a wider audience. Shortly thereafter communication theory became more commonly known as information theory. And eventually it would be recognized as the astounding achievement it was. Shannon's Bell Labs colleagues came to describe it as "one of the great intellectual achievements of the twentieth century"; some years later, Shannon's disciple Bob Lucky wrote, "I know of no greater work of genius in the annals of technological thought."[35] A few observers, in time, would place the work on a par with Einstein's. Shannon himself refused to see the theory in such grand terms; he even urged a magazine journalist, in a genial prepublication letter in the early 1950s, to strip any comparison of his theory to Einstein's ideas on relativity in a story: "Much as I wish it were," he lamented, "communication theory is certainly not in the same league with relativity and quantum mechanics."[36]

In any case, Shannon was ready to move on. He later quipped that at this point in time he was becoming highly interested in "useless" things—games, robots, and small machines he called *automata*. To some observers, Shannon's new interests seemed to suggest that rather than advance his extraordinary insights on communications, he preferred to squander his genius on silly gadgetry or pointless questions, such as whether playing tic-tac-toe on an infinite sheet of graph paper would reveal patterns about tie games. Perhaps this was missing a point. Some of Shannon's more curious pursuits were far less useless than they appeared, but he was growing increasingly reluctant to explain himself. People could see what he was doing, then they could think whatever they wanted.

AFTER THE PUBLICATION of information theory, no one could dare tell Shannon what he should work on. Often he went into his office just like any other Bell Labs employee. But beginning in the early 1950s sometimes he didn't go in at all, or he wandered in late and played chess the rest of the day, much to the consternation, and sometimes the envy, of his colleagues.[37] When he was in the office at Murray Hill he would often work with his door closed, something virtually unheard of at Bell Labs. And yet that breach, too, was permissible for Claude Shannon. "He couldn't have been in any other department successfully," Brock McMillan recalls. "But then, there weren't many other departments where people just sat and thought."

Increasingly, Shannon began to arouse interest not only for the quality of his ideas but for his eccentricity—"for being a very odd man in so many ways," as his math department colleague David Slepian puts it. "He was not an unfriendly person," Slepian adds, "and he was very modest," but those who knew him and extended a hand of friendship realized that he would inevitably fail to reciprocate. They had to seek him out.

And in calling on him (knocking on his door, writing, visiting) one had to penetrate his shyness or elusiveness or—in the case of expecting a reply to a letter—his intractable habits of procrastination and his unwillingness to do anything that bored him.

Shortly after the information paper was published, he did make one significant social connection, marrying a young woman named Betty Moore who worked in the Labs' computational department. Sometimes he and Betty would go to Labs functions—dinner at Jim Fisk's house, where Bill Shockley and his wife would also be guests. But mainly the math department, a social group both in and outside of the office, would extend invitations to Shannon and he would politely refuse.

In the 1940s and 1950s, the members of Bell Labs' math department liked to play a game after lunch. "I don't know who invented this," Brock McMillan recalls, "but it was called 'Convergence in Webster.' Someone was supposed to compose in his head a four-word sentence. And people would try to guess letters and words." As the men would guess, the creator of the mystery sentence would stand before a blackboard filling in blank spaces (as in Hangman) and telling them whether their guesses fell alphabetically before or after his words. Thus they would gradually converge on the right words, as one might home in on a word he was seeking in *Webster's Dictionary*. Fifty years later, the men could still recall their favorites. One time a supervisor dropped in, offered to take a turn as the leader, and drew blanks on the blackboard for what eventually turned out to be *You Are All Fired*.[38] The mathematicians thought it hilarious.

Shannon was in this world but not of it. Just as he had stood in the doorway when he met Norma at the college party—the one where she had thrown popcorn at him to get his attention, but rather than enter her life he had brought her back to his—he would wander by his colleagues' offices some afternoons after lunch, and if he saw "Convergence in Webster" in progress, he would lean against the doorjamb and watch.

In a math department that thrived on its collective intelligence—where members of the staff were encouraged to work on papers together rather than alone—this set him apart. But in some respects his solitude was interesting, too, for it had become a matter of some consideration at the Labs whether the key to invention was a matter of individual genius or collaboration. To those trying to routinize the process of innovation—the lifelong goal of Mervin Kelly, the Labs' leader—there was evidence both for and against the primacy of the group. So many of the wartime and postwar breakthroughs—the Manhattan Project, radar, the transistor—were clearly group efforts, a compilation of the ideas and inventions of individuals bound together with common purposes and complementary talents. And the phone system, with its almost unfathomable complexity, was by definition a group effort. It was also the case, as Shockley would later point out, that by the middle of the twentieth century the process of innovation in electronics had progressed to the point that a vast amount of multidisciplinary expertise was needed to bring any given project to fruition. "Things are much more complex than they were probably when Mendel was breeding peas, in which case you would put them in a pot and collect the fruits, and then cover up the blossoms and have that suffice," Shockley said, referring to the nineteenth-century scientist whose work provided the foundation for modern genetics. An effective solid-state group, for example, required researchers with material processing skills, chemical skills, electrical measurement skills, theoretical physics skills, and so forth. It was exceedingly unlikely to find all those talents in a single person.[39]

And yet Kelly would say at one point, "With all the needed emphasis on leadership, organization and teamwork, the individual has remained supreme—of paramount importance. It is in the mind of a single person that creative ideas and concepts are born."[40] There was an essential truth to this, too—John Bardeen suddenly suggesting to the solid-state group that they should consider working on the hard-to-penetrate surface states on semiconductors, for instance. Or Shockley, mad with envy, sitting in his Chicago hotel room and laying the groundwork for the junction transistor. Or Bill Pfann, who took a nap after lunch and awoke, as if from an edifying dream, with a new method for purifying germanium.

Of course, these two philosophies—that individuals as well as groups were necessary for innovation—weren't mutually exclusive. It was the individual from which all ideas originated, and the group (or the multiple groups) to which the ideas, and eventually the innovation responsibilities, were transferred. The handoffs often proceeded in logical progression: from the scientist pursuing with his colleagues a basic understanding in the research department, to the applied scientist working with a team in the development side, to a group of engineers working on production at Western Electric. What's more, in the right environment, a group or wise colleague could provoke an individual toward an insight, too. In the midst of Shannon's career, some lawyers in the patent department at Bell Labs decided to study whether there was an organizing principle that could explain why certain individuals at the Labs were more productive than others. They discerned only one common thread: Workers with the most patents often shared lunch or breakfast with a Bell Labs electrical engineer named Harry Nyquist. It wasn't the case that Nyquist gave them specific ideas. Rather, as one scientist recalled, "he drew people out, got them thinking." More than anything, Nyquist asked good questions.[41]

Shannon knew Nyquist, too. And though Shannon worked alone, he would later tell an interviewer that the institution of Bell Labs (its intellectual environment, its people, its freedom, and, most important, the Bell System's myriad technical challenges) deserved a fair amount of credit for his information theory.[42]

What the Labs couldn't take credit for, though, was how Shannon, whether by provocation or intuition, seemed to anticipate a different era altogether. His genius was roughly equivalent with prescience. There was little doubt, even by the transistor's inventors, that if Shockley's team at Bell Labs had not gotten to the transistor first, someone else in the United States or in Europe would have soon after. A couple of years, at most.[43] With Shannon's startling ideas on information, it was one of the rare moments in history, an academic would later point out, "where somebody founded a field, stated all the major results, and proved most of them all pretty much at once."[44] Eventually, mathematicians would debate not whether Shannon was ahead of his contemporaries. They would debate whether he was twenty, or thirty, or fifty years ahead.

Eight

MAN AND MACHINE

Early in his Bell Labs career, Shannon had begun to conceive of his employer's system—especially its vast arrangement of relays and switches that automatically connected callers—as more than a communications network. He saw it as an immense computer that was transforming and organizing society. This was not yet a conventional view, though it was one that Shockley, too, would soon adopt. As Shannon put it, the system and its automatic switching mechanisms was "a really beautiful example of a highly complex machine. This is in many ways the most complex machine that man has ever attempted, and in many ways also a most reliable one." He was also intrigued by the fact that the phone system was built to be efficient and tremendously broad in its sweep but was not built to think in depth. In connecting callers, it did innumerable simple tasks over and over and over again. But he knew that other machines could be built for a contrary purpose, to be deep rather than broad, and he began thinking about how to do so. Soon after Shannon's information theory was published, he started working on a computer program and a scientific paper for playing chess. "He was a very good chess player," Shannon's colleague Brock McMillan recalls. "He clobbered all the rest of us." But there seemed little point in studying chess at an industrial lab for communications. While preparing for a radio interview at around that time in Morristown, New Jersey—the town where Shannon and Betty had just moved to an apartment—he scribbled down some responses to the questioner:

Q: *You do other things besides analyze communications, though, don't you, Dr. Shannon? Didn't I see an article in* Scientific American *by you about a chess-playing machine? Don't tell me the Laboratories is interested in chess.*

    Shannon: No—but interested in potentiality of computer machines. They can solve complex problems in a few minutes.

Q: *But how does chess fit into the picture?*

    Shannon: A test to see what we can do with computing machines.

Q: *Would a chess-playing machine of the type you are talking about play perfect chess?*

    Shannon: No, make same kind of errors as human players. . . .

Q: *Would you say that these computing machines are capable of "thinking"?*

    Shannon: Depends on how you define "thinking"—memory, decisions, but must all be programmed into the machine.

Q: *What sort of things will the Telephone Company use the computers for?*

    Shannon: The telephone exchange itself is a type of computer.[1]

At the time, Shannon hadn't yet built a machine; computers of the era, even the simplest ones, were bulky and complex, and so he admitted it would be too expensive to use a computer for "so trivial a problem" as chess. But it's unlikely he actually believed the problem to be trivial, and not long afterward he did in fact build a primitive chess computer. Shannon's first paper on the subject—the one from which *Scientific American* adapted an article—happened to be the first paper ever written on chess programming. Much like his work on cryptography and information, it combined philosophical and mathematical elements, exploring the purpose of a chess machine as well as the logical theory behind its possible mechanisms. It also contained something unusual: an explanation that was meant to clarify why computer chess, a radical notion in 1949, could prove useful. "It is hoped that a satisfactory solution of this problem will act as a wedge in attacking other problems of a similar nature and of greater significance."[2] If you could get a computer to play chess, in other words, you could conceivably get it to route phone calls, or translate a language, or make strategic decisions in military situations. You could build "machines capable of orchestrating a melody," he suggested. And you might be able to construct "machines capable of logical deduction." Such machines could be useful as well as economical, he offered; they could ultimately replace humans in certain automated tasks.

Almost surely, these justifications for the chess program would have placated his bosses at the Labs, worried as they were that some of their researchers could stray too far from the subject of communications. Regulatory anxiety created a background hum for the top managers at Bell Labs: What if government officials became alert to the fact that the phone monopoly was using its customers' dollars to fund the research of a game-playing eccentric? Shannon's offering—a few crumbs of justification—were about as much as he would give. He would acknowledge that building devices like chess-playing machines "might seem a ridiculous waste of time and money. But I think the history of science has shown that valuable consequences often proliferate from simple curiosity."[3] "He never argued his ideas," Brock McMillan says of Shannon. "If people didn't believe in them, he ignored those people."

"WHAT DO YOU GIVE A GUY like Claude for Christmas?" his wife, Betty, asks. "He liked Erector Sets, and so I bought the biggest Erector Set I could find and it was 50 bucks."[4] At first, Shannon began to stay up all night with the Erector Set building small machines. "He built a little turtle that walked around the house," Betty says. "It would bump into something and back off and walk the other way. And then he built the mouse." The mouse was named Theseus. It was a small object, built of wood with copper wire whiskers, which was intended to search for a piece of electronic "cheese" within a maze that Shannon also built. Theseus was named in winking recognition of the mythical Greek hero who found a way to navigate a labyrinth ruled by a deadly minotaur.

Shannon did the project at home—just as he had the information paper—thus making it a surprise to almost everyone on the Labs staff when he brought it in.[5] But in fact the mouse was far less ingenious than the maze, which was about the size of a small kitchen table and which Shannon had fitted with sliding aluminum panels so its pattern could be easily reconfigured. "I decided that my mouse would be basically a bar magnet, moved by means of an electromagnet under the floor of the maze," Shannon explained. "The bar magnet, covered by a mouse-like shell, could be turned, and when it hit a wall of the maze could signal a computing circuit. The computer would then cause the mouse to try a different direction."

Watching the mouse trot through the maze in its first run-through wasn't exactly breathtaking. Theseus would move slowly through the labyrinth until he hit a wall; he would bump against the wall and then turn in various directions until he finally found an open route. He didn't move quickly. But he was unerring in ultimately calculating which way to go. What was interesting about Theseus was not that he could successfully navigate the aluminum labyrinth. What was interesting was that he (or, more precisely, the relays underneath the floor of the maze) could learn while he navigated the maze, and could likewise *remember* the location of walls and routes and whether it made sense to turn north, east, west, or south. Therefore, on his second try, Theseus could make it through the maze much more quickly—perhaps in as little as fifteen seconds. Or Shannon could actually pick up Theseus and place him anywhere in the maze and he could find his way out.

Among the researchers at the Labs the mouse was wildly popular. The legal department was less enthusiastic, seeing little value in the patent they obtained for Shannon on the gadget's circuitry.[6] This was, after all, a communications company—and one that was limited by the federal government to the telecommunications business. But complaints aside, Theseus was a boon for the Labs in making Shannon a minor celebrity in a way that information theory never had. The Labs produced a short movie about the maze and the mouse, hosted by a thin and dapper Shannon, who narrated in his folksy midwestern manner. And soon *Time* magazine came calling, too. "Theseus Mouse is cleverer than Theseus the Greek," the magazine's writer noted, "who could not trust his memory but had to unwind a ball of string to guide him out of the labyrinth."[7] In large part, this observation matched Shannon's intuition that machines would someday be smarter than men in some respects. Shannon often rhapsodized about the human brain and the inimitable processing power of its billions of neurons. But he believed without question that machines had the potential to do calculations and perform logical operations and store numbers with a speed, efficiency, and accuracy that would soon dwarf our own. It was only a matter of time. Often, he said, he was rooting for the machines.

"I wish you could come east some time so I could show you the beautiful new laboratories we have here at Murray Hill, and some of the many new scientific developments in progress," Shannon wrote to Irene Angus, a beloved high school science teacher of his, just after the *Time* story appeared in 1952. "Probably the most important is work on the transistor, the small germanium device which competes with the vacuum tube. I consider it very likely the most important invention of the last fifty years. As soon as the final manufacturing 'bugs' are out and large scale production starts, electronic devices that we only dream about today will become realities." Shannon told his teacher that the electronic mouse *Time* described was "a demonstration device to make vivid the ability of a machine to solve, by trial and error, a problem, and remember the solution." His fondest dream, he explained, was to now try and build a machine that actually thinks, learns, and communicates with humans.

He added, "I am currently trying to design a machine that will be able to repair itself."[8]

BY THE EARLY 1950S, Shannon's admirers from around the world began to seek him out. They wrote to the oracle at Bell Labs to ask about computers or chess or information theory, and then tried to tease out what he was thinking and why he was thinking it. Requests also came through official channels: In London in 1950, during one of his periodic swings through Europe, Mervin Kelly wrote to Ralph Bown at Murray Hill to say that a British contingent at the Imperial College of Science and Engineering was insisting on hosting Shannon for a visit. "They hope to build a Communication Theory Conference around Shannon," Kelly said. "His work has made a great impression here."[9] The following year, the director of the Central Intelligence Agency, Walter Bedell Smith, contacted Kelly for help with cryptography. "We urgently need the assistance of Dr. Claude E. Shannon of your Company who, we are informed on the best authority, is the most eminently qualified scientist in the particular field concerned."[10]

Still, most of the letters to Shannon came from academics or chess enthusiasts or weekend tinkerers, schoolchildren or hobbyists wanting to know more about Theseus. On occasion Shannon would answer the letters; more often than not, he would let them languish in piles and folders in his office. He would frequently receive letters from some of the most notable scientists in the world. And these, too, would languish. David Slepian recalls that the letters would eventually get herded into a folder he had labeled "Letters I've procrastinated in answering for too long." On rare occasions when Shannon did reply to someone whose original query he had pushed aside, he would begin, *I am sorry to be so slow in returning this, but.* . . . It seemed lost on Shannon that the scientist who had declared that any message could be sent through any noisy channel with almost perfect fidelity was now himself a proven exception. Transmissions could reach Claude Shannon. But then they would fail to go any farther.

Information theory, in the meantime, was getting ready for the masses. In 1953, one of the premier science journalists of the era, Francis Bello of *Fortune* magazine, profiled Shannon along with Norbert Wiener, an MIT mathematician who was putting forward theories on the command and control of machines, a discipline closely related to Shannon's work on information. Wiener called his work *cybernetics.* "Within the last five years a new theory has appeared that seems to bear some of the same hallmarks of greatness," Bello wrote. "The new theory, still almost unknown to the general public, goes under either of two names: communication theory or information theory. Whether or not it will ultimately rank with the enduring great is a question now being resolved in a score of major laboratories here and abroad."[11] Bello didn't say whether Shannon was working toward any kind of resolution, and in fact he wasn't. At the Labs, Shannon had continued to work on various aspects of information theory—on coding, for instance—but he was increasingly drawn to computing. Often his ideas were incorporated into the machines he was building—machines that were constructed for research, for amusement, or for both. Just as he had challenged people to think of information as a word bereft of meaning, he seemed to be challenging people to see whether the things he was building had any deeper significance. Game-playing machines "may seem at first an entertaining pastime rather than a serious scientific study," he said at the time; he noted that there was "a serious side and significant purpose to such work, and at least four or five universities and research laboratories have instituted projects along this line."[12] Shannon didn't say, however, whether the things he was building on occasion—one was a large desk calculator, known as THROBAC, that did calculations only in roman numerals[13]— had any deeper purpose apart from curiosity or wit. Often they were merely responses to somewhat mundane questions: *Could it be built? How would you build it?* Shannon's self-proclaimed "ultimate machine," for instance, seemed a jesting commentary on the subject of the meaning of his tinkerings. It was a wooden box with a single switch. A user hit the switch to turn it on, the box top opened and a mechanical hand reached out and turned the switch off, then the hand retreated into the box and the top closed.

In his speeches in the 1950s, Shannon seemed to make the point that he was not necessarily interested in automated machines per se. He was interested in how machines interact with other machines (as in the telephone switching system) and how they interact with human operators (as in a chess machine). In the latter instance, there was a psychological aspect that seemed curious to him: "We hope that research in the design of game playing machines will lead to insights in the manner of operation of the human brain."[14] To his colleagues at Bell Labs, who had firsthand experience of Shannon's electronic endeavors, this rang true. Shannon's games and machines from those years weren't only about engaging opponents in matches where a computer was pitted against man. Sometimes they were about engaging opponents in games that depended not only on outwitting them but in subtly deceiving them. (Or making fun of them: One game of Shannon's had a computer make sarcastic comments after each move.) Many of these were games in which the unsuspecting player would invariably lose—and thus Shannon, as the games' creator, would win.

His friend David Slepian recalls one of Shannon's electronic games that was peculiar on two accounts.[15] First, the computer was programmed to take an absurd and arbitrary amount of time to calculate its next move, all to give its human opponent a false impression of formidable strategizing. Second, the design of the game board was done so artfully, and with such mathematical precision, that the opponent didn't realize it was created so that the computer would have extra squares on which to move its pieces. "The person playing it from the other side couldn't notice that one side, from one direction, was smaller," Slepian says. Slepian played Shannon's computer several times. He couldn't win. No one could. And obviously that was interesting as well as pleasing to Shannon. "My characterization of his smartness is that he would have been the world's best con man if he had taken a turn in that direction," Slepian says.

In the early 1950s, one of the few people at the Labs whom Shannon would actually seek out at lunchtime was David Hagelbarger, a self-described "tall, skinny kid" from Ohio who went to the University of Michigan and got a PhD under Robert Millikan at Caltech. Hagelbarger always wore a bow tie, in part because he liked to work in the Labs' machine shop: So dressed, he wouldn't have to worry, while working a lathe or drill press, about the safety hazard of a dangling cravat. His mechanical skills were in part what drew Shannon to him. "He would just come around about lunchtime," Hagelbarger recalls. "He didn't make a formal appointment." Usually the two would talk about ideas, both serious and frivolous, some of which Hagelbarger would build for the two of them in the shop. Around 1954, Hagelbarger on his own built a machine with electromechanical relays that would guess whether a human player had chosen heads or tails on a coin. By taking into account a human player's tendency to fall into various patterns of guessing, the machine could beat a player about 53 percent of the time, a success rate that Bob Lucky, a Bell Labs executive, later noted, "by chance alone would happen with probability less than one in 10 billion" over the course of about 10,000 trials.[16] Shannon was fascinated by his friend's machine, so he built his own—a simplified version with a smaller memory but greater calculating speed. "After considerable discussion concerning which of these two machines would win over the other we decided to put the matter to an experimental test," Shannon recalled. The men built a third "umpire machine" to pass information between the two competing machines and keep score. Shannon recalled, "The three machines were plugged together and allowed to run for a few hours to the accompaniment of small side bets and large cheering."

For Shannon, it was an ideal experiment, with an almost optimal range of psychological and technological combinations. Here were machines working with (and against) machines; here were men working with (and against) machines. And here were men working with (and against) each other. Even with the perspective of a half century, it is hard to say whether, or how much, the contest meant to the evolution of computing. But it certainly meant something to Shannon. His machine won, by an "outguessing" ratio of about 55 to 45. The victory thrilled him.

. . .

NOT ALL MACHINES HAD TO BE ELECTRONIC or built with complicated relay circuits to earn Shannon's devotion. By his own admission, he appreciated movement the way an aesthete appreciates beauty.

One year, Betty gave him a unicycle as a gift. Shannon quickly began riding; then he began building his own unicycles, challenging himself to see how small he could make one that could still be ridden. One evening after dinner at home in Morristown, Claude began spontaneously to juggle three balls, and his efforts soon won him some encouragement from the young kids in the apartment complex. There was no reason, as far as Shannon could see, why he shouldn't pursue his two new interests, unicycling and juggling, at Bell Labs, too. Nor was there any reason not to pursue them *simultaneously*. When he was in the office, Shannon would take a break from work to ride his unicycle up and down the long hallways, usually at night when the building wasn't so busy. He would nod to passersby, unless he was juggling as he rode. Then he would be lost in concentration. When he got a pogo stick, he would go up and down the hall on that, too.

Here, then, was a picture of Claude Shannon, circa 1955: a man— slender, agile, handsome, abstracted—who rarely showed up on time for work; who often played chess or fiddled with amusing machines all day; who frequently went down the halls juggling or pogoing; and who didn't seem to care, really, what anyone thought of him or of his pursuits. He did what was interesting. He was categorized, still, as a scientist. But it seemed obvious that he had the temperament and sensibility of an artist.

IN 1956, Shannon was invited to spend a semester at MIT as a visiting professor. Almost immediately, he and Betty felt rejuvenated by the intellectual life of Cambridge—the plays, concerts, lectures, bookstores, libraries, and the like. "There is an active structure of university life that tends to overcome monotony and boredom," Shannon explained. "The

new classes, the vacations, the various academic exercises add considerable variety to the life here."[17] Years before, Betty says, she had lamented
the Labs' move to the New Jersey suburbs from Manhattan: "At West
Street when we went to lunch we went to the Village or a bookstore. At
Murray Hill, we would go to the cafeteria and then get right back to
work." Cambridge reminded her more of the old days.

Not long after he arrived at MIT, the school offered Shannon a permanent job. Bell Labs made a counteroffer, and Shannon wrestled with
this painful dilemma. In October, he wrote his supervisor at the Labs,
Hendrik Bode, to say, "I have finally decided to have a go at academic
life." Like a mathematician balancing out a complex equation, he drew
for Bode a dichotomy between life at Bell Labs and life in Cambridge.
"With regard to personnel, I feel Bell Labs is at least equal in caliber to
the general level in academic circles. In some of their specialties Bell Labs
is certainly stronger." But the intellectual range of the university—and
the long summer vacations—appealed to him more. "It always seemed
to me that the freedom I took [at the Labs] was something of a special
favor," he told Bode. At MIT, it seemed to him, freedom "in hours of
work" would be less unusual.

At around this point in his career, Shannon was beginning to publish
less. Perhaps it would have been impossible to keep up the extraordinary
run he'd had in the 1940s; perhaps, too, a torrent of ideas still rushed
through his mind but he was less interested, as he later conceded, in
writing any of them down. "We've got boxes full of unfinished papers,"
Betty would remark to visitors.[18] Many years later Shannon would leave
behind these half-written papers along with scraps of ideas and mathematical scribbles that were titled "good problems"—but with no indication as to whether he had ever found it worth his time to discover good
answers.[19] Family life, by the time he left Bell Labs for MIT, was in any
event often taking precedence over work. He and Betty had three children. After the visiting professorship at MIT, Shannon decided to take a
one-year fellowship in California, at the Center for Advanced Study in
the Behavioral Sciences, before returning to Massachusetts for the next

phase of his career. Rather than flying, he bought a Volkswagen microbus and drove the family out west.[20]

In Shannon's old Bell Labs office, his machines and his folders of unanswered letters were packed up and sent to Massachusetts. "When all these advantages and disadvantages are added up it seems to me that Bell Labs and academic life are roughly on par," Shannon had told Bode in his resignation letter, "but having spent fifteen years at Bell Labs I felt myself getting a little stale and unproductive." It was unclear precisely what he was going to do at MIT. His initial project involved hosting a popular lecture series on different topics—information, communication, computers—for students and faculty. Meanwhile, he was also involved in several projects with the National Security Agency, a secret government organization within the Department of Defense that focused on cryptography and the deciphering of international communications.[21] But he still intended to pursue his personal curiosities. He and Betty bought a grand house by a lake in Winchester, Massachusetts; soon after, Shannon purchased a Massachusetts Transport Authority bus so he could gut it and reconfigure the inside into a perfect camping vehicle—with a stove, bunk beds, and folding tables. In his free time—or was it all free time?—he experimented with a rocket-powered Frisbee, a gasoline-powered pogo stick, and his various unicycles. He also started to build intricate juggling machines where balls and rings weren't actually juggled by mechanical figurines but were instead moved on hidden guidewires. All the while he began thinking about outfitting a special room in the big house with mirrors—on the floor, ceiling, walls—to create the illusion of an infinite stretch of rooms where none existed.

"Mostly he worked at home," Betty says. Students would come to the house to visit him, mainly to seek advice on their research but also to hear about his newest enthusiasms or experience his latest electromechanical gadget. He quit smoking and began running two or three miles in the morning. Letters kept arriving, meanwhile—letters he still rarely answered—asking him questions both trivial and profound. There were invitations to give speeches or lectures in far flung locales, or congratulations on the bestowment of honorary degrees and awards. Sometimes letter writers asked, would Dr. Shannon consider contributing to a new book? And sometimes he would actually send back a typewritten response. "Unfortunately," he would say, "I am currently completely snowed under with a google of other jobs."[22] When he did answer a letter it was more likely to be about juggling than mathematics.

At first, Shannon visited Murray Hill occasionally. He would come to talk with his old colleagues like Slepian and Hagelbarger about his most recent ideas, and to hear about theirs and perhaps offer a suggestion or two. Shannon and his supervisors at Bell Labs agreed that he would stay on the Labs payroll as a part-time employee, so he still maintained an office there. But eventually the visits to Murray Hill grew less and less frequent. And then, finally, he would not come at all. Shannon's office, its nameplate burnished and its door always closed, stood in wait. He remained in the Bell Laboratories telephone book. Those who phoned his office discovered they would instead be directed to a Bell Labs secretary, who would inform them that no, no, unfortunately, Dr. Shannon wasn't in today.[23]

Nine

FORMULA

I n the late winter of 1950, Mervin Kelly embarked on a whirlwind trip through France, Switzerland, Sweden, Holland, Belgium, and England. Kelly had gone to Europe just a few years before—in 1948, not long after the invention of the transistor, for a monthlong excursion. At that time, he had made his way through a continent that was broken not only socially and politically, but technologically as well. Repairs to the havoc wrought by World War II were only beginning. Now, two years later, Kelly was keenly interested in seeing how much Europe's laboratories and communications systems had rebounded.

But he had come across the Atlantic with another purpose as well.[1] Over the past year he had prepared a lengthy presentation about Bell Labs, and in Europe he intended to lecture on an organization that he now described as "an example of an institute of creative technology." He was a missionary with a gospel to preach. On the same trip during which British engineers were urging Kelly to send Claude Shannon to Europe to talk about Shannon's new theory of information, Kelly was meanwhile offering them a broader context into which they could place Shannon's work, or for that matter Bill Shockley's. Kelly wanted to explain how their efforts were enmeshed within the vast machinery of ideas at Bell

Labs. Listeners who hadn't before seen or heard Kelly reacted to his habitual manner—speedy, impatient, assertive—with surprise. "The remark was made that never before were so many words said in 1.5 hours in England," Kelly wrote to Ralph Bown, his deputy, about one of his speeches. "So I apparently was talking at my home speed."[2]

In London, late in the afternoon on March 23, 1950, Kelly gave a polished version of the lecture about Bell Labs in front of the Royal Society, making every effort to speak more slowly. Even sixty years after the fact, it is worth pausing to consider what Kelly was trying to do in the London speech, for he not only tried to explain the empire he was building, but why he was building it. Only good manners kept him from suggesting to a packed auditorium that Bell Labs was the world's foremost example of a place where scientists pursued creative technology. Echoing Shannon's ideas on the subject, Kelly told his audience in London that "the telephone system of the United States could be viewed as a single, integrated, highly technical machine in which electrical currents that are very small and complex in wave form are sent from any one of more than 40 million points to any one of all the others."[3] Bell Labs helped maintain and improve that system, he said, by creating an organization that could be divided into three groups. The first group was research, where scientists and engineers provided "the reservoir of completely new knowledge, principles, materials, methods and art." The second group was in systems engineering, a discipline started by the Labs, where engineers kept one eye on the reservoir of new knowledge and another on the existing phone system and analyzed how to integrate the two. In other words, the systems engineers considered whether new applications were possible, plausible, necessary, and economical. That's when the third group came in. These were the engineers who developed and designed new devices, switches, and transmissions systems. In Kelly's sketch, ideas usually moved from (1) discovery, to (2) development, to (3) manufacture.

To look at the process in more concrete terms, Kelly explained, one might imagine the kernel of new knowledge that arose from the Labs' solid-state research team just two years before The transistor moved from there to development and manufacture. At the same time, the systems engineers began to consider how to insert it into the phone system. Though its impact on technology was still only speculative, the transistor validated in Kelly's mind everything he had done in his management career, along with everything he thought America's industries (and now Europe's, too) should be doing in the years to come.

In truth, the handoff between the three departments at Bell Labs was often (and intentionally) quite casual. Part of what seemed to make the Labs "a living organism," Kelly explained, were social and professional exchanges that moved back and forth, in all directions, between the pure researchers on one side and the applied engineers on the other. These were formal talks and informal chats, and they were always encouraged, both as a matter of policy and by the inventive design of the Murray Hill building. Researchers and engineers would find themselves discussing their respective problems in the halls, over lunch, or they might be paired together on a project, either at their own request or by managers. Or a staffer with a question would casually seek out an expert, "whether he be a mathematician, a metallurgist, an organic chemist, an electromagnetic propagation physicist, or an electron device specialist." At the Labs this was sometimes known as going to "the guy who wrote the book." And it was often literally true. The guy who wrote the definitive book on a subject—Shockley on semiconductors, John Tukey on statistics, Claude Shannon on information, and so forth—was often just down the hall. Saddled with a difficult problem, a new hire at Bell Labs, a stuttering nobody, was regularly directed by a supervisor toward one of these men. Some young employees would quake when they were told to go ask Shannon or Shockley a question. Still, Labs policy stated that they could not be turned away.

Physical proximity, in Kelly's view, was everything. People had to be near one another. Phone calls alone wouldn't do. Kelly had even gone so far as to create "branch laboratories" at Western Electric factories so that Bell Labs scientists could get more closely involved in the transition of their work from development to manufacture.

Kelly never mentioned the word "innovation" in his speech. It would be a few more years before the executives at Bell Labs—especially Jack Morton, the head of transistor development—began using the word regularly.[4] What he went on to describe in London, though, was a systematized approach to innovation, the fruit of three decades of consideration at the Labs. To Kelly, inventing the future wasn't just a matter of inventing things for the future; it also entailed inventing ways to invent those things. In London, Kelly seemed to be saying that Bell Labs' experience over the past few years demonstrated that the process of innovation could now be professionally fostered and managed with a large degree of success—and even, perhaps, with predictability. Industrial science was now working on a scale, and embracing a complexity, that Edison could never have imagined. Please listen, Kelly was telling the Europeans. He had a formula.

IN TECHNOLOGY, the odds of making something truly new and popular have always tilted toward failure. That was why Kelly let many members of his research department roam free, sometimes without concrete goals, for years on end. He knew they would fail far more often than not. To gather new knowledge in real time, as Kelly had done all his life—even back to those early days in his career, when he counted oil drops in Millikan's lab, or when he shared an office on West Street with the slow-moving and specter-thin Clinton Davisson—was to see how difficult and faltering the process was. So how was one to make something truly innovative every year? "There is a certain logic in the reasoning that methods which have produced much new knowledge are likely to be the best to produce more new knowledge," the science historians Ernest Braun and Stuart Macdonald wrote some years after Kelly's 1950 speech. "Though there is also something paradoxical in the thought that there should be established methods of creating the revolutionary."[5]

Here, then, was the dilemma: Just because you had made something new and wondrous didn't mean you would make something else new and wondrous. But Bell Labs had the advantage of necessity; its new inventions, as one of Kelly's deputies, Harald Friis, once said, "always originated because of a definite need." In Kelly's view, the members of the technical staff had the great advantage of working to improve a system where there were always problems, always needs.

Sometimes innovations sprang from economic needs—making something cheaper, for instance, such as a long-distance phone call, by efficiently combining many different conversations on a single cable, interleaving them at different frequencies or at different time intervals (or both), and then pulling apart those conversations at the receiving end. Sometimes innovations sprang from operational needs—making something that worked better and faster, such as direct dialing so subscribers wouldn't need to use an operator to complete a phone call. Sometimes, apparently, innovations sprang from cultural necessity—making something that appealed to an evolving society, such as a cross-country phone link or, by 1950, a new Bell product like the car phone. And sometimes they sprang from military necessity—an invention such as radar or automatic gun controllers, which were urgent for national defense.

To innovate, Kelly would agree, an institute of creative technology required the best people, Shockleys and Shannons, for instance—and it needed a lot of them, so many, as the people at the Labs used to say (borrowing a catchphrase from nuclear physics), that departments could have a "critical mass" to foster explosive ideas. What's more, the institute of creative technology should take it upon itself to further the education and abilities of its promising but less accomplished employees, not for reasons of altruism but because industrial science and engineering had become so complex that it now required men and women who were trained beyond the level that America's graduate schools could attain. In 1948, Bell Labs began conducting a series of unaccredited but highly challenging graduate-level courses for employees known as the Communications Development Training Program, or CDT. But nobody at Bell Labs really called it CDT. The program was informally known—much to Kelly's discomfort—as "Kelly College," because that's what it was.

An institute of creative technology needed to house its critical mass close to one another so they could exchange ideas; it also needed to give them all the tools they needed. Some of these tools took the form of expensive machinery or furnaces for the laboratories. Some of them were human, however. Bell Labs employed thousands of full-time technical assistants who could put the most dedicated graduate students to shame. Such assistants sometimes had only a high school diploma but were dexterous enough, mentally and physically, that PhDs would often speak of them with the same respect they gave their most acclaimed colleagues. The TAs, as they were known, formed a large subculture—a stratum parallel to the one formed by the Labs' esteemed scientists—where they would exchange valuable information among themselves over lunch. "They were the keepers of practical information," John Rowell, an experimental physicist, recalls.[6] "They knew secrets, tricks. And they knew all this lore about what had been done in the early days." The best of the assistants had the same talents that Walter Brattain and other physicists would idealize, a natural ability to take apart car engines or radios and put them back together, an ability that at Bell Labs might translate into a gift for growing crystals, preparing the surface of a metal for a contact, or constructing experiments.[7]

An institute of creative technology required a stable stream of dollars. "Never underestimate the importance of money," the physicist Phil Anderson says—and it was true.[8] Thanks to the local phone companies, AT&T, and Western Electric, Bell Labs had ample and dedicated funding. Plans could thus be made for the near term as well as for the far future— five, ten, and even twenty years away.

Perhaps most important, the institute of creative technology needed markets for its products. In the case of Bell Labs, there were markets for consumers (that is, telephone subscribers) as well as for manufacturing (with Western Electric). There was no precise explanation as to why this was such an effective goad, but even for researchers in pursuit of pure scientific understanding rather than new things, it was obvious that their work, if successful, would ultimately be used. Working in an environment of applied science, as one Bell Labs researcher noted years later, "doesn't destroy a kernel of genius—it focuses the mind."[9]

Finally, something else seemed important. "A new device or a new invention," Kelly once remarked, "stimulates and frequently demands other new devices and inventions for its proper use."[10] Just as the invention of the telephone had led to countless developments in switching and transmission, an invention like the transistor seemed to point to even more developments in switching, transmission, and computer systems. Or to put it another way, the solution to a technological problem invariably created other problems that needed solutions. So making something truly new seemed to ensure that you would be making something else truly new before too long. The only trouble was, this rule suggested that your competitors—that is, if you weren't a regulated monopoly like the American Telephone and Telegraph company, and you actually had competitors—could do the same.

HE HAD ALWAYS BEEN in a rush, ever since his Missouri childhood, but in the late 1940s and early 1950s Kelly became even busier, a blur of a man. Robert Oppenheimer and several other scientists remarked that he seemed to be working himself to death, but that at least he didn't look quite as exhausted as he did during the war.[11]

His days were long. "He would read at night until 12 o'clock," Kelly's wife would recall.[12] But in the mornings, at 5 a.m., Kelly would rise and dress and make his way down the stairs of his big Dutch colonial house and out into the backyard toward the garden. His neighborhood in Short Hills, New Jersey, was a leafy maze of streets flanked by the homes of the wealthy and the very wealthy. Kelly's house was among the more modest—six bedrooms instead of ten, his lot comprising a single landscaped acre as opposed to neighbors who had two or three.[13] It was his backyard gardens—ornate, multitiered, shrieking with color—that might be called extravagant. They were a private indulgence. Each year Kelly supervised the arrangement of tens of thousands of tulip and daffodil bulbs, some of which he ordered from Holland—"1,000 bulbs every year just to keep it going," his wife recalled—but most of which he would store during the winter in the corners of a basement room, secreted under piles of sand and sorted according to a complex color classification system of his own devising.[14] For a hobby, it was almost absurd in its meticulousness. Then again, this was Kelly. In the yard he would turn the dirt himself, impatiently, before the gardeners arrived, working methodically in the cool near-dark.

When he was finished, he would shower and eat breakfast and dress for work.[15] The uniform was almost always the same: a pinstriped double-breasted suit, white shirt, and patterned tie; his dark hair, slightly gray now, combed straight back; his round-rimmed glasses softening the severity of his face and giving him a vaguely scholarly air. Kelly worked out of two offices, one at Murray Hill and one at the old West Street building in Manhattan. His rush began on the way to work. "He had one official driver for his car, a company car," Brock McMillan, the Bell Labs mathematician, says. "And he would beat on this guy—'drive faster, drive faster, get going, get going.'" When Kelly once hectored the driver so intently that he hit a car pulling out of the company lot, Kelly left the wreck without pause. He walked back to the office to get another car.

"You get paid for the seven and a half hours a day you put in here," Kelly often told new Bell Labs employees in his speech to them on their first day, "but you get your raises and promotions on what you do in the other sixteen and a half hours."[16] He seemed to live by his own advice. In 1950, Kelly was still the executive vice president of the Labs, serving as Oliver Buckley's deputy, but it had been arranged that Kelly would succeed Buckley upon his retirement in 1951.[17] Buckley had Parkinson's disease. The fact was not publicly known.[18] Ostensibly, Buckley was in charge; in truth, Kelly was. And with his 1950 speech in London, Kelly began to move from manager to statesman, an emissary of industrial science who took every opportunity to consider, in speeches to academic audiences and professional groups all over the United States, how Bell Labs' work fit into the future of American science. His pace was grueling, and the frenetic schedule sometimes resulted in fits of distemper. "Twice he submitted his resignation to the president of AT&T, stating that important work at Bell Laboratories was not being adequately funded," a colleague would recall. "In each case, he got the funds."[19] The constant travel and constant meetings and constant speaking engagements—and almost certainly, too, his constant chain-smoking—sometimes resulted in bouts of utter exhaustion, requiring him to take time off and convalesce near his tulip gardens.[20] But within a week or two he would come roaring back.

By 1950, too, Kelly was involved in military and government affairs to such a degree that it required half of his working hours.[21] He now served as a scientific consultant to the United States Air Force and as a frequent advisor on government science commissions; in turn, he enjoyed the same level of security clearance as the head of the CIA. This was in large part a consequence of Bell Labs' work on radar and gun control during World War II, and on the Labs' electronics breakthroughs in the years since: The success of the work had thrust Kelly, willingly, into a shadow society of wise men—people like Frank Jewett, or Vannevar Bush—whose scientific training and large social networks allowed them to move smoothly between the elite circles of industry, academia, military intelligence, and public policy. Truman advisor William Golden visited Kelly and Oliver Buckley in 1950 and early 1951 seeking advice on who might serve as a science advisor to President Truman because Kelly and Buckley were on a short list of the elite. ("While Mervin Kelly was courageous," Golden pointed out later, reaffirming his belief that Kelly was his first choice, "Buckley was timorous.")[22] But Kelly wasn't interested in the job, preferring instead to move into the presidency of the Labs after Buckley's retirement. He directed Golden to friends of his in the scientific elites: Lee DuBridge (president of Caltech), James Conant (president of Harvard), James Killian (president of MIT), and Robert Oppenheimer, now at the Institute for Advanced Study in Princeton, who had successfully managed the Manhattan Project.[23]

Why was an office in the White House so unappealing to Kelly? For one thing, he was already immensely influential at the highest military and policy levels. The tightening alignment between a handful of the largest American corporations and the armed forces—"the huge industrial and military machinery of defense," as President Dwight D. Eisenhower would call it when he left office a decade later—had already become an enormous business for AT&T, which entrusted its Bell Laboratories and manufacturing divisions at Western Electric to design and manufacture a vast array of secret equipment for the Army, Navy, and Air Force. Most of the industrial work orders related to radar and communications equipment; these were considered vital for national defense.

These contracts earned AT&T more than revenue; they gave the company strong allies within the government that the company would need as the twentieth century reached its midpoint. In 1949, Thomas Clark, Harry Truman's attorney general, filed a complaint against AT&T alleging that it and Western Electric, the phone company's equipment manufacturing arm, had "unlawfully restrained and monopolized trade and commerce in the manufacture, distribution, sale and installation of telephone equipment."[24] In effect, the government sought to break the bond between Ma Bell and its factories—cleaving the companies in two and then again cleaving Western Electric into three separate businesses, so that AT&T could buy phone equipment more cheaply through a competitive bidding process. Clark's belief, shared by many in the government, was that telephone costs were being inflated by the cozy arrangement between AT&T and Western. It may well have been true, but the data and accounting records were extremely difficult to penetrate. A countervailing belief, however, little noted at the time but discussed privately among military leaders and AT&T executives—and eventually with Attorney General Clark and President Truman—was that a company that the U.S. government depended upon to help build up its military during the cold war was arguably worth far more intact than apart.[25] In a private letter, Leroy Wilson, the president of AT&T, pointed out the contradiction. "We are concerned by the fact that the anti-trust suit brought by the Department of Justice last January seeks to terminate the very same Western-Electric–Bell Laboratories–Bell System relationship which gives our organization [its] unique qualifications." The Attorney General's office, in other words, seemed to be fighting to break up AT&T at the same time the Department of Defense was moving to capitalize on its broad expertise. If that was in fact true, then Wilson—and Kelly, too—realized they had some leverage. They could make AT&T indispensable in the affairs of government. Kelly had long been willing to do anything necessary to preserve Bell Labs' existing structure, size, and influence. If he had to work even harder to do so, he would.

THE CHIEF, the passenger train that rumbled southwest from Chicago through Kansas City and on to the Pacific, brought Mervin Kelly and Jim Fisk to Albuquerque, New Mexico, on March 6, 1949. The trip had been Fisk's idea.[26] On leave from Bell Labs, the physicist Kelly had hired in 1939 and had put in charge of the radar magnetron work at Bell Labs was doing a stint as the research director for the Atomic Energy Commission in Washington. Earlier that year, Fisk had been informed that the University of California, which had been running the government's Sandia Labs in New Mexico, wanted to stop managing the facility.[27] Sandia was Los Alamos' less glamorous sister. Whereas Los Alamos' famous scientists were charged with researching and developing the nuclear components inside America's missiles and bombs, Sandia's fifteen hundred employees built all the non-nuclear components of those weapons. Sandia's scientists and engineers tested new ballistic shapes and designed sophisticated fuses for detonation. They also trained the troops who would ultimately handle the weapons.

Managing Sandia required extraordinary expertise in research, development, and manufacturing. And some in the military felt the job was beyond the capabilities of any university. Fisk had proposed a solution to his superiors at the Atomic Energy Commission: If the University of California could no longer manage the lab, some other organization would have to take charge. Fisk suggested that Mervin Kelly would be an excellent person to assess Sandia and advise the commission about possible replacements for Cal. Kelly could visit the lab, gather information, and then make an informal report to the AEC about how to improve its operations and administration.

The commission readily accepted Fisk's suggestion and Kelly traveled twice to Sandia that year. Hour after hour, he sat in meetings, eyes closed, as was his habit, listening to managers explain their work. When he made a lengthy report in early May 1949, Kelly unsurprisingly concluded that the AEC should place Sandia under the management of "an industrial contractor with experience, professional know-how, and a sense of public responsibility."[28] By the middle of the month, the AEC had determined that Bell Labs and AT&T would be the best contractor for Sandia. "This operation, which is a vital segment of the atomic weapons program, is of extreme importance and urgency in the national defense, and should have the best possible technical direction," President Truman wrote to AT&T president Leroy Wilson.[29] He urged Wilson and Bell Labs president Oliver Buckley to take on the job (Kelly had apparently recused himself from the negotiations, owing to the fact that he had been hired as an impartial assessor). In early June, following a meeting at Wilson's home with AEC chairman David Lilienthal, the two parties sealed the deal, on the condition that AT&T would not profit from the management of Sandia. In July, Lilienthal wrote Kelly an effusive note of thanks for his work. "It was a splendid job," he noted, "and a real contribution to the atomic energy program."[30]

Despite its distance from New Jersey, Sandia soon became a frequent stopover for Bell Labs managers moving up through the executive ranks—a place where they could be rotated in or out, like a pitcher on a minor league baseball team, depending on the needs of the parent organization. With its focus on the development of missiles and bombs, Sandia fit into the Labs' expanding portfolio of military work. In the final days of World War II, for instance, the Army's Ordnance Department, along with the Air Force, had selected the Labs "to determine the practicability of developing a ground based guided-missile system." The results—a concerted effort of the Army, Air Force, Bell Labs, Western Electric, and the Douglas Aircraft Company—were code-named Nike, after the Greek goddess of victory, and put into operation in 1953. "Essentially a defensive weapon," the Bell Laboratories Record explained, "the Nike system will provide defended areas with a far greater degree of anti-aircraft protection than was ever before possible with the more limited ranges and altitudes of conventional anti-aircraft guns."[31] Nike "systems," essentially clusters of missiles poised for flight, were sited on the outskirts of major U.S. cities and near strategic locations, including Bell Labs' Murray Hill offices. The first missiles were known as Nike-Ajax; each was twenty feet long and about a foot in diameter, with a serration of sharp fins surrounding the white tube containing the propellant fuel and explosives. Ajax missiles were not nuclear. But the next iteration of larger Nike rockets—the Nike-Hercules, which in the late 1950s offered "increased lethality"—were. Later still came the even more sophisticated Nike-Zeus.

What made the Labs essential to the Nike program was an expertise in radar and communications. "Telephone technology has much in common with that of new weapons systems," Kelly remarked as the Nike installations were being built.[32] The new missiles, outfitted with several antennas, were guided by a complex control system, both in the air and on the ground, that involved radio detection and guidance and required, according to one assessment, approximately 1.5 million parts. Though nuclear arms and communications were often perceived as distinct phenomena—one was military, the other was civilian; one was deadly, the other benign—it was becoming increasingly difficult to separate the atomic age from the information age. Indeed, at the military's request, Bell Labs and Western Electric also began designing and building a string of remote radar installations in the frozen wastes north of the Arctic Circle from Canada's Baffin Island to Alaska's northwestern coast; these installations, "the arctic eye that never sleeps" (as the *Bell Laboratories Record* put it), were meant to warn North America of a Soviet nuclear attack. Named the DEW—for Distant Early Warning—line and made possible by a string of nonmilitary discoveries years earlier at the Labs regarding microwave communications,[33] the defensive systems were sister projects to the Labs' military work that included BMEWS (Ballistic Missile Early Warning System) and White Alice, which connected radio sensors in Alaska to Air Force command headquarters in Colorado.

Sandia, Nike, DEW—"All that is part of our good citizenship and, I think, fully meets the obligation imposed by the unique place that we have in our society," Kelly said.[34] He wanted to limit the Labs' military contracts so that they would not get in the way of its communications business, yet he harbored no apparent qualms about such endeavors. All were either strategically or financially important to the phone company; all were potentially useful in keeping at bay the antitrust regulators, who still sought to break up the Bell System. The military work could easily be construed as part of the implicit pact between the phone company and the government that allowed it a monopoly.

To counter communist intransigence, Kelly remarked, would require a "two-front defense," each as important as the other. Americans "are faced with maintaining a military strength adequate to deter the Russians from a general war, while at the same time maintaining a civilian economy that provides our people with an increasingly abundant life." Both pursuits were to him necessary, and so he decided to split his lab, and his career, between the two.

Ten

SILICON

O n the civilian front at Bell Labs, there was still the business of semiconductors. Slowly, in the five years since the unveiling of Bardeen,
Brattain, and Shockley's discoveries, Jack Morton, the transistor's development chief, had shepherded the device through the Labs' development
process to the point that it had begun to infiltrate the mainstream economy. It had also moved outside of Ma Bell. The company's executives—
wary of the regulatory implications of hoarding the technology to itself,
and also convinced that production costs of transistors would decrease
much faster if the semiconductor industry was large and competitive—
had licensed its patents to a number of other companies, including Ray-
theon, RCA, and GE. They were poised to join Western Electric in the
transistor business. The year 1953, *Fortune* magazine proclaimed, would
be "the year of the transistor," when the "pea-sized time bomb," fashioned
from a sliver of purified germanium, finally went into volume production
and thus began to erode the electronics industry's dependence on the
vacuum tube.[1] The doubts that had dogged the invention after its unveiling had since vanished. The transistor, Francis Bello wrote in *Fortune*, in
what seemed an uncanny echo of Mervin Kelly's own thoughts, "will
almost certainly stimulate greater changes in commerce and industry

than reaction motors, synthetic fibers, or even, perhaps, atomic energy." The new devices were compact, reliable, and used so little power they could "lift information handling and computing machines—the nub of the second industrial revolution now upon us—to any imaginable degree of complexity." "In the transistor and the new solid-state electronics," Bello concluded, "man may hope to find a brain to match atomic energy's muscle."

In comparison to the vacuum tube, the transistor was still expensive. It had been helped along commercially during five years of incubation in large part by military contracts. For the armed forces, price was often less important than utility; the transistor's size and low power requirements made it ideal for deployment on ships and planes (and in the Nike systems, too), where every ounce and every fraction of a watt—it used as little as one-hundred-thousandth of the power a vacuum tube required—made a difference.[2]

Within the consumer electronics industry, there seemed to be general accord that the transistor's greatest value would be in computers and communications devices. But so far very few transistors had been integrated into the phone system, and those that had—to generate pulses for nationwide dialing in an office in Englewood, New Jersey, and to help route phone calls automatically in an office in Pittsburgh[3]—were more like demonstration projects than actual technological overhauls. Long ago, the dream of an electronic switch had prompted Kelly's initial push on semiconductors. As the *Fortune* story pointed out, a switching office with 65,000 electromechanical relays could do "slightly less than 1,000 switching operations a second." Transistors—using a fraction of the power and lasting far longer—could potentially do a million.

So what was the Bell System waiting for? Kelly acknowledged that the phone company would capitalize on the transistor long after "other fields of application" such as the home entertainment industries.[4] The recent Justice Department antitrust suit, which was now moving forward, was a stark reminder why: The phone company was a regulated monopoly and not a private company; it had no competitors pushing it to move forward faster. What's more, it was obliged to balance costs against service quality in the most cautious way possible. "Everything that we design must go through the judgment of lots of people as to its ability to replace the old," Kelly told an audience of phone executives in October 1951. "It must do the job better, or cheaper, or both."[5] Any element within the system was designed (by Bell Labs) and built (by Western Electric) to last thirty or forty years. Junking a functional part before its time had to be economically justifiable. And if it wasn't justifiable on economic grounds, it had to be justifiable on technological grounds.

The transistor could not be justified on technological grounds—at least not yet. The awesome intricacy of the phone system was not hospitable to sudden changes, even when they allowed for stark improvements. In time, Kelly remarked, the traffic and data needs of the system would require replacing tubes and switches with transistors.[6] The Labs' managers had already begun planning for new transistorized phones and an electronic switching station that, as it turned out, would take nearly twenty years to fully deploy. In the meantime, the system was working reliably and was giving customers a reasonably good product for their money. AT&T shareholders, who now numbered more than a million, making AT&T not only the world's largest corporation but the most widely owned, also liked its steady profits and sizable dividends. On Wall Street, brokers called the dependable AT&T a "widows-and-orphans" stock; if you couldn't rely on anyone else, you could rely on Ma Bell. The paradox, of course, was that a parent corporation so dull, so cautious, so predictable was also in custody of a lab so innovative. "Few companies are more conservative," *Time* magazine said about AT&T, "none are more creative."[7]

PERHAPS, as Kelly realized, one need not rush the phone system's evolution. But the business of communications was different than the science of communications, and in the science, Kelly's employees could do whatever they liked to push ahead. By the mid-1950s, one of the essential questions facing researchers was whether the transistor of the future, and therefore the future of electronics, would be fashioned out of germanium or silicon. All of the transistors so far had been made of germanium. But there were a number of reasons to favor silicon.[8] Germanium is far rarer than silicon, which can be derived from sand. If the transistor industry were potentially as enormous as *Fortune* magazine envisioned, germanium's scarcity (and its high price) could at some point limit the industry's growth. More fundamentally, germanium had performance issues. For reasons owing to the behavior of its electrons, as germanium transistors got hotter, they became less reliable. In a very warm environment—150 degrees Fahrenheit and over—they were often useless.

In late 1952, a young chemist with a PhD from Princeton named Morris Tanenbaum joined the Bell Labs research department. As was typically the case with new recruits, Tanenbaum was encouraged by his supervisors to look around the Murray Hill complex—literally to drop in on neighboring laboratories for a few days—and see what kind of research interested him before he settled into a particular project. "The transistor was just a few years old then," Tanenbaum recalls, "and there was still a lot of awful good work going on with germanium crystals. The question was: Are there better semiconductors than germanium?"[9] Tanenbaum found this question intriguing. He began his search by obtaining a variety of materials—aluminum, gallium, indium, and so forth—in the purest states he could find. Usually they came from DuPont, and were sold as powders or in small chunks. He or his assistant would melt them into crystals. When Tanenbaum began his tests, he tried a rare element called tellurium; soon after, following some work he had heard about in the Siemens labs in Germany, he began working with a compound of indium and antimony. At that point, Shockley got involved.

"Bill said, 'Look, germanium has a number of properties that really aren't very good,'" Tanenbaum recalls, "'so let's really look at silicon.'" Several years before, Bardeen and Brattain and Brattain's lab mate Gerald Pearson had tried to make silicon transistors but had been discouraged by the results. Shockley wanted to try again. "I was asked if I would like to work with him to do that," Tanenbaum recalls. "I had heard about Shockley's reputation," he says, noting that it almost scared him away from the project. "Shockley had an ego substantially larger than a house—and he deserved that. He was a very brilliant guy. But it turned out I never had any problems. Being a physicist with Shockley, well then, you had better be very, very good or you're going to have a hell of a time." But being a chemist, with knowledge outside of Shockley's sphere of expertise, put Tanenbaum beyond the reach of his bullying.

Working with silicon, as Tanenbaum soon discovered, was far more frustrating than working with Shockley. Still, the rewards of success seemed obvious to most people in the solid-state area. "If anyone could actually build a silicon transistor," Tanenbaum remarks, "then we knew it would even work in boiling water." DuPont was already selling what it called "pure silicon" for semiconductor devices; according to the *Wall Street Journal*, the company was then charging $430 a pound.[10] The product, which arrived at Bell Labs as a powder, was a useful starting point. But it was only a starting point. Silicon has an extraordinarily high melting point, about 2,500 degrees Fahrenheit, and is easily tainted in the "melt" by all sorts of other unwelcome elements. The residue from the crucible that it has been melted in, for instance, can easily ruin its potential as an electronic device. Indeed, while it was true that the metallurgists at Bell Labs wanted their silicon to have small amounts of impurities, they only wanted certain *kinds* of impurities, so as to affect the conduction in useful ways. They wanted a few atoms of one specific type of element to transform the silicon into the negative n-type and a few atoms of another specific type of element to transform the silicon into the positive p-type. Then they wanted to join the two. The junction between the p-type and n-type was where the movements of electrons and holes resulted in transistor action.

Success, to a certain degree, came quickly. After several months of work, Tanenbaum, with the help of his lab technician, Ernie Buehler, "grew" a long crystal of silicon through a complex process that involved varying the rate at which the crystal was being "pulled" up from the molten silicon. By varying the rate, the men could alternate the amount of n-type and p-type impurities that were incorporated into the crystal. When they were done, this long crystal—about four and a half inches long and three-quarters of an inch wide—had dozens of tiny n-p-n sandwiches all stacked up, giving it the appearance of a thin rod made up of tiny gray slices piled on top of one another. After slicing one of the n-p-n portions from the tiny stack, the men fashioned, in January 1954, the world's first working silicon transistor. A few months later, Gordon Teal, a former Bell Labs metallurgist who had pioneered techniques for making germanium crystals before joining a tiny semiconductor company called Texas Instruments, unveiled his own silicon transistor. But neither of these developments were cause for celebration. One of the drawbacks of Buehler and Tanenbaum's silicon transistor was their complicated fabrication method, which seemed unsuitable for mass production. To actually change the course of an entire industry, a silicon transistor would have to be reliable and easy to make.

For nearly a year afterward, Tanenbaum kept working with silicon. He was in Building 2 at Murray Hill, with Shockley's research team; just down the hall was Jack Morton's transistor development group. In Building 1, across a courtyard, Tanenbaum had a chemistry colleague named Cal Fuller. Like many of his fellow researchers at the Labs, Fuller had a background that should never have led him into a distinguished life in science. A slim and scholarly-looking man, the son of a bookkeeper who was raised in a poor family in Chicago, Fuller as a teenager had experimented diligently with his wireless radio and home chemistry set. He never imagined he would go to college. But then his high school physics teacher thought otherwise. She was Mabel Walbridge, "a very demanding teacher," Fuller recalled, "a woman in her fifties or early sixties" who had taken courses earlier in her life from Robert Millikan. She knew that the University of Chicago offered exams to high school students in science and math. She also knew that for those who passed, the university "provided full tuition for the first year and, if you were among the top twenty-five students in your class at the university, for the following three years." Walbridge demanded that Fuller pursue the scholarship. "She tutored me nights on her own time," Fuller said, still astounded by the fact sixty years later, "so that when I took the exam I was prepared for almost every question." Fuller went on to an undergraduate degree at Chicago and a PhD in chemistry. He put himself through graduate school by working the four-o'clock-to-midnight shift at the *Chicago Tribune*.[11]

His early work at Bell Labs had been on plastics and rubber. After Mervin Kelly reorganized the Labs research department after the war, however, Fuller begun working on semiconductors such as germanium and silicon, with a special interest in how infinitesimal impurities affected them. He had noticed that a germanium crystal could be rendered impure if someone touched it after handling a brass doorknob,[12] and he wondered if there was a way to take advantage of the remarkable sensitivity of these crystals to impurities. By the time Morry Tanenbaum got to know him, Fuller was using a technique called diffusion that suggested a way to manipulate the concentrations of impurities in silicon with remarkable precision.

In diffusion, a long silicon crystal (it looked about the size of a pretzel rod, says Tanenbaum) is cut into thin, round slices; the slices are then placed in a furnace. In the furnace, the silicon slices are exposed to a gas containing an impurity, such as aluminum. "At high temperatures, these impurity atoms bombard the crystal surface and slowly force their way into the interior," the *Bell Laboratories Record* explained. Or to put it another way, the impurities in the furnace atmosphere, depending on the type, could create, on top of, say, an n-type silicon slice, exceedingly thin layers of p-type and n-type silicon stacked on top of one another. One might envision a crystal disc the size of a dime that emerges from a furnace with two thin coatings on top, each less than a thousandth of an inch thick. These are, respectively, the p-type and n-type layers.

Now that he had the right materials, Tanenbaum faced the challenge of actually making an electrical contact with the thin middle p-layer within the diffused silicon disc. That middle layer was far finer than a human hair. He spent weeks on the problem, trying to grind the dime-sized crystal on an angle to attach a wire, and attempting all sorts of other tricks. "Finally, one night, I went back into the lab because my wife was having a bridge game," he recalls, and by trying a blunt method—in his lab notebook he wrote, *will try direct approach*—he melted an aluminum wire "through" the thin top layer. He made a good contact. It was late on the evening of March 17, 1955. When he took some instrument readings, he was shocked to see that the device performed better than any germanium transistor then in existence. In his notebook he wrote, *This looks like the transistor we've been waiting for. It should be a cinch to make.* "Right away," he recalls, "I knew that this would be *very* manufacturable." He drove home like a demon to tell his wife. He could barely sleep, wondering if he had imagined the whole thing, and rushed back to the Labs in the morning to test it. Almost immediately the supervisors were called in, including Jim Fisk, who had returned from the Atomic Energy Commission and was now Bell Labs' chief of research.

Jack Morton was in Europe, Tanenbaum recalls, but cut his trip short and flew home when he was told the news. Right away Morton—whose opinions on transistor innovation were akin to a final judgment at Bell Labs—understood the potential value. Even Kelly, too busy with management to properly assess the technical details that entered into Morton's calculus, would defer to him on such matters. If Morton was on board with the diffused silicon transistor, Kelly was, too. That meant the future would be silicon.

Diffused silicon had another use, too.

Fifteen years had passed since the day Walter Brattain had been ushered into Mervin Kelly's office to regard a strange piece of silicon that had been discovered down in Holmdel, New Jersey. The men had shone a light on the blackened chunk and the resulting electric charge had stunned them. In later years it came to be understood that this chunk of silicon contained a naturally occurring p-n junction where two types of silicon met. The junction is extremely photosensitive. In very general terms, the photons in light are hitting the semiconductor crystal and "splitting off" electrons from their normal location in the crystal; the process, if properly captured, can create a flow of electrons, that is, a flow of electricity.[13] Kelly and Brattain and Ohl didn't know it at the time, but in Kelly's office the men had been looking at the world's first crude silicon solar cell.

In the early 1950s, Cal Fuller, who had made the diffused silicon that Morris Tanenbaum used to fashion his silicon transistor, was also working at Murray Hill with the experimental physicist Gerald Pearson. A genial presence at the Labs, slim and handsome and neatly kept, with his dark hair always combed straight back from his forehead, Pearson was Walter Brattain's old laboratory mate, the same G. L. Pearson who had signed Brattain's lab notebook on that fateful Christmas Eve in 1947 after Brattain and Bardeen had demonstrated the transistor for the Bell Labs brass. Now Fuller and Pearson were trying to build something with diffused silicon called a silicon power rectifier. In the course of the work on the device, Pearson noticed it was highly sensitive to light.[14] Pearson had an old college friend at the Labs named Daryl Chapin, who he knew was trying to develop power sources for remote telephone installations. Often these remote installations—places where phone repeaters might be located, for instance—used diesel generators or dry cell batteries. The batteries had problems in humid weather. Pearson wondered if Chapin might be able to use the power of the sun.

With Pearson as the go-between, the three men—Fuller, Pearson, and Chapin—created over the course of a few months what the Labs eventually called a silicon solar battery. These were thin strips of specially diffused silicon, connected to a circuit, that in sunlight could generate a steady electric voltage. The "battery" was not the first solar cell; functional ones had been made before from the element selenium, for instance. But this was, by Bell Labs' calculations, "at least fifteen times more efficient than the best previous solar energy converter."[15] That made it the first truly usable solar power device. The batteries promised to last forever, since they had no moving parts. What was striking but almost always overlooked about its invention, Fuller later recalled, was that all three inventors of the device were working in different buildings. "The solar cell just sort of happened," he said. It was not "team research" in the traditional sense, but it was made possible "because the Labs policy did not require us to get the permission of our bosses to cooperate—at the Laboratories one could go directly to the person who could help."[16]

The silicon solar cell generated a hurricane of publicity when it was unveiled. "The subsequent attention," Fuller recalled, "which exceeded that of the announcement of the transistor, was unbelievable."[17] By the front-page newspaper headlines,[18] one might easily imagine that the cells' ability to effectively harness the sun meant modern society had reached a pivotal juncture, and that soon enough the world's energy supplies would be clean and inexhaustible. For those who knew anything about transistors, which had extremely low power requirements and could thus be a perfect match for the new solar technology, the news seemed yet another fantastical bit of Bell Labs augury. On October 4, 1955, a test project for the solar cells was set up by the Labs' engineers at a remote rural phone installation in Americus, Georgia, 135 miles south of Atlanta. For six months, they powered equipment at the installation, and hinted at the remarkable future where power could be generated anywhere the sun shone.

And then the great excitement of the solar breakthrough dimmed. As Pearson would later recall, the installation was "a huge technical success, but a financial failure." The solar battery could power the remote telephone equipment with ease. But for the power they generated, the solar cells, at several hundred dollars per watt, simply cost too much.[19] In 1956, Daryl Chapin figured that it would cost the average homeowner nearly $1.5 million to buy enough Bell solar cells to power his house.[20] By one of Kelly's fundamental dictums of innovation—something that could do a job "better, or cheaper, or both"—the cost of the cells and the results in Georgia suggested solar power was not going to be a marketable innovation anytime soon. Sometimes, in describing a new invention that seemed technically brilliant but impractical, industrial scientists would quip that they had found "a solution looking for a problem." The silicon solar cell needed a problem, as yet unimagined, to appear.

THE PUBLICITY AROUND inventions like the solar cell tended to distort public perceptions about the actual work being done inside the Labs. Kelly would often point out that the Labs workforce—including PhDs, lab technicians, and clerical staff—by the early 1950s totaled around nine thousand.[21] Only 20 percent of those nine thousand worked in basic and applied research, however. Another 20 percent worked on military matters. Meanwhile, the rest of the Labs' scientists and engineers—the majority—toiled on the never-ending job of planning and developing the system. Their work was arguably less glamorous. The research scientists at the Labs were thinking ahead to a glorious future that was ten or even twenty years away. The development and systems engineers were thinking about what they could do in the next year or two or three. And yet the projects undertaken by the latter group during Kelly's presidency were in many ways just as ambitious as those done in research; one might see that they were logistically more difficult. In development, mistakes were not excusable. Building a new product or invention, and then putting it into the working telephone system, demanded perfection.

Systems engineers—the ones who looked at new ideas and decided whether they could improve the system—lived by Kelly's rule: *Better, or cheaper, or both*. In the years immediately following the war, one idea that met with their approval involved a project whereby the Labs, working in conjunction with AT&T's Long Lines Department, could ease the congestion on the long-distance phone network. The plan was to create a new, long-distance corridor that would forsake cables altogether—in effect, it would be wireless. At the time, all long-distance calls, along with some coast-to-coast TV transmissions, were carried on thick underground coaxial cables that crisscrossed the country. But it was believed that a nationwide chain of microwave relay antennas, linked to one another in a relatively straight line, could efficiently move calls and programs great distances. Indeed, the microwave links would liberate the phone company from the onus of buying and burying or stringing more of its expensive cables. To test the idea, "an eight-hop route," as the *Bell Laboratories Record* described it—with seven antenna stations along a 220-mile corridor—was built in the late 1940s. The test route linked New York with Boston, via eight microwave towers, some built from concrete blocks and others from steel girders, most located on hillsides (or in some cases on tall urban buildings), topped by special horn chaped antennas, vaguely resembling megaphones, which had been invented at

Bell Labs largely under the direction of Harald Friis, the head of Bell Labs' Holmdel, New Jersey, research office.[22] Usually, two horn-shaped antennas on the towers would receive calls; a repeater apparatus inside the tower would amplify them; and then two other horn-shaped antennas, facing the opposite way, would instantly relay them to the next tower in the phone link. The height of these towers was crucial: Transmissions by microwaves traveled in straight lines and required a clear line of sight. Any interruptions—buildings, trees, mountains—would affect the signal.

When the local test proved successful, AT&T and Western Electric built a national system of microwave links, requiring the construction of 107 towers across the United States, or about one every thirty miles. A phone call could now be handed off automatically—that is, received and immediately resent—between towers in some of the most remote locations imaginable. For instance, a long-distance call made from a Manhattan advertising agency could be switched through the network to a microwave tower atop a New York skyscraper; from there it could move from east to west at nearly the speed of light, passing from station to station—through a 406-foot steel tower in Des Moines, for instance, and another installation high atop Mount Rose, in Nevada—before ultimately arriving at a receiving station perched high in the Oakland hills east of San Francisco, where it would be routed to a switching center and tied into the local phone exchange.

"Mark, how are you?" an AT&T vice president in New York asked Mark Sullivan, president of Pacific Bell, in San Francisco, during a call on the system's opening day in August 1951.

"It's nice to hear your voice," Sullivan said. "I'm fine, thank you."[23]

This exchange of dull pleasantries seemed fitting for the occasion. In contrast to the opening of the first transcontinental line some thirty-five years before, the relay system produced only a slight riffle of excitement, suggesting that the public had come to take for granted easy coast-to-coast communications. Microwave towers would shape the future of telecommunications, as well as the fate of Bell Laboratories. But at this point, nobody could see how.

Eleven

EMPIRE

One project under Kelly's direct supervision captured the public's imagination, and expanded the system's reach, in a way that microwave towers could not. The project was given the name TAT-1; it was the first transatlantic phone cable, a joint project of AT&T and the British Post Office, that was intended to carry thirty-six phone conversations at any given time from the tiny village of Clarenville in Newfoundland, Canada, to the city of Oban, Scotland. Actually, TAT-1 was two cables that would be laid side by side. One cable would carry voices to Europe, the other would carry responses back.

Sending a message over land had always been easier than doing so over or under the water. Engineers had first tried to connect North America to Europe in the 1850s, when successive attempts were made to lay down a telegraph cable on the floor of the North Atlantic. "It was a mathematical impossibility to submerge the cable successfully at so great a depth, and if it were possible, no signals could be transmitted through so great a length," the British royal astronomer predicted at the time.[1] Indeed, the first two tries had ended in failure; the cable layings, done by way of a sailing ship outfitted with a giant spool of copper wire in its hold, were deemed perilous, multimillion-dollar disasters. Cables would snap, snag, kink, and leak; ocean storms would batter the crews and equipment; and in the end, transmissions on the line might work for a couple of weeks before going dead for no apparent reason. But in 1866, a cable—made of better materials, and laid down with more care and expertise—finally succeeded in carrying dots and dashes back and forth between Canada and Ireland. And in the decades after, engineers figured ingenious ways to increase the speed and capacity of other submarine cables. By the early 1900s, overseas telegraph communications had become a lucrative business.

Human voices were different than telegraph signals. Carried by copper wire, the telephone signals were more complex and delicate; moreover, they would attenuate—that is, fade out—after a hundred miles or so. Years before, in the early part of the century, the same challenge had vexed Frank Jewett as he contemplated the idea of a telephone line that connected New York and San Francisco. "The crux of the problem," Jewett wrote at the time, "was a satisfactory telephone repeater or amplifier." Harold Arnold, Jewett's research chief, had answered his prayers by developing a vacuum tube that could amplify signals and relay them forward. But as difficult as that task had been, stringing a cross-country line meant working on hard ground and in dry air. To put tubes and repeaters in a cable located miles under the sea, and then provide a constant electrical current for them, and then make sure the cable covering never leaked, that sea worms never gnawed through the covering, and that the components never failed? "If such a complex and technically difficult system could be made operational," Kelly would later reflect, that would not automatically make it viable. The cable would still have to prove that it was economically justified. "In the event of a failure in midocean," Kelly pointed out, "a cable ship would have to lift the cable at that point and make the necessary repairs. In winter, this might well remove the cable from service for two or so months and be a costly operation."[2] By Kelly's calculations, the cable would have to work for at least twenty years without a single problem in order to pay for itself.

For several decades, the safe answer had been to forsake the idea of an ocean cable altogether. Some of the early pioneers of radio transmission, including the Italian inventor Guglielmo Marconi, had proved that it was possible to send radio signals across the ocean by bouncing them off the earth's outer atmosphere. In the late 1920s, as it happened, Kelly had worked with a Bell Labs team stationed at the end of Long Island that had built a transatlantic transmitting station to create regular radio-telephone service between the United States and Europe. By the 1950s there were sixteen radio channels operating between the continents. These were relatively cheap to operate (and certainly less expensive than a 2,200-mile undersea cable, which was projected to cost about $42 million, or roughly $340 million in today's dollars). But overseas radio had one ineradicable fault: Weather and atmospheric conditions could wreak havoc on the transmissions. "When conditions were good," Arthur C. Clarke wrote in his history of overseas communications, "transatlantic speech was of excellent quality, with little distortion or interference. But all too often the radio beams picked up most peculiar noises, like the sounds of cosmic frying-pans. These were usually no more than annoying, but sometimes they could obliterate the signal."[3] In fact, there could be periods of days when weather conditions made an overseas radio call impossible. In the early 1950s, Kelly noted, "No way has been found to provide day-to-day continuity and reliability comparable to that of good wire lines."[4]

If you were building the best communications system in the world, as Kelly liked to imagine he was, this was not acceptable. At the same time, he noted, there was an obvious and unique answer: Figure out a way to make an undersea cable with repeaters spaced every forty miles or so, and then figure out a way to make it work for twenty years without leaks or interruption. To a large degree, then, the cable was a challenge of engineering rather than science, but even Kelly seemed humbled by "the magnitude of the task, its tremendous scope, the necessity for establishing that the designs of each and every item meet not only functional requirements but also the twenty-year no-failure requirement."[5] Fortunately, by the time Kelly began planning an undersea cable with British phone engineers in 1953, a small library already existed about what would and would not work. In the years following World War II, the Labs had tested various designs for undersea repeaters on shorter routes—notably a submarine cable that successfully connected Key West with Havana, Cuba. In the course of this project, the Labs had created a "flexible" repeater containing three vacuum tubes that could be inserted within the cable at forty-mile intervals and could be wound around a horizontal spool on a cable ship. In the case of the TAT-1, the planners would use a British vessel called the *Monarch*, the world's largest cable ship.

Even to the untrained eye, the repeater in the transatlantic line was not hard to spot. The final Bell Labs designs specified a cable 2,250 miles long and about one and a half inches thick—"such a precise piece of construction," as *Reader's Digest* put it, "that communications engineers speak of it almost with awe." Unpeeling the ten-layer cable (an impossible task without power tools, actually, owing to its formidable strength), one might see that at its center were copper wires, surrounded by polyethylene insulation, and then another layer of copper conducting tape. These copper wires and tape were the coaxial cable used to carry the calls. Then there were six outer layers, designed with the knowledge gained from the late 1800s onward, for protection and strength. First came a copper layer to protect against undersea teredo worms, then successive layers of cotton, jute, steel, and more jute. At forty-mile intervals, where a repeater would be inserted, the thickness of the cable gently bulged to about three inches in diameter for a length of about twenty-eight feet before graduating back to one and a half inches again—making the cable look like a snake that had just had a modest meal, as Arthur Clarke put it.

The engineers at the Labs had spent years drawing up an exact architecture for the repeater system so it could, as Kelly explained, "withstand the shocks of laying and recovery and also the pressure of water encountered." The floor of the North Atlantic wasn't smooth. Between Newfoundland and Scotland there existed a series of huge underwater peaks and troughs, meaning the cable would descend in places to well over two miles before climbing mile-long underwater inclines to less punishing depths. The engineers also designed the repeater sections with the cable ship in mind—so that it could "without damage to itself or to the cable be passed over the drum and sheaves of the cable ship and paid out without slowing up or stopping the vessel."[6] Clarke described the laying of the cable as a spider spinning out its thread.

Kelly had spent years considering how the cable's repeaters could operate flawlessly. It was the kind of question that suited the Labs statistical experts, the same ones who had invented the disciplines of quality control and quality assurance for the Western Electric factories. For something to be "better, or cheaper, or both," as Kelly insisted, it had to last and last. It was not unusual at Bell Labs to run an experiment on, say, telephone poles—burying them in a swamp, or exposing them to harsh temperatures[7]—lasting twenty-five years and keeping a meticulous record of the results. In Kelly's negotiations with the British engineers, there had been a fair number of disagreements over which country's technology to use for the cable. "It was realized that compromise was essential but innovation and relatively untried methods were too risky," an internal history of the Labs would put it.[8] In fact, by the time the first cable was ready for laying down in 1955 (the return cable would go down in 1956), Kelly had made sure that the undersea project, in sum a great innovation, used the least innovative components. A transistor for the repeater system was out of the question; the technology was too new, and there was no telling how long it would last underwater. Only simple vacuum tubes that had been designed in the late 1930s, back when Kelly was still the Labs' head of research, were to be considered—and these would be assembled from the best possible materials in a specially designed clean room at a new factory, built expressly for the purpose of the undersea project, in Hillside, New Jersey. The Labs' engineers had reams of data on how well these tubes would work. Some had been running continually in a testing lab for sixteen years.

A hurricane blew through the North Atlantic while the *Monarch* was laying down the first cable; otherwise, the operation went smoothly. The opening of the cable, on September 25, 1956, was a gala event at three main sites of the Labs—West Street, Murray Hill, and Whippany. On the occasion, AT&T chairman Cleo Craig called Charles Hill, Britain's postmaster general. Technically speaking, Craig's first words—"This is

Cleo Craig in New York calling Dr. Hill in London"—were carried by a phone cable from New York to Portland, Maine; a radio relay system in Portland broadcast it by antenna to Sydney Mines, Nova Scotia. From Sydney Mines, the voice signals then entered a shallow-water cable running to Clarenville, Newfoundland. At Clarenville, the signals entered the deepwater portion of the newly-completed transatlantic cable. All along the bottom of the ocean, the message flashed through fifty-two repeaters over 2,250 miles, before emerging in Oban, Scotland. At Oban the transmission was directed by cable to London, where Hill was poised to respond to Craig. It all took less than a tenth of a second.

Occasionally, fishing trawlers near the shore would cause breaks in phone service via the first transatlantic cable. But for twenty-two years after it was first activated, its technology never failed once.

BY THE TIME the transatlantic cable came online, only two of the four men most closely associated with the transistor—Mervin Kelly and Walter Brattain—were still working at the Labs. John Bardeen had settled in as a physics professor at the University of Illinios; Bill Shockley had left to form a transistor company—Shockley Semiconductor—in Palo Alto, California. He had received some help from Kelly, who had introduced him to some wealthy investors. The assistance was almost certainly not selfless. Judging Shockley as unsuited for upper management, Kelly had refused to promote him—and as if to prove Kelly's judgment correct, Shockley had in 1953 and 1954 sent his boss peevish memos written as "pinks," the nickname for informal notes, complaining about the Labs and Kelly himself.[9] Perhaps it was merely evidence of Shockley's broader frustration. According to Ian Ross, a colleague of Shockley's who would later become president of the Labs, Shockley simply felt he was not getting the rewards he deserved at Bell Laboratories. By the early 1950s, he was a department head overseeing a small group of scientists. He thought he should be higher in the organization and paid much more. Thwarted by Kelly, Shockley appealed to people even more powerful. In one instance, he even went to the president of AT&T. When he got nowhere,

Ross recalls, "he said, the hell with that, I'll go set up my own business, I'll make a million dollars that way. And by the way, I'll do it out in California."[10]

As yet there wasn't much in the way of technology out in Palo Alto. Mostly it was apricot orchards and undeveloped land, but it had been Shockley's hometown for most of his childhood. Also, there was Stanford University, where he had a booster named Frederick Terman, the school's provost. Come to the Valley, Terman had told Shockley, and he could help him find an office in a new industrial park for young, innovative companies.[11]

Shockley tried to lure several of his Labs colleagues out west with him. His notebooks of the era are filled with potential candidates for jobs. For those he liked, he promised grand adventures in a new industry and, at least to the silicon transistor inventor Morris Tanenbaum, twice his Bell Labs salary. Shockley only managed to woo one person from Bell Labs. Mostly, he located and hired some promising young scientists from other companies—most notably Gordon Moore, Robert Noyce, Jean Hoerni, and Eugene Kleiner, all four of whom would do much to put Silicon Valley on the map. None of the recruits seemed particularly aware of Shockley's shortcomings as a manager. And even so, the attraction of working for his new company was obvious. Robert Noyce famously described what it was like for a young solid-state physicist, toiling in obscurity, to discover that Bill Shockley was calling him: "It was like picking up the phone and talking to God."[12]

Even as Shockley was preparing to leave the Labs, Kelly, as a foreign member of the Swedish Academy of Sciences, had lobbied for the transistor team, though now disbanded, to be awarded the Nobel Prize.[13] For years there had been rumors they were in the running, and Brattain, Bardeen, and Shockley heard on November 2, 1956, that they would indeed share a Nobel Prize. Brattain refused to attribute the honor to his own genius. "It was really only a stroke of luck," he later explained, that he and Bardeen, rather than someone else, had gotten there first. Everyone at Bell Labs knew the prize reflected well on Kelly. Even Shockley was moved, in his own way, to share some credit. "It seems to me that this is a suitable occasion for me to repeat what I told you over the telephone when resigning from Bell Telephone Laboratories," he wrote to Kelly from California upon the receipt of the Nobel. "It is hard for me to see how a research director and vice president in your position could have proceeded more effectively to get a transistor out of a solid state physicist like myself."[14]

To Kelly, however, what was looking to be the most memorable thing about a year of extraordinary events was not the Nobel. The 1949 Justice Department lawsuit to break Western Electric off from AT&T had also been settled in 1956. AT&T's public relations department circulated an internal memo to staffers in the wake of the decision, explaining that the "decree *does not* affect the general relationship among AT&T, Western Electric, Bell Telephone Laboratories," and the local phone companies. That much was true; the government had agreed that the phone monopoly could remain intact, as long as its business was limited either to publicly regulated communications services or to military work like the kind being done on the Nike missiles and the Distant Early Warning line.[15] Some years later, it would be revealed in congressional testimony that just prior to the 1956 consent agreement, Herbert Brownell, President Eisenhower's attorney general, had quietly given to AT&T's general counsel "a little friendly tip" to analyze its operations and identify "practices" that it could compromise on without harming its business.[16]

The government had hinted that in exchange for a fig leaf of compromise from AT&T, it was inclined to drop the suit and allow the company to maintain its monopoly. AT&T offered two fig leaves. The first was its agreement not to enter the computer or consumer electronics markets. The second concession, at least on its face, seemed far more dramatic: The phone company agreed to license its present and future U.S. patents to all American applicants, "with no limit as to time or the use to which they may be put." In other words, eighty-six hundred or so of AT&T's U.S. patents "issued prior to January 24, 1956 are in almost all cases to be licensed royalty-free to all applicants." (All future patents, meanwhile, would be licensed for a small fee.) But this largesse didn't worry AT&T executives, whose confidence in the strength of their business had now been restored. The patent giveaway was in fact deceptive. So what if entrepreneurs all over the country now had essentially free access to transistors, microwave long-distance systems, underwater repeaters, solar cells, coaxial cables, and thousands of other devices and industrial processes? The Bell System remained a monopoly. Competitors trying to gain a foothold in the telephone equipment business still had no way in.[17]

THE CAKE WAS SHAPED like a transistor, which had just turned ten years old. Walter Brattain—now older, now grayer—came down from his office at Murray Hill to cut the first slice. Kelly looked on, wearing a dark suit and a tie patterned with the atomic energy symbol, three electrons running ovals around a nucleus. It was a daylong program for journalists all over the country—demonstrations on fabrication techniques and exhibitions exploring the transistor's military and commercial applications, as well as its potential uses in electronic switching, pulse code modulation, and the like. Kelly made a keynote speech to the crowd. In the previous year, he noted, 30 million transistors had been manufactured and semiconductors were now a $100 million industry. The transistor would infiltrate the industries created by vacuum tubes and would be the driving force in transforming the world in ways "yet undreamed."[18]

One striking proof of this assertion was imminent. The capacity of the two-year-old transatlantic cables would soon be increased thanks to a transistorized technology developed at the Labs known as TASI, or Time Assignment Speech Interpolation. How this worked seemed a marvel. Bell Labs technicians had long known that when someone is talking on a telephone line, they send out signals only about 35 percent of the time. Between all the spoken words there are gaps, pauses, hesitations. The TASI machinery would "inspect" the voices coming through the channels two thousand times per second and would switch back and forth as the talkers' speech paused and resumed, paused and resumed. "During pauses or listening times," one Bell Labs manager told the audience at the tenth-anniversary gathering, "a talker is disconnected from the channel so that it may be used for another; and, upon his resumption of speech, he is automatically connected to another channel in a millionth of a second." TASI could save both the money and effort of installing another cable. TASI made enough room on the cable to double the traffic from thirty-six channels to seventy-two. It was better; it was cheaper.

If there was an air of victory at Bell Labs in the late 1950s, it was in the dawning recognition of an empire built. Though not without its surprises, the future seemed to be weaving itself in a way Kelly had foreseen many years before. He had predicted grand vistas for the postwar electronics industry even before the transistor; he had also insisted that basic scientific research could translate into astounding computer and military applications, as well as miracles within the communications systems—"a telephone system of the future," as he had said in 1951, "much more like the biological systems of man's brain and nervous system."

In addition to the ocean cable and the military defense systems and the Nobel Prize, Kelly's management had been validated in *The Organization Man*, William Whyte's influential 1956 book that analyzed the conformity of America's corporate culture and the merits of creative thinking. "If ever there were proof of the virtues of free research, General Electric and Bell Labs provide it," Whyte wrote, pointing in particular to the achievements of thinkers like Claude Shannon. "Of all corporations' research groups these two have been the two outstandingly profitable ones . . . of all corporation research groups these two have consistently attracted the most brilliant men. Why? The third fact explains the other two. Of all corporation research groups these two are precisely the two that believe in 'idle curiosity.'"[19]

Even more flattering was an extremely detailed November 1958 *Fortune* story that christened Kelly's shop as "The World's Greatest Industrial Lab." Francis Bello, who had written on Shannon's information theory and the transistor, had spent months at Murray Hill chronicling almost every aspect of its research and development. "The preeminent discovery of the twentieth century is the power of organized scientific research," Bello began. "The industrial enterprise that has carried out this mobilization most brilliantly in the U.S.—and indeed the world—is Bell Telephone Laboratories, Inc."[20] The story opened with Kelly and Fisk together, noting that Kelly's retirement as president was just a few months away. Fisk looked to be the heir apparent. Usually in photographs Kelly was grim and unsmiling; clutching his cigarette, he would appear as if trapped in a moment of forced repose, ready to burst with kinetic relief when the shutter clicked. But the *Fortune* portrait showed another man entirely. Kelly had the look of a man kicking back at the end of the day. On his face was what might have been a grin.

Bello highlighted the combined power of the transistor and Shannon's information theory to create the future. It's likely that at the time of their breakthroughs, Shannon and Shockley did not see their work as being linked.[21] But within a decade their ideas had become intertwined. As Bello wrote, "The transistor pointed the way to tiny, inexpensive, and indefinitely lived devices—requiring little power—that could be used in very large numbers to implement the teachings of Shannon's theory."

The executives at Bell Labs would no doubt agree with this observation. But some time later, a number of Labs veterans would also reflect that Shannon's and Shockley's work had fostered something potentially self-defeating. One of these scientists was Max Mathews, who had joined the acoustics department at the Labs in 1955 and eventually rose to department head of its acoustic and behavioral research unit. In many respects, says Mathews, a phone monopoly in the early part of the twentieth century made perfect sense. Analog signals—the waves that carry phone calls—are very fragile. "If you're going to send sound a long way, you have to send it through fifty amplifiers," he explains, just as the transatlantic cable did. "The only thing that would work is if all the amplifiers in the path were designed and controlled by one entity, being the AT&T company. That was a natural monopoly. The whole system—an analog system—wouldn't work if it was done by a myriad of companies."[22]

But when Shannon explained how all messages could be classified as information, and all information could be digitally coded, it hinted at the end of this necessary monopoly. Digital information as Shannon envisioned it was durable and portable. In time, any company could code and send a message digitally, and any company could uncode it. And with transistors, which were increasingly cheap and essential to digital transmission, the process would get easier by the year. Mathews argued that Shannon's theorem "was the mathematical basis for breaking up the Bell System." If that was so, then perhaps Shockley's work would be the technical basis for a breakup. The patents, after all, were now there for the taking. And depending on how it played out, one might attach a corollary to Kelly's loose formula for innovation—namely, that in any company's greatest achievements one might, with the clarity of hindsight, locate the beginnings of its own demise.

Part 2

Twelve

AN INSTIGATOR

J ohn Robinson Pierce's first book was entitled *How to Build and Fly Gliders*; it was published in 1929 and cost one dollar. It was an attempt to cash in on the fad among Pierce's teenage friends in Southern California of building large glider planes that could be launched from a high dune and piloted by a single man, circling around, up, and down—sometimes dangerously so, like a berserk kite—in the air currents wafting in from the Pacific. "In those first days," Pierce later noted, "it was practiced not by wealthy men or sportsmen, not by those who knew something of aviation, but by the ignorant, the odd, the impecunious. Those who had a romantic turn and plenty of time and scraped together a hundred dollars or so built gliders."[1] A glider pilot like Pierce sat on a wooden bench called a "skid"; there was nothing else between him and the earth, hundreds of feet below.[2] Often the glider and its pilot came hurtling down, their descent attributable to an equipment malfunction or a sudden gust or an error in judgment. Sometimes, too—it had happened to several of Pierce's friends—a pilot died on the beach. His comrades would pull his body, bloodied and limp, from a heap of splintered wood and torn sailcloth.

Pierce's book was ninety-three pages long. Since he knew only a modest amount about gliding and even less about writing, he composed only forty of those pages himself. The remaining text was a montage of government reports on aeronautics that he obtained and stuffed into the manuscript to make it publishable. *Gliders* was, by Pierce's later admission, "atrocious." On the other hand, the title and the rushed nature of the endeavor said a lot about the man Pierce was becoming. For one thing, it demonstrated that he enjoyed pushing ideas and projects on people—"dropping them on our desks like an egg," as one of his colleagues would later put it.[3] For another, the book suggested that even as a young man Pierce did things quickly and then moved on, excited by another new idea and rarely polishing a finished product after the idea took shape. When he happened to reflect back on the glider book many years later, for instance, he wondered if his work had done actual harm: "Because of me, did human beings build crazy gliders without benefit of engineering, and kill themselves therewith? I wouldn't be a bit surprised." He seemed less troubled by that prospect than by the poor quality of his prose.

Standing about five foot ten and a skeletal 126 pounds,[4] with wispy blond hair that had receded dramatically by his late twenties, Pierce impressed people mainly with his nervous energy, droll demeanor, and a tendency to become easily bored. He was so thin and slight that even in late middle age he could fit into the slender office lockers that staffers used to stow their jackets or lab coats. "People think you can't fit into these, but you really can," Pierce said one day to Henry Landau, a Bell mathematician, who looked up from his desk to see Pierce walk unannounced into his office, squeeze himself into Landau's locker, close the locker door, open it, squeeze himself out, and then exit the room.[5] It was quite common for Pierce to suddenly enter or leave a conversation or a meal halfway through. Sometimes it was involuntary—something in his makeup, "in the way his mind would click on and click off," as his colleague Bob Lucky, who later succeeded Pierce in his management position at Bell Labs, describes it. Other times it was by design. Lucky recalls that during a phone call Pierce might suddenly hang up in the middle of his own sentence, leaving the person on the other end with the impression that a technical glitch had ended the call. No one could imagine that he would hang up on himself.

There was no obvious explanation for his peculiar social rhythms. Pierce had grown up happily in Iowa and Minnesota; he admired his parents and was especially close with his mother, in part because (as he once told an interviewer) she had a sharper mind than his father.[6] His father sold women's hats to clothing stores, a job that often took him out of town for weeks on end, leaving Pierce's mother in charge at home. He was always interested in technical things, even before he could understand them. Before he learned how to read, he would ask his mother to get him library books on electromotive force; as he grew older, he and his friends played with electric motors and steam engines, crystal radio sets and vacuum tube receivers. In the mid-1920s, Pierce, an only child, moved with his parents to California. At his high school in Long Beach, he discovered that algebra came easily to him. Then he discovered that geometry came easily to him. Then he discovered that chemistry came easily to him. He graduated first in his class. Later he would say that during these years he saw "a glimmer of the dawning of the idea that things can be understood, and that learning, in science at least, is understanding." His forays with his new gliding friends were important, too. "This"—he said about constructing gliders—"was the first time I built something complicated that really worked, that was a practical realization of purpose rather than mere tinkering."[7]

He was moving, unwittingly, along a path similar to that of other men—a brotherhood of radio enthusiasts still unbeknownst to him—who were on their way to Bell Labs as scientists and engineers. Pierce's parents hadn't gone to college, but they encouraged their son to take an entrance examination for Caltech, the science school in nearby Pasadena. "They were bothered when I was young that I seemed to be not very capable of dealing with the world," he said later, noting that it was clear to both parents that he would never succeed as a hat salesman like his father.[8] When Pierce passed the examination he decided to go. Caltech was now Robert Millikan's school, a fact that put Pierce, also unknowingly, in the center of a web of contacts that led east—to Frank Jewett, Bell Labs' president, for instance, who was still close with Millikan. Caltech was also home to several other physicists destined for the Labs, like Bill Shockley, who had come through Caltech just a few years earlier, and Dean Woolridge, a classmate of Pierce's.

Pierce wasn't drawn to physics, though. At first he thought he might try chemistry, a science of unforgiving precision, which proved a misstep for a young man with a fascination for electronics and a hummingbird-like tendency to flutter from idea to idea. He was befuddled by his introductory chemistry lectures; even worse, his experiments—"I dropped things, spilled things," he recalled—were disasters.[9] Aeronautical engineering seemed a sensible next step, especially for a young man who had built and flown gliders, and had even written a book on the subject, but he was soon bored with that subject, too, claiming later that his instructors made him draw pictures of rivets over and over again. That left electrical engineering, and here Pierce finally felt at home. He quickly demonstrated his expertise in electronics classes and was soon asked to help teach; a professor who had a side job as a legal witness even hired Pierce to dissect radio circuits for him in an effort to prove patent violations.[10] Years later, Pierce would avoid giving reporters, as well as readers of his own voluminous writings, the sense that he was calculating or unusually accomplished, sometimes offering the impression instead that his life at Caltech, and the successes that followed, were mostly serendipitous. Though he was eventually the subject of a long profile in the *New Yorker*, Pierce was inclined to describe his career as a procession of fortunate events rather than the product of his own restless intellect.[11] He described himself, depending on the circumstances, as either lucky, lazy, stumbling, or shambling. And sometimes he purported to be all of those things at once.

No one made that mistake at Caltech. "Even then, everyone knew who John Pierce was," Chuck Elmendorf, Pierce's friend at Caltech, recalls. In part it was Pierce's eccentricity. But mainly it was his dazzling quickness of mind. As Elmendorf, an electrical engineer, recalls, "Everyone *knew* John Pierce was going to be a great man."[12]

ELMENDORF ARRIVED in New York City during a sweltering week in the summer of 1936. He had been hired to work at Bell Labs' West Street building in Greenwich Village, and after finishing his master's degree at Caltech in engineering, he had driven with a friend from California to Chicago. In Chicago he had hopped on a train to New York. "It was one of the hottest days in New York history," he recalls, pegging the exact date of his arrival in Manhattan as July 9, 1936. "I went down to West Street and met with a guy named Ernie Waters. Ernie was the personnel manager, and he gave me 275 dollars." The money, a substantial amount in a country just emerging from the Great Depression, was ostensibly meant to reimburse him for his first-class train ticket.

"Ernie, I didn't travel first class on the train," Elmendorf admitted.

Waters looked up at Elmendorf. "If we thought you'd have gone first class," he replied, "then we wouldn't have hired you."

Waters suggested to Elmendorf that he go to the YMCA on 34th Street, where rooms cost fifty cents a night, and Elmendorf immediately took his advice. A few weeks later he moved up to a different YMCA, "a nicer one," as he recalls, that was located around 56th Street. One day in August at his new digs, he bumped into John Pierce, who had also been hired in the Labs' research department. The two men hadn't seen each other in several months. After finishing his PhD at Caltech, Pierce had taken some time before his arrival in New York to travel alone, on a trip financed by his parents, through Europe. Mostly he had bicycled around England, but he had also taken a quick tour of the Continent. "I went to Paris, and to the north of Italy, to Viareggio," he later recalled. "I went to Munich and went to the opera." It was three years after Hitler's ascension to power, but Pierce tended to be oblivious to things that didn't hold his interest. He conceded later he was almost completely unaware of Europe's political crosscurrents, save for the fact that in one open-air café, after some Italians drank a toast to Roosevelt in honor of the American visitor, he returned the favor by toasting Mussolini. "That was the right answer, apparently," he said later.[13] Often when Pierce didn't know what to do, he made a good choice anyway.

Pierce and Elmendorf decided to move out of the Y and find a place together. As Elmendorf recalls, they rented a one-room apartment on the first floor of the London Terrace apartment complex at 23rd Street and 9th Avenue. The apartment had one Murphy bed. "As a PhD John was getting $33 a week," Elmendorf recalls. "As a master's they were paying me $29 a week. So he got the bed." At the apartment, Elmendorf did the cooking and Pierce did the washing up. The two men were within walking distance of work—fifteen minutes south, at a brisk pace, was all it took.

At Bell Labs, the two men were not on an equal footing. Elmendorf was placed in what he calls "the salt mines," a lowly department where he was tasked with designing equipment for the phone system, a terrible letdown for an engineer who aspired to research complex circuits. Years later, after a career in which he had worked his way up to a high executive position with AT&T, Elmendorf discovered that his first job was due partly to his friend Pierce, who when asked about his colleague from Caltech had informed the Bell Labs management—not with malice but in keeping with his habit of saying whatever might be on his mind—that Elmendorf "wasn't that smart."[14] In relation to Pierce, of course, few people were. Pierce himself was assigned to work in the research department on vacuum tubes, where he was given free rein to pursue any ideas he might have. He considered the experience equivalent to being cast adrift without a compass. "Too much freedom is horrible," he would say in describing his first few months at the Labs. Indeed, he eventually came to believe that freedom in research was similar to food; it was necessary, but moderation was usually preferable to excess. It was almost certainly true, however, that Pierce's freedom was a reflection of his supervisors' belief in the recruit.

Soon after his arrival at the Labs, Pierce met Bill Shockley—"a new employee who was making the rounds of various parts of the research department," he recalled.[15] Shockley patiently explained the physics of vacuum tubes to Pierce, and the men quickly became friends. And soon afterward, Pierce began to find his way. He and Shockley worked together researching various devices and writing several papers during the late 1930s. But Pierce began to show a knack for inventing things on his own, too, putting together a variety of new types of highly complex and specialized vacuum devices—electron multipliers, they were called, and later reflex klystrons, which were useful for radar—that the Labs quickly patented.

Pierce always gravitated toward the smartest people in the building, just as he had gravitated toward his mother rather than his father. It was also true that the smartest people gravitated toward him. Surprisingly, he wasn't especially intimidated by the likes of Shockley. And Pierce's awkward manner and social peculiarities did not seem to preclude him from fitting in; on the contrary, he developed a wide circle of devoted admirers, charmed by his wit and his lively mind. Not long after he'd started at West Street, the young engineer was stopped in the hallway at Bell Labs by an unfamiliar manager—a man who seemed every bit as brusque as Pierce.

"What are you doing here?" the man asked him.

Pierce looked at him with suspicion. "Who are you?" he demanded in response.

"I'm Mervin Kelly," the man said.

It was one of Pierce's favorite stories, one that demonstrated both his obliviousness and his audacity. In time, Pierce would consider Mervin Kelly his great hero—one of two he had at Bell Labs. A close colleague of both men would later marvel that Kelly and Pierce didn't obliterate each other with their matching energies and enthusiasms.[16] But in many ways their mutual admiration made sense. Those who worked for Pierce, just as those who worked for Kelly, were often frightened of him—frightened of the way he raked through their ideas, ruthlessly and without regard to their feelings, in an attempt to separate the good from the bad. "You took your life in your hands every time you went into his office," recalls one of his former employees. Pierce, like Kelly, was a man of action, a man of strong opinions, and above all things a pragmatist in regard to science and innovation. Pierce, like Kelly, ran up and down staircases. He needed to get where he was going as fast as possible.

Pierce's other hero at Bell Labs was none of those things. He was a dreamer as well as an unrepentant futurist; and he was someone with no aspirations whatsoever to manage people or wield power. That would be Pierce's good friend in the mathematics department, Claude Shannon.

AN INSTIGATOR is different from a genius, but just as uncommon. An instigator is different, too, from the most skillful manager, someone able to wrest excellence out of people who might otherwise fall short. Somewhere between Shannon (the genius) and Kelly (the manager), Pierce steered a course for himself at Bell Labs as an instigator.

"I tried to get other people to do things, I'm lazy," Pierce once told an interviewer.

"Do you think this has helped your career?" the interviewer asked.

"Well, it was my career," Pierce replied.[17]

It was probably more accurate to say that Pierce had too many ideas to actually pursue on his own, and too many interests—airplanes, electronics, acoustics, telephony, psychology, philosophy, computers, music, language, writing, art—to focus on any single pursuit. Also, as he readily admitted, organization was not his strong suit. "John is not the typical inventor," Pierce's good friend Harald Friis, the director of the Labs' microwave research lab in Holmdel, New Jersey, told the *New Yorker* writer Calvin Tomkins. "His mind goes off at too many different angles and sees too many different possibilities in everything. He is like a child in that, but a very mature child."[18] Pierce's real talent, according to Friis and Pierce himself, was in getting people interested in something that hadn't really occurred to them before.

"You should do something on that," Pierce would suggest to Shannon about some idea or another. "Should? What does should mean?" Shannon would reply. Then both men would laugh. It was Shannon's way of saying he was never going to follow Pierce's suggestion, just as he never followed anyone's.[19]

But when most people heard Pierce float an idea, they didn't laugh. According to Pierce, he had made certain suggestions that led to Jack Morton's early work on something that came to be known as the Morton triode—a powerful vacuum device that would be essential for the vast national system of microwave relay towers. A few years later, Pierce had been the third collaborator, with Shannon and another member of the technical staff, Barney Oliver, who was destined to run Hewlett-Packard's research laboratories, in the first definitive paper on pulse code modulation, or PCM, the new method for sending communications that would rely on pulses representing digital information rather than on electrical waves.

Around that same time, Walter Brattain had bumped into Pierce in the hallway at Murray Hill and invited him into his office. Brattain was looking for a name for the new device he and John Bardeen had built, and Pierce had a reputation around the Labs as a wordsmith. As usual, Pierce didn't hesitate before tossing out a suggestion: How about calling it a transistor?[20] That his suggestion was eventually adopted after a vote was, Pierce would say, "my one claim to eternal fame."[21]

He seemed to have a knack for involving himself—or being right nearby—in situations that made historical ripples. Pierce was apparently the last person at Bell Labs that Bill Shockley spoke with—as recorded in a cryptic entry in Shockley's diary, for noon on December 31, 1947—before he began working around the clock on his idea for the junction transistor.[22] Pierce had suggested that his friend Arthur C. Clarke compose a history of overseas communication,[23] and he demonstrated for Clarke, during a visit to Murray Hill, a computer rendition, created by a group of scientists working in acoustics at Bell Labs, of "A Bicycle Built for Two." This rendition eventually found its way into the film *2001: A Space Odyssey*.[24] Pierce had also been the supervisor who came by when the mathematicians were playing their guessing game on the blackboard, "Convergence in Webster," and scratched out the infamous sentence, *You Are All Fired*.[25]

Essentially, Pierce's job was coming up with ideas for the next generation of communications technology. He was vain about his intelligence, but not so vain as to consider his own ideas automatically superior to those of his colleagues. In looking for what was interesting and doable, he could be an able scavenger. One day in 1943, while Pierce was working on vacuum devices for the military—like everyone else at the Labs, he had been engaged in engineering work for World War II—he read a British report on new developments in the field. Something caught his eye. A man working at Oxford, an Austrian immigrant named Rudi Kompfner, had created a new vacuum tube that seemed decidedly different from anything Pierce had seen before. It was called the traveling wave tube, and was sometimes abbreviated as TWT. Although Pierce later liked to say that the device was simple, the mechanism within was quite complex: The tube was pencil-thin and around a foot long, with an interior consisting of a long wire wound in a helix pattern that resembled a stretched screen-door spring.[26] A communications signal—a traveling wave—was sent along the curling spring of wire; meanwhile a focused beam of electrons was sent through the center of the spring. The two interacted, and the result was a considerable transfer of energy from the beam of electrons to the spiraling signal wave. In plain terms, the merits of the device were twofold: (1) a spectacular amplification of the signal; and (2) a remarkable capacity, or *bandwidth*, meaning it could amplify a vast number of signals at one time. Sitting in his Bell Labs office in the West Village—the office with a five-drawer file cabinet that he had labeled "bottom drawer," "next-to bottom drawer," "middle drawer," "next-to top drawer," and "top drawer"[27]—Pierce saw in the report something potentially invaluable to the phone system. "It was interesting because it seemed to have great prospects, but it had all sorts of problems," he remarked. Such problems were at once theoretical—how did the traveling wave tube work?—and developmental—could Bell Labs make the device reliable enough for the system?

In November 1944 Pierce traveled to England, on a trip sponsored jointly by Bell Labs and the U.S. Navy. Over the course of an exhausting four-week stretch he visited twenty different British industrial labs and government research installations. Mostly he was trying to assess for the United States the state of the art of tube development in England.[28] In the midst of the trip he sought out Kompfner. He was searching out new technologies, not new friendships, but here he encountered a man who was perhaps closer to himself in sensibility, quickness, and intelligence than anyone else he would ever meet. Kompfner made an indelible impression on everyone. He had begun his career as a trained architect and had drifted toward physics and inventing only because he found it interesting. He was less eccentric than Pierce, and more worldly. Buffed with a European polish, Kompfner was blessed with an uncommon wit and elegance. He was a world-class storyteller and a superb skier. But like Pierce, he could be blunt and dismissive of ideas or people he considered ignorant. He always spoke his mind to Pierce, who would in time seek Kompfner's advice on almost all of his ideas.

Pierce later liked to say that "Rudi invented the traveling wave tube and I discovered it"—a quip Kompfner enjoyed and never disputed. Pierce brought Kompfner's design back to Bell Labs and started working on a prototype. At the unveiling of the device to the press in 1946, the traveling wave tube was hailed as an invention that could make possible a coast-to-coast network "over which 10,000 telephone conversations may go simultaneously, or all the television programs needed for all the video stations likely to be operating in this country in years to come." The *New York Times* report quoted an unattributed source at Bell Labs, quite possibly Pierce, noting that perhaps for the first time, "radio men will have created a device that eventually may provide the means of setting up more channels for long-distance communications than they will know what to do with."[29]

In fact, the new tube wasn't quite ready for that. And as it turned out, Pierce—along with a team of engineers and scientists that included Kompfner, who joined Bell Labs in 1951—spent the next thirteen years investigating, testing, and perfecting various kinds of traveling wave tubes. Pierce had been correct in some respects about the traveling wave tube's potential. But as he came to understand, inventions don't necessarily evolve into the innovations one might at first foresee. Humans all suffered from a terrible habit of shoving new ideas into old paradigms. "Everyone faces the future with their eyes firmly on the past," Pierce said, "and they don't see what's going to happen next."[30]

WE KNOW SO MUCH about John Pierce's opinions on Bell Labs and innovation because his career as a technologist was complemented, almost from the start, by his career as a writer. He had made great strides after the publication of his 1929 book on gliders. By the late 1940s, as he was becoming immersed in his work on traveling wave tubes, he was turning out essays, science fiction stories, lectures, and books. This work was in addition to his lengthy technical memoranda (distributed internally at Bell Labs) and his papers (published in various engineering journals) describing his vacuum tube research. Some of Pierce's writing was published under his own name; some of it, however, was done under the pseudonyms John Roberts and j.j. coupling, the latter of which he borrowed on a whim from the physics literature (a j-j coupling described the spin and orbital functions of electrons), so that he would not have to seek clearance from the Bell Labs publications department, a sometimes formidable obstacle, each time he wanted to disseminate one of his new ideas.[31] Pierce would later describe himself as a "writerholic," and indeed his literary output was prodigious. It was also done in haste. "I am rather alarmed at your habit of sending finished articles before we have had a chance to discuss them," an editor at *Scientific American* told him in 1949.[32] Manuscripts were frequently rejected by publishers and, in the case of Pierce's nonfiction work, riddled with minor errors.[33]

The rejections of his stories must have stung him terribly. A publicly unsentimental man, Pierce nevertheless kept in his files until his death what appeared to be every rejection letter he'd ever received. There were hundreds. His factual mistakes, however, were not something he worried much about. In 1950 he wrote a book on traveling wave tubes, pouring "everything I knew about the subject" into the text. He made many small errors, he later conceded, but had gotten the gist of things right. And that was what he thought most important. With what seemed like amusement, he noted that the corrections amounted to a stack of papers a quarter of an inch thick. "There was," he acknowledged, "about one correction for every page of the book."[34]

Pierce's closest friends recognized that his wry, skeptical, and crusty exterior concealed the warm inner core of a romantic. Sometimes a person visiting Pierce in his office would find him reading poetry; sometimes they would find him reading poetry in another language. A devotee of verse—in college he had organized a group of students and then convinced a professor to read all of *Paradise Lost* to them aloud—he often tried to write poems, usually with stilted and unsuccessful results. His science fiction, meanwhile, was imaginative and promising, and something he worked on far more diligently. He and Shannon often discussed their favorite stories and traded books. Pierce often admitted that he perceived in writing and publishing—and in writers themselves—a kind of glamour. In fact, while he was in England in 1944 visiting Rudi Kompfner, he decided in London to look up one of his literary idols, the author H. G. Wells. Pierce didn't know Wells, or anybody who knew him. But he explained in a note to the author that he and his traveling companion, an accomplished Bell Labs engineer named Homer Hagstrum, were important American scientists visiting London—at the time an exaggeration—and would like to meet. Wells invited the two for tea in his house, and Pierce and Hagstrum complied.

It was a dreary November afternoon, thick with London fog. "He sort of appeared at the door, in a huge dressing gown, with the collar turned up around the neck," Pierce wrote in notes he jotted down just after the meeting. Wells was sick, with a bad cough, but he was hospitable. The men sat down in the dining room, at a large table laid with doilies, and a tea service arrived by dumbwaiter. Pierce jumped up to pour the tea, only to realize that "I didn't know how to go about it." Did one put the milk in first, or the tea? Wells gave him instructions, and the men settled down to eat toast and cakes and cucumber sandwiches. Pierce let Wells know that one of his science fiction concepts—an atomic bomb—was coming true: America was building one. He had deduced this from the way most of the country's good physicists were disappearing and being directed to secret laboratories around the country. Pierce told Wells that he and his fellow engineers joked that promising scientists had been "body snatched." But Wells was largely uninterested in what Pierce was saying. He wanted to talk about politics—among other things, Churchill, Roosevelt, and race in America.[35] Pierce was happy to talk about whatever Wells preferred. Afterward, when Pierce and Hagstrum left the house and headed home in the fog, he concluded that he was "feeling rather bad about the whole imposition." Wells had struck him as being both old and tired.[36]

The meeting didn't diminish Pierce's desire to write, however. In the 1930s and 1940s, Karl Darrow, a gifted writer and explainer employed by Bell Labs, had translated in various journals the new ideas of quantum mechanics for a younger generation of physicists. In many respects, Pierce picked up in the 1950s where Darrow left off, but instead of technical journals his venue was books and magazines for a general audience. Often Pierce's books on technology weren't nearly as accessible as he imagined; frequently they were disorganized and dense with mathematics. Nevertheless, in addition to his book on traveling wave tubes, he wrote general-interest books, some of them fairly successful, about cathode-ray tubes, electromagnetic waves, acoustics, communications infrastructure, information theory, and transistors. He would say that instead of naming the transistor, he wished he had actually invented it. He also remarked that rather than writing about information theory, he wished that he, rather than Shannon, had thought of it.

He thought of something else, though. In 1952, Pierce wrote a nonfiction essay for *Astounding Science Fiction* entitled "Don't Write: Telegraph." The piece explored the idea of sending messages to and from the moon. Pierce later noted, "What struck me then was how much easier it would be to communicate between the moon and the earth than across the United States—if only one could put the apparatus in place." Microwaves travel vast distances in straight lines, which is ideal in space. Here on earth, however, they can't follow the curvature of the planet. To send microwave signals across the United States, therefore, as the phone company was now doing with its nationwide system of relay towers, required stations every thirty miles or so. "The truth," Pierce wrote, "is that you could order equipment for an Earth-Moon link from any of several manufacturers."

Slowly, the idea evolved in his mind. In October 1954, he was invited to give a talk about space in Princeton at a convention of the Institute of Radio Engineers. Pierce decided he would discuss an idea he had for communications satellites—that is, orbiting unmanned spaceships that could relay communications (radio, telephone, television, or the like) from one great distance to another. A terrestrial signal could be directed toward the orbiting satellite in space; the satellite, much like a mirror, could in turn direct the signal to another part of the globe. Pierce didn't consider himself the inventor of this idea; it was, he would later say, "in the air." In fact, unbeknownst to Pierce, Arthur Clarke had written an obscure paper about ten years before suggesting that a small number of satellites, orbiting the earth at a height of about 22,300 miles, could connect the continents. Clarke never developed the idea any further and quickly lost interest in it. "There seemed nothing more that could be said until technical developments had validated (or invalidated) the basic concept," he later wrote. In Pierce's talk, however, he made some detailed calculations about satellites. He concluded that orbiting relays might not be financially viable over land; in the United States, the Bell System already had an intricate system of coaxial cables and microwave links. The oceans were a different story. The new cable that Bell Labs was planning for the Atlantic crossing in 1954 would carry only thirty-six telephone channels at tremendous expense and tremendous risk of mechanical failure. A satellite could satisfy the need for more connections without laying more cable.

One academic in the audience that day in Princeton suggested to Pierce that he publish his talk, which he soon did in the journal *Jet Propulsion*. "But what could be done about satellite communications in a practical way?" Pierce wondered. "At the time, nothing." He questioned whether he had fallen into a trap of speculation, something a self-styled pragmatist like Pierce despised. There were no satellites yet of any kind, and there were apparently no rockets capable of launching such devices. It was doubtful, moreover, whether the proper technology even existed yet to operate a useful communications satellite. As Pierce often observed ruefully, "We do what we can, not what we think we should or what we want to do."

Thirteen

ON CRAWFORD HILL

I deas may come to us out of order in point of time," the first director of the Rockefeller Institute for Medical Research, Simon Flexner, once remarked. "We may discover a detail of the façade before we know too much about the foundation. But in the end all knowledge has its place."[1] Flexner was speaking of the biological sciences; he had seen how the individual fruits of research might have little use—and little clarity—but could accrue over time to create a grand idea. The same might be said about any branch of the sciences, or about many of the large projects in the planning stages at Bell Labs. The transatlantic cable, for instance, which had been on the drawing boards for several decades until a variety of developments made it technologically feasible as well as cost-effective, was a good example. For communications satellites, the main idea had come too early. But was Pierce decades early, or just a few years?

Already, by the time he had given his talk in Princeton, some of the elements needed to create a satellite were available. The transistor—rugged and miserly in its power requirements—was in wide production. It could undoubtedly be useful in an orbiting satellite, Pierce realized. The traveling wave tube, now moving out of its development stage, could likewise be valuable, since it could amplify a multitude of telephone or television channels simultaneously. Much of the challenge in creating satellite communications lay not in the satellites themselves but in building an adequate system for transmitting and receiving signals from the ground, as well as a system for tracking the satellite as it moved across the sky. In this regard, a third existing technology appeared vital. It was the horn antenna, which had been designed by Bell Labs' Harald Friis at the rural Holmdel lab in southern New Jersey. Horn antennas were already a crucial component in microwave towers across the country: They allowed for the reception of signals in a focused manner that greatly reduced surrounding noise and interference. There was no reason to think they couldn't be adapted for satellite communications.

By late 1956, about two years after Pierce's talk at Princeton, some other necessary elements for his idea had sprung into existence. Pierce realized that the batteries in an orbiting satellite would wear out unless the satellite could tap into a dependable and inexhaustible power source. Bell Labs' silicon solar cell had been celebrated until the silicon strips had proven too expensive and too inefficient for use in rural phone installations. But satellites seemed to be the perfect problem for this solution. Since there was no other way to recharge a satellite's batteries in space, the expense or inefficiency of solar cells was no hindrance.

Just as important was another development, which originated mainly from the work of a former colleague of Pierce's. Charles Townes had come to Bell Labs from Caltech in 1939, just after Pierce; Townes was the curious recruit who had used the Labs' travel advance that year as a means to fund an adventurous bus tour for himself, taking him south from California through Mexico, and then up again through Texas, North Carolina, and ultimately the Bell Labs offices on West Street. Townes was an experimental physicist. Regarded as brilliant by his colleagues at Bell Labs, he had been reluctant to spend his career in an industrial setting, and had left Bell for a position at Columbia University a few years after World War II ended. In the late 1940s, Townes's work was concentrated in the field of microwave spectroscopy—analyzing the "spectra" (or energy emissions) of excited molecules as a means of studying their makeup. By the early 1950s, he had moved on to something known as "stimulated emission." Townes believed that under the right circumstances, and inside a complex apparatus he'd spent years building in his Columbia laboratory, he could (1) take a gas such as ammonia and direct its molecules through a slit to form a narrow stream; (2) move the narrow stream through an electric field; and then (3) have the stream finally enter something called a "resonant chamber." In the chamber, these ammonia molecules, now in a highly excited state, would become increasingly excited—to the point where they could generate a beam of highly focused energy.[2]

Townes faced skepticism for his project, even from his boss at Columbia, the Nobel laureate I. I. Rabi. But then his machine actually started working. When it came time to name the device, he and his colleagues dubbed it the maser, which stood for "microwave amplification by stimulated emission of radiation." Within a few years, physicists (including Townes himself) would try building similar devices that emitted energy in the form of light rather than microwaves. These devices were often called optical masers, until that name fell out of fashion. Afterward, they were simply known as lasers.

Townes's maser became operational in 1954. By 1957, a team at Bell Labs had built a more sophisticated maser that appeared to have some use in communications.[3] Its functional part looked like a huge metal test tube, about eight inches in diameter and a couple of feet long; a crystal about the size of a fingernail (rather than ammonia gas) was placed in a cavity at the bottom of the tube that was cooled to about minus 456 degrees Fahrenheit by a bath of liquid helium. The crystal was then bombarded with electrical energy from a power source called a "pump."[4] Pierce hadn't recognized that the maser might be useful for his satellite project, but one day his old friend Rudi Kompfner, home sick in bed, realized that the maser could be ideal in amplifying the faint signals coming from a satellite in orbit. Its sensitivity and fidelity would surpass any other known device. "As soon as I could get out of bed," Kompfner said, "I rushed over to tell John, and he agreed right away that the maser would make all the difference."[5]

So now there were transistors, the horn antenna, the traveling wave tube, solar cells, and the maser. Even with the right electronic components, though, communications satellites weren't going anywhere yet. There was still no proof that aeronautical engineers had developed rockets that could propel the idea into space. Proof arrived dramatically in October 1957 when the Soviet Union launched its Sputnik satellite.[6] A year later the United States, fearful of losing ground to the opposing superpower, launched its first satellite, Explorer I. The event was followed soon after by the National Aeronautics and Space Act, which created a new national space agency called NASA. With the technology as well as the money now available, the ideas Pierce had been writing about for the past few years were now plausibly real. When someone asked him for his reaction to the Sputnik launch, Pierce said, "It's like a writer of detective stories going home and finding a body in his living room."[7]

AS THEY WAITED for the gun to fire, the communications engineers poised at the starting line of the satellite era had two choices before them. Pierce had proposed that they could first explore the frontiers of space by launching either a "passive" satellite or an "active" one. A passive satellite would merely be a reflector of sorts that circled the earth in a low orbit, perhaps a thousand miles up, that engineers could bounce signals to and from. A transmission from California, for instance, could be directed at an angle toward the traveling satellite, which in turn could *passively* reflect the signal at a different angle back down to earth, perhaps to a receiving station in New Jersey. An active satellite, on the other hand, contained batteries, transistors, antennas, and vacuum tubes, with which it could amplify a signal received from earth before sending it back down. Essentially, it followed the same principles as the microwave relay towers already stationed around the country—except this relay was orbiting in space.

At least in theory, active satellites were better than passive ones. The signals directed from earth to a passive satellite are reflected out in all directions, for instance, so that those received at any one point on the ground might be so faint—perhaps a millionth of a billionth of the originally transmitted signal—that the task demanded extraordinarily sensitive equipment (huge horn antennas, expensive masers, and the like) for receiving even a single voice transmission. That wouldn't be the case with an active satellite, which could broadcast a broad band of strong, accessible signals and likewise receive instructions from ground stations on what signals to carry. (An active satellite could also carry television signals far more easily than a passive one.) On the other hand, Pierce considered himself conservative about any satellite gambit. He wasn't certain that Bell engineers knew enough yet to build a foolproof and durable active satellite—one that could operate for more than a few weeks or months. He also knew that the small research department at Bell Labs, unlike the huge development department, lacked the manpower and budget necessary for an active project. "There's a difference, you see, in thinking idly about something, and in setting out to do something," he explained to an interviewer in the early 1960s. "You begin to see what the problems are when you set out to do things, and that's why we thought [passive] would be a good idea." A passive satellite, he added, probably wouldn't be useful in terms of the business of communications. But it tested the possibility of orbiting relays before they were developed into something more. A passive satellite, in other words, was an *experiment*. Ten years earlier, Pierce had witnessed how problems with the transistor didn't show up until the device entered the development and production stage. Here, too, was a relatively low-risk opportunity to confront and solve the practical challenges of this new technology—"to get one's hands dirty"—before making a slew of big and expensive mistakes.[8]

In Pierce's caution one could discern some of the management wisdom of his mentor, Harald Friis, the pipe-smoking, Danish-born chief of Bell Labs' small Holmdel lab, where most of the Labs' microwave research was conducted. Friis often asked his young scientists, "Are you sure this isn't too big a ball of wax?"[9] He had a gentle way of helping them see if they could sharpen their vision and perceive the difference between what was doable and what was not. The passive satellite was certainly a big ball of wax, but the active satellite, in Pierce's view, was too big a ball of wax. Indeed, Pierce soon found out that the military, through their Advanced Research Projects Administration (ARPA), were thinking about building an extremely expensive active satellite called Advent. "The tendency of ARPA has been to project elaborate and complicated schemes," Pierce wrote derisively in a memo to Jack Morton after visiting with ARPA's directors.[10] Pierce thought the project was too much, too soon. He believed—correctly, it turned out—that Advent was hurtling toward failure.[11]

All through 1958, Pierce's main advocate for building a passive satellite at Bell Labs was his old friend Rudi Kompfner, the traveling wave tube inventor he had first met in England during the war. At the start, the two men had no precise idea of how they would get such a project to happen; nor did they have any sense yet whether Mervin Kelly, then in the last year of his Labs presidency, would agree to put several million dollars behind the idea. But various aspects of the project fell into place with surprising speed. Early that year, for instance, Pierce and Kompfner found out that a government engineer named William O'Sullivan, stationed at Langley Field in Virginia, was proposing various atmospheric tests by putting a huge aluminum-clad Mylar balloon—100 feet in diameter, but weighing only 136 pounds—into orbit about 800 or 1,000 miles above the earth. It struck Pierce as an ideal design for a passive satellite, and when he and Kompfner actually obtained samples of the balloon, they discovered, much to their satisfaction, that it would reflect 98 percent of the radio waves directed at it.

O'Sullivan couldn't get anyone to launch his balloon. But in the summer of 1958 Pierce and Kompfner began discussions with William Pickering, the director of the Jet Propulsion Laboratory in California, who was intrigued by the project. As Pierce recalled, Pickering "agreed to supply a west-coast ground station" that would communicate with another ground station in New Jersey "if we could get the balloon launched."[12] The men still didn't have a way to fire the balloon into space. But when NASA officially took over the Jet Propulsion Laboratory, just a few months later,

NASA's new chief, Keith Glennan, put his support behind the passive satellite launch, too. That problem now seemed solved.

Still, there had been the challenge of garnering support at Bell Labs during 1958. Bill Baker, a chemist who had become the head of research, supported Pierce's satellite idea. So did Baker's boss, Jim Fisk, now the executive vice president of the Labs. But during the summer, Mervin Kelly, relying partly on the mathematical analysis of Claude Shannon's old friend and office mate Brock McMillan, told Pierce one day during a lunch meeting that it was a nonstarter. McMillan's calculations had been woven into a three-page memo on the project, written by one of Kelly's deputies, that was hostile to both the economics and technology of the satellite project.[13] Whether Kelly's opinions were the starting point for the memo or were merely reinforced by it wasn't wholly clear. Pierce later would say that Kelly had a negative view of the satellite project from the very start.

Kelly may have been swayed by arguments within the Labs that satellite communications were too expensive to justify, or he may have worried that Bell Labs' entrance into the satellite business might upset the delicate state of relations between AT&T and the federal government. AT&T, he believed, should not be in the space business. But all of these concerns may have been magnified by Kelly's opposition to the kind of innovation that might later be described as "discontinuous."[14] Bell Labs had just completed the successful transatlantic cable; the future of communications to Europe and beyond appeared to reside in new and better cables. These would be incremental innovations. In such a vision of the future, orbiting satellites weren't only a risky and unproven technology; they were also—at least to a telephone executive with a well-defined, step-by-step ten-year plan for improving the system—a strange sideways leap.

When Kelly said something was dead, it was dead. Except to Pierce. He and Kompfner prepared a long memo in early October 1958 that made their case to Kelly yet again. Pierce showed the memo to his boss, Baker, but it was not sent on to Kelly, almost certainly on Baker's wise suggestion.[15] The strategy was to wait. "Kelly was a great hero of mine,"

Pierce later reflected, "and a great leader of Bell Laboratories. Even great men can be wrong." When Kelly retired at the end of 1958, Fisk took charge of the Labs. Pierce and Kompfner then circulated another proposal on January 6, 1959, suggesting that the cost for the satellite would be $1 million during 1959 "and more in subsequent years." It seemed possible, they noted, "that satellites could provide very broad-band transoceanic channels more cheaply than cable or other means." The two men also suggested that if Bell Labs didn't act, someone else would. "It is important for the Bell Laboratories to assume a leading role in research on satellite communications. This would enable us to evaluate satellite communication technically and to exploit it expeditiously if it proves advantageous to us."[16] This time, Pierce's proposal was authorized by AT&T as Case 20564. It was also given an official name: Project Echo. NASA, in turn, scheduled a tentative launch for the fall of 1959.[17]

Pierce and Kompfner asked Bill Jakes, a young electrical engineer, to take charge of the project. Jakes was wiry and athletic and quietly commanding; after growing up in Evanston, Illinois, he had spent most of World War II in charge of the radio room on an aircraft carrier in the South Pacific. Jakes's team on Echo quickly swelled to several dozen men, and to Pierce's relief Jakes soon took almost complete control. As Pierce would later say, Jakes's word was law. At times the younger man would remark to Pierce, empathetically, that it must have bothered him that he didn't get to run Echo's day-to-day operations. "I don't know what I said, but it certainly wasn't the full truth," Pierce later wrote. "The Echo ground terminal would never have got built or worked had I been the project engineer. Through a mixture of ineptitude or boredom, I would have flubbed." For Pierce, it was plenty of work just being an idea man. Jakes was the kind of guy who would actually make ideas happen.

AT THE TIME Echo was authorized, Pierce held a high position in Bell Labs' research department. His office was located at the Labs' Murray Hill location, the large campus of brick buildings twenty-five miles west of New York City where Shockley and Shannon had also worked before they made their respective moves to California and Massachusetts. Every day at noon at Murray Hill, he stretched out on a table in his office and took a nap. "I tried to call him one day," Bill Jakes recalls, "and his secretary said, 'Oh, you can't talk to him now, he's sleeping.'" At Murray Hill, Pierce, with several hundred employees reporting to him, held the title of executive director of communications research, which in organizational terms put him one step below Bill Baker, the Labs' vice president for research. (Baker, in turn, was one of several vice presidents who reported to Labs president Jim Fisk.) The satellite work, however, was centered not on Pierce and Baker and Fisk in Murray Hill but on Jakes in Holmdel, thirty miles south. Holmdel was where Pierce's friend Harald Friis had worked and where he had invented the horn antenna. And it was where Pierce's other good friend Rudi Kompfner, the traveling wave tube inventor, had now taken charge upon Friis's retirement in 1958.

Holmdel was a world unto itself. The laboratory and its environs comprised hundreds of acres of open land, assembled from three farms purchased by Bell Labs in 1929, where several dozen men did research on microwave transmissions. They worked just a few minutes from Sandy Hook beach and the Atlantic Ocean. "You drove off the main road," Bill Jakes recalls of his commute to work, "and it would be about a half mile until you came to the laboratory, with nothing but beautiful mown fields on both sides." In the 1940s, the vast and isolated property had been used for some of the early research on silicon and germanium. But mostly Holmdel was the Labs' central facility for the experimental installation of large antennas and to test emerging technologies on wireless microwave transmissions. The families of the men who worked at Holmdel would congregate there on the weekends for group picnics and to throw boomerangs in the empty fields; their children could play with an ancient water pump and explore an abandoned Dutch farmhouse on the grounds dating from the early 1700s. Years later some men, John Pierce included, would be moved nearly to tears as they recalled this vanished outpost. Its main laboratory building was a modest one-story wooden clapboard structure with three wings radiating from a peaked central core. The wings had so many windows that the building resem-

bled an exceedingly long glassed-in porch. Some of the men affection-
ately called the building the Turkey Shed, after a traveling poultry-feed
salesman, seeing the old farmhouse and the clapboard laboratory, mis-
took it for a working farm during a sales call. To work in Holmdel was to
work in a place that was breezy and bucolic and hushed in a country way.
"Lord, it was just *beautiful*," Jakes's wife, Mary, says—so beautiful that
you couldn't possibly think it was on the vanguard of radio research,
though in fact it was. Holmdel's appearance simply didn't jibe with any
conventional notions of laboratory science.[18]

Rudi Kompfner's deputy at Holmdel was named C. Chapin Cutler.
Technically speaking, Bill Jakes reported to both men, but on matters of
Echo they formed a triumvirate with Jakes in the lead. Around the time
Echo was authorized, Kompfner, Cutler, and Jakes went about buying
the equipment for building their ground receiving station. About a mile
north of the Holmdel lab was a flat-topped rise known as Crawford Hill,
about five hundred feet above sea level, with an expansive view to the
ocean just a few miles away, that the Holmdel men used for research on
antennas.[19] In the summer of 1959 plows arrived and construction began
atop the hill. Several small prefabricated buildings were set up to house
men and equipment. More important, two circles were cleared for the
installation of two large antennas.

One of the novel aspects of Echo was the intention to use sophisticated computer systems to track the satellite as it moved across the sky
and then automatically align the antennas in accordance with the balloon's trajectory. For about $230,000, Jakes's Echo crew purchased a
huge dish antenna, sixty feet in diameter, that would transmit signals to
the orbiting satellite—signals that the balloon would reflect to a ground
site being built in tandem by the engineers at the Jet Propulsion Laboratory in Goldstone, California, just south of Death Valley. The Echo team,
meanwhile, had also designed and requisitioned an immense horn antenna (the cost was about $128,000) resembling a kind of huge empty
tobacco pipe, which would *receive* signals sent from the Jet Propulsion
station in California as they bounced off the orbiting satellite.[20] It was,
in essence, a simple two-way conversation, but one made possible through

THE FIRST IDEA FACTORY: Western Electric's former headquarters on West Street in lower Manhattan, which in 1925 became the original home of Bell Labs. Kelly, Shockley, Pierce, Fisk, and Shannon all began their careers at the building; after the war, the men, along with the Labs' most important research projects, moved to New Jersey. *Courtesy of AT&T Archives and History Center*

Bettmann/Corbis

Courtesy of AT&T Archives and History Center

AIP Emilio Segre Visual Archives
Physics Today *Collection*

BELL LABS' EARLY ARCHITECTS: Robert Millikan (top left) served as a friend and advisor to Bell Labs' first president, Frank Jewett (top right), pictured in 1938, as well as a mentor to Mervin Kelly (right), photographed here soon after his arrival at Western Electric in 1917. After making a name for himself at the University of Chicago, Millikan became president of Caltech.

In 1915, Theodore Vail, AT&T's chairman, listens in on the first transcontinental phone conversation—connecting New York with San Francisco—from his vacation home on Jekyll Island, Georgia.

Courtesy of AT&T Archives and History Center

AIP Emilio Segre Visual Archives, Physics Today Collection   Photograph by Parker Studio, Courtesy of AIP Emilio Segre
Visual Archives

TWO YOUNG TURKS: Portraits of a young William Shockley (top left) and his friend from MIT, Jim Fisk (top right). "If that man gets hired," Shockley once said of Fisk, "we'll all be working for him in ten years." His prediction proved correct. BELOW: Building 1 on the Murray Hill, New Jersey, campus, just after World War II. Kelly considered the design of the building—where research scientists and development engineers from all disciplines were housed together in close proximity—yet another grand Bell Labs invention.

Courtesy of AT&T Archives and History Center

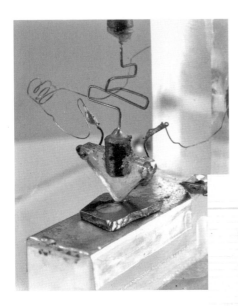

TRIUMPHS OF A MAGIC MONTH: A crucial page from the December 24, 1947, entry in Walter Brattain's notebook, the ur-text of the transistor's genesis; next to it, the first transistor. The germanium metal slab under the arrowhead is about one quarter the size of a penny.

Courtesy of AT&T Archives and History Center

DATE Dec 24 1947
CASE No. 38139-7

We obtained the following A.C. values at 1000 cycles

$E_g = .015$ R.M.S. volts $E_p = 1.5$ R.M.S volts

$P_g = \dfrac{}{5.4 \times 10^{-7} \text{ watts}}$ $P_p = 2.25 \times 10^{-5}$

Voltage gain 100 Power gain 40
Current less $\frac{1}{2.5}$

This unit was then connected in the following circuit.

[circuit diagram]

This circuit was actually spoken over and by switching the device in and out a distinct gain in speech level could be heard and seen on the scope presentation with no noticeable change in quality. By measurements at a fixed frequency

The solid-state triumvirate in Brattain's cluttered Murray Hill laboratory: from left to right, Walter Brattain, William Shockley, and John Bardeen.

Courtesy of AT&T Archives and History Center

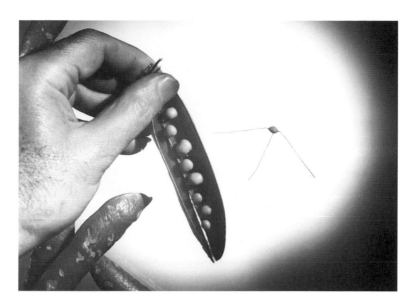

THE TECHNOLOGY OF ENVY: TOP: An early junction transistor, a breakthrough device based on a 1948 theory by Bill Shockley that was built in 1951. Bell Labs' publicity department was fond of comparing the technology to a pea or a kernel of corn. Shockley began formulating the idea as he dealt with the blow of not being one of the transistor's original inventors. "I did not want to be left behind on this one," he said. BOTTOM: Gerald Pearson, Daryl Chapin, and Cal Fuller, the creators of the first silicon solar cell. The circumstances of the innovation were serendipitous; all the men worked in different buildings.

Courtesy of AT&T Archives and History Center

BEHOLD, THESEUS: Top right: Claude Shannon, with the mouse-machine that could find its way through a maze and learn from its mistakes. The mouse's logic circuits were beneath the floor of the maze. Bell Labs' patent department was unimpressed with Theseus; Shannon's colleagues, on the other hand, thought it brilliant. Left: One of 107 microwave towers that served as links in a long-distance coast-to-coast network that opened in 1951. Right: John Pierce and Rudi Kompfner, circa 1951, standing before an array of traveling wave tubes.

READY FOR LAUNCH: TOP: The Echo satellite balloon was made from a thin film of Mylar and stood about ten stories high. BOTTOM: Telstar was the first active communications satellite; in addition to sending and receiving phone and television signals, it collected a trove of data about radiation in space. The black plates on the satellite face are solar cells. *Courtesy of AT&T Archives and History Center*

A LOST WORLD: The small corps of scientists and engineers at the Holmdel lab, in southern New Jersey, circa 1933. The men spent their days researching antenna technology in the "turkey shed" building. Just as often, they worked outdoors, amid hundreds of acres of mown lawns. "Lord, it was just beautiful," one visitor recalled. *Courtesy of AT&T Archives and History Center*

THE BLACK BOX: An austere new laboratory, viewed from outside (top) and from within its atrium, (bottom), that was designed by the architect Eero Saarinen. The building, opened in 1961, replaced the old lab in Holmdel, New Jersey, and served as a dramatic contrast to the Murray Hill buildings, thirty miles to the north. "Gone completely are the old claustrophobic, dreary, prison-like corridors," Saarinen remarked with pride. *Ezra Stoller©Esto*

AN INSTIGATOR: John Pierce in 1961, around the time of the Echo launch. "You took your life in your hands every time you went into his office," a colleague recalls. But Pierce's close friends realized that his tough and skeptical exterior hid the warm inner core of a romantic. *Yale Joel/Time & Life Pictures/Getty Images*

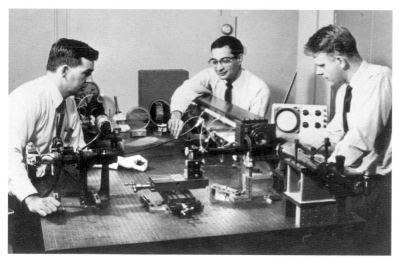

FAINT AND DISTANT SIGNALS: TOP: The large horn antenna atop Crawford Hill—located a few miles from the Black Box in Holmdel—that was built for the Echo satellite experiment and later served to collect data that proved the existence of the Big Bang. ABOVE AND AT RIGHT: Two teams that pioneered lightwave communications. In 1961, Donald Herriot, Ali Javan, and William Bennett work on the first gas laser; around the same time, Arthur Schawlow—a coinventor of the laser theory—works with C. G. B. Garrett (foreground) on a solid-state laser technology.

THE BOSS: Mervin Kelly, around the time of his 1958 retirement from Bell Labs. "He is most certainly an empire builder," a White House advisor wrote in a private memo just after World War II, when Kelly turned down the offer of becoming the U.S. president's first science advisor. To John Pierce, Kelly was "an almost supernatural force." *Courtesy of AT&T Archives and History Center*

A GREAT MISTAKE: The Picturephone model that debuted at the 1964 World's Fair. At the fair, a survey conducted by AT&T indicated that a majority of those who tried the device perceived a need for Picturephones in their business. A near majority said they perceived a need for Picturephones in their home. The rollout of the device proved disastrous.

Courtesy of AT&T Archives and History Center

THE CHEMIST: Bill Baker, in his beloved research labs. His mystique, perhaps even more than his intelligence, separated Baker from his colleagues. His work in Washington and on military affairs was shrouded by secrecy. "Nobody knows what I do," he would sometimes lament to friends and family. *Courtesy of AT&T Archives and History Center*

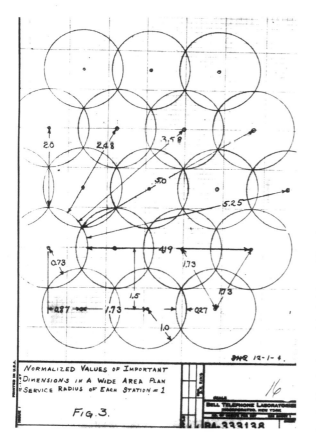

NORMALIZED VALUES OF IMPORTANT
DIMENSIONS IN A WIDE AREA PLAN
SERVICE RADIUS OF EACH STATION = 1

FIG.3.

BELL TELEPHONE LABORATORIES

WITHOUT WIRES: A page from Doug Ring's landmark 1947 memo (left) outlining the ideas behind a nationwide mobile phone system; by the early 1970s, the Bell System mobile phone plan—as described in the company's explanatory literature (below)—had assumed the more sophisticated design of today's cellular telephone and wireless data infrastructure.

Courtesy of AT&T Archives and History Center

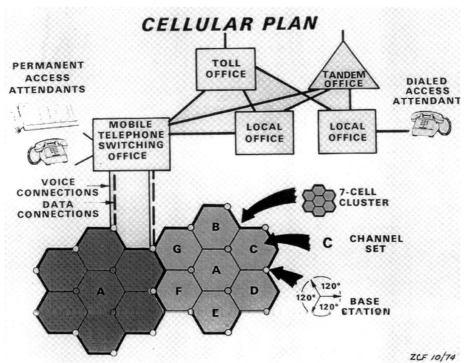

CELLULAR PLAN

PERMANENT ACCESS ATTENDANTS

TOLL OFFICE

TANDEM OFFICE

DIALED ACCESS ATTENDANT

MOBILE TELEPHONE SWITCHING OFFICE

LOCAL OFFICE

LOCAL OFFICE

VOICE CONNECTIONS
DATA CONNECTIONS

7-CELL CLUSTER

C CHANNEL SET

120°
120°
120°
BASE STATION

ZCF 10/74

THE INFORMATIONIST: Claude Shannon, at home in Massachusetts in 1962. "I think that this present century in a sense will see a great upsurge and development of this whole information business," Shannon had remarked two years before. The future, he predicted, would depend on "the business of collecting information and the business of transmitting it from one point to another."

Henri Cartier-Bresson/Magnum Photos

an immensely complex electronic infrastructure on each edge of the continent. The horn antenna on Crawford Hill would be "steerable." That is to say, it would be mounted on a circular track so that it could swivel according to the predicted path of the satellite. Near the base of the horn antenna, the Echo team would install the supercooled maser that would amplify the faint signals.

It was true, to a large extent, that the satellite project was about technology. But it was also about a vast logistical coordination—worked out on numerous trips to Washington and California—between Bell Labs, the Jet Propulsion Lab, NASA, and several other scientific affiliates (such as Douglas Aircraft and MIT's Lincoln Labs) that were providing technological support. In mid-June 1959, the teams convened at NASA headquarters in Washington to offer progress reports. A series of tests were agreed upon in preparation. NASA would try out a new Delta-Thor rocket it intended to use for the satellite launch and would begin overseeing "shotput" experiments—abbreviated launches to heights of a few hundred miles—to investigate whether the huge reflective balloon, folded within a round canister only two feet in diameter, could be deployed and inflated correctly.[21] The teams met at NASA headquarters again in late October. The date for the fall launch had apparently been scrapped, with hopes resting on a balloon launch during the spring of 1960. The team at the Jet Propulsion Lab and Jakes's team in Holmdel meanwhile planned a series of prelaunch communication tests. Each would direct a radio signal toward the moon while the station on the opposite coast would try to receive the reflection. The first dates were set for late November 1959.

Moonbounces, they were called.

THE ECHO TEAM on Crawford Hill now ceased working normal hours.

"Poor *Mary*," Bill Jakes says of his wife. "I'd go out there at ten o'clock at night. And come home at three o'clock. And we'd do this night after night after night after night to see how things were going."

There were about a dozen moonbounces, some more successful than others, during the fall of 1959 and the winter of 1960. By the spring the equipment of all the various teams around the country appeared to be working satisfactorily. NASA's hopes for a balloon launch from Florida's Cape Canaveral in March were eventually pushed back to the morning of May 13, 1960. The space agency issued a lengthy press release that morning explaining the equipment and the launch procedure. The rocket's payload—the thin Echo balloon, "about half the thickness of the cellophane on a cigarette package"—would be filled with about thirty pounds of "sublimating powders" (a combination of benzoic acid and anthraquinone) and folded in an accordion fashion into a twenty-six-and-a-half-inch magnesium container, which was put into the Delta rocket's third stage. After separating from the rocket, the magnesium container would be split open by an explosive charge and the balloon would release. The powders inside, warmed by the sun, would turn to gas and help inflate the balloon. Moving in a southeasterly direction, Echo would then circle the earth at a height of about 1,000 miles once every two hours at an estimated speed of 16,000 miles per hour. It would be about as bright in the night sky as the star Vega. NASA predicted that Holmdel and Goldstone would have their first chance to connect during the satellite's first pass, 150 minutes later. At that point, both teams would have the silvery balloon in their direct line of sight.[22]

There were four small buildings on Crawford Hill. One was a command center where Jakes and his team monitored the transmission and receiving equipment. Jakes would park himself in front of a laboratory bench loaded with equipment that showed whether the satellite dishes and horn antenna were tracking properly. He would watch the dials constantly. On May 12, Jakes and his colleagues spent the night there following the countdown from Cape Canaveral, which was sent to them by Teletype. Their families gathered nearby to wait. Pierce, Kompfner, and other Bell Labs executives convened in a small support building, perhaps fifteen by twenty feet, located near the command center.

"Here we all were," Mary Jakes recalls. "All of us waiting, sitting on a grassy hill. And here was this building with the brains inside and the women and children outside. All of this stuff had gone into this launch. And it went up and it was a failure. Everyone else felt that a member of the family had died."[23] A bulletin from Bell Labs that afternoon explained that the third stage of the Delta rocket did not achieve orbit. The balloon payload burned up in the atmosphere.

What followed for the distraught members of the Echo team were several more months of tests—more moonbounces, more computer tracking experiments, more waiting. NASA scheduled a second launch for early August, but that, too, faced delays due to mechanical complications and weather. Jakes and his crew and Pierce showed up at the command shack around midnight on August 9–10 only to have their hopes dashed. And then, forecasting a night of clear weather on August 11, NASA gave the go-ahead for an early-morning August 12 launch.[24] To allow his crew to concentrate on the task at hand, Jakes ordered everybody in the control shack—Pierce and Kompfner included—to another building. "I just wanted them the hell out of there," he says, somewhat surprised and amused that his bosses actually followed his order.[25] The countdown came from Florida over the Teletype machine. And at 5:39 a.m., the Delta rocket with the balloon payload went up.

The men on Crawford Hill waited anxiously. Around the world, several tracking stations had been enlisted to follow Echo's progress, both by telescope and by monitoring its small solar-powered radio beacon, as it sped forth at 16,000 miles per hour. A station in Trinidad got it first. Then Johannesburg confirmed a signal. At 7:05 word came to Jakes from Woomera, a station in Australia—"Woomera has acquired it!" he announced over the loudspeaker—which virtually assured the men that the balloon was functioning and on the correct path. Pierce, drinking coffee and eating a doughnut in the support building, began jumping up and down; his glasses bounced on his nose. It was a startling sight to colleagues who had never before seen him visibly happy. "It's in orbit," he yelled. "Echo is in orbit." He telephoned his wife to tell her.[26]

Within about thirty minutes the balloon was in the sights of both Goldstone and Holmdel and their antennas began their automated tracking movements. Goldstone intended to broadcast to Holmdel a recorded message from President Eisenhower prepared especially for the occasion. At 7:41 Jakes called California and asked them to start playing the message: A microwave signal was beamed from Goldstone's antennas to the orbiting balloon, reflected down, and then sucked in by the giant horn in New Jersey, where it was amplified four thousand times by the maser in the horn's base. Eisenhower's voice came through the loudspeakers on Crawford Hill so clearly—the president lauded Echo as "one more significant step in the United States' program of space research and exploration"—that Pierce and other visitors didn't even realize at first it was coming from the other coast. When they did, a cheer went up.

Pierce quickly regained his composure. Soon afterward, he talked with reporters and answered questions about the experiment before departing for his house in Berkeley Heights, New Jersey. He'd been up all night. At home, he took a short rest, then went outside to paint his garage.[27]

It was possible that Pierce, the most restless thinker, was already on to the next idea. For Jakes, meanwhile, the work on Echo was actually just starting. During the following days and weeks he and his counterparts in California did a series of successful experiments, mostly at night—the first-ever two-way phone conversations over a communications satellite, for instance, and the first-ever satellite fax transmissions. Jakes and Pierce, now minor celebrities—Echo was on the front page of virtually every newspaper in the country—also began receiving piles of fan mail and crank calls. ("Your balloon made me pregnant," one woman told Jakes.) In the course of its first year, Echo went around the earth 4,481 times; the ten-story balloon meanwhile began to wrinkle slightly as it was pelted with micrometeors and its interior gases leaked slowly out, giving it a twinkling effect to the naked eye.[28] By then it had become an international curiosity, though, a bright dot streaking regularly across the night sky. Soon after, the Bell System set up a phone service in New York City to tell callers at which times Echo would be visible. A schedule was published in various newspapers as well. At Jones Beach, not far from New York City, the bandleader Guy Lombardo would sometimes look at his watch and momentarily stop his orchestra's summer evening concert. Then he would suggest that everyone in the audience pause to look up and see Echo, at one point just a curious idea, and now something more, pass overhead.[29]

NOT LONG AFTER the Echo experiment, Pierce was walking down the hall in Bell Labs' Murray Hill office when he bumped into a big, intimidating executive named Frederick Kappel. Kappel, a plainspoken Minnesotan, had begun his career at AT&T in the 1920s digging holes for telephone poles at $25 a week; he had risen through the ranks over a long career to become the company's president. He and Pierce liked each other. They'd had an amicable encounter several years before, when an executive at AT&T had complained that the psychological research being done at Bell Labs under Pierce was too distant from communications science to deserve funding. Kappel headed a committee that looked into the matter, and after a review he let Pierce's group proceed. Now, on the day when Kappel saw Pierce at Murray Hill, the executive was in the midst of a series of speeches extolling the potential of a global satellite communications system, run in large part by AT&T. "Look what you've got me into, John," Kappel joked.[30]

His point was that satellites, only a year after Echo, had progressed from an intriguing experiment to a cutthroat business. A number of multinational corporations—RCA, General Electric, ITT, among others—had already made clear their intentions, as *Fortune* magazine put it in July 1961, to install "the great cable in space." Satellites offered an alternative to the increasingly burdened underseas cables; more important, they could carry live television, which the current underseas cables could not. Within a decade, some economists predicted, orbiting relays could be a billion-dollar-a-year business.[31] "We think the best practical system would use about 50 satellites in space," Kappel explained to *U.S. News & World Report*. Would AT&T spend $25 million to do it? "I'm talking, really, about a system costing more than that," Kappel remarked. He put the cost at $200 million. Even at such an expense, he said, he believed that it could eventually make transatlantic calling cheaper.[32]

Pierce would later observe that Project Echo proceeded quickly and smoothly in part because it was considered eccentric: Few people in the business community perceived its practical importance, and as a result Pierce and his crew on Crawford Hill were largely left alone. In truth, apart from Mervin Kelly, there had been little skepticism about satellites at Bell Labs. Even as Echo and its ground stations were being planned in late 1959 and early 1960, Pierce was attending meetings about building a more complex *active* satellite, rather than a passive one like Echo.[33] In simplest terms, the active satellite—it was soon approved and given the name Telstar—would be able to receive a radio signal beamed from the ground, amplify it ten billion times, and then retransmit it, on another frequency, back to earth. It could carry hundreds of phone calls or several television programs. (Television required far more bandwidth than phone calls.) Kappel, in other words, didn't want to spend $200 million of AT&T's money on big silver balloons. He wanted to spend it on an array of active satellites.

Within Bell Labs, the engineers began debating various possibilities. One basic question was how high the orbit of an active satellite, or a group of active satellites, should be. Among the companies vying to control space communications, one possibility was to launch several satellites that circled the earth at about 22,300 miles, which was the height at which their orbits would keep pace with the earth's rotation. Such "geo-stationary" satellites would stay fixed in the sky above a particular point on earth, and could theoretically be available for transmission at any time. The disadvantage here was twofold, however. Rocket technology had advanced tremendously since Pierce first laid out a rough plan for communications satellites in 1954. But NASA still couldn't put a relay so high up in space. The second problem, as Bell Labs researchers soon found out, was that a satellite at such heights would necessitate such a long path for a phone call—at least 44,600 miles—that a delay of about six-tenths of a second would result. This was tolerable for a television signal, yet it posed challenges for a two-way conversation. Indeed, in psychological experiments at Murray Hill, test subjects reacted to the long conversational delay (and to a related problem of echo) with annoyance and frustration. And so, for the moment at least, the Bell engineers decided that an active satellite orbiting at no more than 3,000 miles high was the better option.

Telstar moved fast. "We were all working on an impossible schedule and nobody was going to speak up and say, 'I can't do that,'" says Ian Ross, who worked on the transistor and tube elements of the satellite. "There was pell-mell competition to see who would get done first," Eugene O'Neill, Telstar's project engineer, would recall, noting that the tight deadlines and financial pressures were a departure for Bell engineers accustomed to working on a more orderly schedule and with a focus on quality and durability rather than speed.[34] Telstar, if it worked, could show the world that AT&T and Bell Labs could do it all. During the course of the work, John Pierce offered O'Neill suggestions, but mostly he was relegated to the status of a wise and distant uncle. It didn't rankle Pierce. The point of a successful research project like Echo, after all, was to hand it off to the Labs' development group. And the contrasts between the research and development approaches were substantial. Echo was done on a shoestring, at a cost of less than $2 million, with a staff of about three dozen men. Telstar—the rocket launch alone, billed by NASA to AT&T, cost $3 million—was a development project that required the work of more than five hundred Bell Labs scientists and engineers. Echo was a big shiny balloon with a small radio beacon; Telstar had fifteen thousand parts.

The engineers' objective was not to build a satellite for customer phone calls and generating revenue. It was still too early for that. More precisely, Telstar was meant to demonstrate that Bell engineers could design, develop, and deploy an active satellite, and that in doing so they could flag potential problems, especially those of reliability, that might befall AT&T's plans for a large-scale relay business. The complexity of the project was easily on par with the undersea cable, but the time frame for development was much faster, making the task even more difficult. Nothing in the satellite could be allowed to fail, moreover, for as hard as it was to repair an undersea cable, it would be impossible, in a world that had yet to send a man into space, to fix a satellite.

O'Neill, the project manager, recalled that every one of Telstar's fifteen thousand parts was tested exhaustively: "We shook them and racked them in every way imaginable and stressed them electrically and physically until hell wouldn't have it to satisfy ourselves that it was going to be rugged enough to withstand the vibration of a rocket launch and survive in outer space."[35] The engineers also attempted to work into Telstar a number of specialized parts—special semiconductor diodes, for instance—that were intended not for communication tests but to make scientific measurements on temperature and radiation in space. The information would be transmitted to Bell scientists on the ground. Space was as yet an utter mystery that the satellite could help unravel by showing how the intense radiation of the Van Allen Belt surrounding the earth, or an atmospheric vacuum, or the extraordinary temperature changes—hundreds of degrees in a single orbit—might affect a working spaceship.

Telstar was slightly bigger than a beach ball—about three feet in diameter—and as heavy as a man—170 pounds. After it was assembled in a laboratory in Hillside, New Jersey, then tested at Murray Hill and Bell Labs' Whippany, New Jersey, facility, it was transported to Cape Canaveral, Florida, for a Delta-Thor rocket launch scheduled for the second week of July 1962. Though it was spherical in shape, Telstar's surface had seventy-two flat facets, giving it the appearance of a large, bizarrely decorated gemstone. In the end, though, it served as an almost perfect example of Pierce's contention that innovations tend to happen when the time is right. Indeed, Telstar was not one invention but rather a synchronous use of sixteen inventions patented at the Labs over the course of twenty-five years. "None of the inventions was made specifically for space purposes," the *New York Times* pointed out. On the other hand, only all of them together allowed for the deployment of an active space satellite.

Some of the ideas used for Telstar had to do with transmission—the maser to amplify faint signals, for instance, and a noise-reducing circuit (a "frequency-modulation feedback receiver") that had been patented by a Labs engineer named Joe Chaffee in 1937 and was considered largely useless at the time but had been dusted off with spectacular success for Echo and would be used again for Telstar. Other inventions, meanwhile, had to do with semiconductors. On the surface of Telstar were thirty-six hundred solar cells to provide the power that allowed the satellite to function; there were also thousands of transistors and diodes, many of them used to make radiation measurements. The essential amplifying component of the satellite, finally, was a pencil-thin, footlong traveling wave tube—a modification of the design that Pierce had salvaged from Rudi Kompfner during the war nearly twenty years before. This particular tube had been patented by Pierce and Kompfner and Jack Morton, the development chief who had helped move the transistor into mass production.

On the night of Telstar's launch, Pierce had dinner at a restaurant with a friend. Then he went up to Crawford Hill to relax with Rudi Kompfner and see whether the active satellite would succeed. Pierce was confident—one of his hallmark traits. He found it continually astonishing that a complex apparatus such as the phone system even worked, but on individual projects he rarely doubted the capability of his fellow Bell engineers.[36] As it happened, the crowd he joined at Crawford Hill was using their horn antenna only as voyeurs. Officially, Telstar was being tracked and guided at a new installation, built on a thousand acres purchased by AT&T in a remote area of Maine. Dubbed "Space Hill," the Maine installation was located in a bowl of mountains that minimized radio interference. The centerpiece of the site was a mammoth horn antenna, 177 feet long, far larger than the one at Crawford Hill, that sat atop a rotating base. To make sure that the Maine weather would not get in the way, the horn had been placed inside an inflatable dome—the largest inflatable structure ever built at the time—made of Dacron and synthetic rubber.

At 4:35 on the afternoon of June 10, 1962, the rocket carrying Telstar lifted off. By 7:30 p.m. the satellite had been released into orbit and was making its sixth pass around the earth. It was then that Fred Kappel, the AT&T chief, placed a call via Telstar to Vice President Lyndon B. Johnson. The two chatted briefly without any hitches. Within the hour, a French ground station that had also been involved in the Telstar project began receiving a live, clear television signal: It was the image of an American flag waving in the breeze at the receiving station in Andover,

Maine. The following morning a British ground station broadcast a live show to American audiences. To millions—by one measure, a full 82 percent of British residents knew Telstar by name—the technology seemed positively startling. "It pleased us that the satellite did no more nor less than just exactly what it was designed to do, but we expected that," Pierce later remarked. What did surprise him, however, was Telstar's "human impact." On the day after the launch, the *New York Times* called it a communications feat "regarded as rivaling in significance the first telegraphed transmission by Samuel F. B. Morse more than a century ago."[37] Soon after, a British band called the Tornados released an instrumental homage to the satellite, aptly named "Telstar," that became a number one hit in both the United States and England. Pierce was especially charmed that six months after its launch, on Christmas Day 1962, Queen Elizabeth of England referred to Telstar as "the invisible focus of a million eyes."

By that point, though, the euphoria for satellites at Bell Labs had largely disappeared. Congress and the Kennedy administration, concerned about ceding the control of space communications entirely to the private sector and worried, too, about AT&T's immense size and aggressiveness, had already pushed all private companies out of the international satellite business. Pierce, in turn, had fallen into something resembling a tailspin.

JOHN PIERCE DIDN'T HAVE pronounced political leanings; in fact, he never even registered to vote.[38] But the Communications Satellite Act of 1962, which created a new government-authorized communications corporation called COMSAT and effectively barred AT&T from the international satellite communications business, radicalized him in a way. "I took that hard, you know," he remarked to an interviewer some years later. "I'd just got into communications satellites and covered my name with glory— deserved or not deserved. I felt thrust out into the cold."[39] Pierce had long been frustrated by Washington's bureaucracy. (NASA officials, in turn, had also been frustrated by AT&T, which they believed had unfairly monopolized credit for Echo and Telstar.)[40] But for all the accusations that the Bell System could be slow to implement its newest technologies, Pierce held an unshakable belief that his organization was the world's supreme example of technical competence. He doubted satellite engineering could advance at the same rate without Bell Labs or private competition.[41] And he was even more astonished that the enormous engineering successes of Echo and Telstar had led to a decisive business failure. Perhaps, Pierce later mused, Mervin Kelly had been right to insist that the Bell System should stay out of the space race entirely.

Pierce occasionally entertained offers to leave Bell Laboratories. Despite his quirks, or perhaps because of them, he had a devoted circle of friends and admirers in academia, government, and industry. His acquaintances included many of the world's most influential scientists. William Shockley, for instance, when setting up his semiconductor company in California, had sounded out his old friend about a move to the West Coast. "He says he's happy at BTL," Shockley noted about Pierce in a personal journal in the mid-1950s.[42] Pierce had also remained close with Shannon, who often invited him to visit at his house in Massachusetts.[43] Even the collapse of the Bell System satellite business didn't change Pierce's views on his employer, however. As he later reflected, "I liked Bell Laboratories better than I liked satellites."[44] For a while, at least, he was staying.

He still had plenty to do. All during his work on satellites, for instance, Pierce had become more and more involved with electronic and computer-generated music. Along with his colleague Max Mathews, Pierce and some Labs researchers had compiled an album of computer-programmed music, released by Decca Records, that they'd created on a primitive IBM 7090 computer. The music was intriguing and nearly unlistenable—beeps and blips, mainly, interspersed with shards of classical melodies and eccentric diversions. The Labs scientists called it *Music from Mathematics*. Pierce sent unsolicited copies of the record, along with an enthusiastic cover letter, to the composers Leonard Bernstein and Aaron Copland.[45]

Not all his work was so tangential to his company's business. Pierce was still thinking about the state of the phone system, too, and especially what might lie beyond. In speeches and television appearances during the years defined mostly by Echo and Telstar, he sometimes impressed people as a kind of philosopher of communications. In his view, it wasn't so much that technologies were changing society; rather, a new web of instantaneous information exchanges, made possible largely by the technologies of Bell Labs, was changing society. Pierce was also coming to the realization that other advances—data transmission, home computers, electronic mail, lightwave communications—might soon define the culture far more radically than the Bell System already had. "It is clear that more and more the Bell System will be concerned with sending digital signals, both to enable machines to talk to one another and to enable people to hear distant sounds or to see distant scenes," he remarked in 1956. By the late 1960s, when he sat down to talk about the future with the CBS news anchor Walter Cronkite, he made the point, just as his friend Claude Shannon had done in a less accessible manner, that all electronic exchanges—letters, calls, data, television—were likely to merge.

> *Cronkite:* You're talking now of some sort of a central communications panel in the home, a single switchboard from which all of this can be done?

> *Pierce:* I think that it's very important to realize that communication is a general function. The wires that will carry telephony will also carry teletypewriter, and a circuit that's capable of handling television will handle high-speed data and many other things as well. So that once you have the transmission facilities available, they can be used for everything interchangeably.[46]

Among other things, he had long been pushing at Bell Labs for more research into mobile phones. If the Federal Communications Commission would grant Bell Labs and AT&T permission to operate such phones in a large frequency range, Pierce believed the business would explode in popularity. (Without a broad range of frequencies, however, there wasn't much logic in moving forward with new technologies, since a mobile system's capacity, and thus its potential pool of customers, would be small.) Was there any way of "prying mobile allocations out of the FCC?" he wrote in a 1957 memo. At the time, Pierce was actually thinking ahead by a decade or two, and wondering where Bell Labs should be with mobile phone research by that point.[47] He perceived that the transistor might create entirely new possibilities for mobile communications.

As it happened, it took almost a decade until the Labs could move forward with mobile technology. Project Echo, the huge balloon that defined Pierce's career, stayed aloft for nearly eight years, disappearing somewhere off the western coast of South America on the evening of May 23, 1968.[48] At that point Bill Jakes, Echo's project engineer, was still working at Holmdel under Pierce. One day, around the time of Echo's demise, Jakes looked up to see his boss come into his office. "Jakes, why don't you do something about mobile communications?"[49] Pierce asked. And then, before waiting for an answer, Pierce walked out.

Fourteen

FUTURES, REAL AND IMAGINED

A year after John Pierce celebrated the success of his Echo satellite balloon, and just around the time that work on Telstar was going full bore, Pierce and a Bell Labs colleague named Edward David sat down to compose a memo. It was the summer of 1961, and the two men were making an effort to compile a list of ideas for a Bell System exhibit at the 1964 New York World's Fair. AT&T executives had already established a planning group and were soliciting advice from some of the company's farthest-thinking scientists. They hoped that the planning team could instill upon the fair's visitors a sense of the phone system's complexity along with a glimpse of how communications might evolve in the coming years. That was precisely where Pierce came in.

For corporations, world's fairs are public relations opportunities. But the fair was not an exercise in cynicism. It was a legitimate chance to display some of the ideas in the company's technological pipeline, a pipeline usually clogged with more inventions and ideas than the business side of the phone company could ever hope to implement.[1] Pierce was thinking about the New York fair around the same time that a modest display of Bell Labs innovations was being demonstrated at Seattle's Century 21 Exposition, which was being marked by the construction of a huge "space needle" on the city's fairgrounds. At the Seattle fair visitors could ride a monorail to a Bell exhibit intimating a future of startling convenience: phones with speedy touch-tone buttons (which would soon replace dials), direct long-distance calling (which would soon replace operators), and rapid electronic switching (which would soon be powered by transistors). A visitor could also try something called a portable "pager," a big, blocky device that could alert doctors and other busy professionals when they received urgent calls.[2]

New York's fair would dwarf Seattle's. The crowds were expected to be immense—probably somewhere around 50 or 60 million people in total. Pierce and David's 1961 memo recommended a number of exhibits: "personal hand-carried telephones," "business letters in machine-readable form, transmitted by wire," "information retrieval from a distant computer-automated library," and "satellite and space communications." By the time the fair opened in April 1964, though, the Bell System exhibits, housed in a huge white cantilevered building nicknamed the "floating wing," described a more conservative future than the one Pierce and David had envisioned. The exhibit was primarily explanatory. Visitors could get a sense of how quality control worked at Western Electric factories, or how researchers at Bell Labs grew pure crystals necessary for transistors. They could experience push-button "dialing," or they could consider how Telstar and satellite communications functioned. Perhaps the only surprise was an open demonstration—available to anyone who visited the floating wing—of what Pierce and David had in their memo termed "visiphone." By 1964, these devices, created by the engineers at Bell Labs over the course of a decade, had a trademarked name. They were known as Picturephones, and they seemed the very essence of the future.

At the fair, a visitor who wanted to try a Picturephone would enter one of seven booths and sit before what was called a "picture unit." The device was a long oval tube, measuring about one foot wide and seven inches high and about a foot in depth. Set within the oval face was a small camera and a rectangular video screen, measuring four and three-eighths inches by five and three-quarter inches. The picture unit was cabled to a touch-tone telephone handset with a line of buttons to control the screen. If you wanted to make a Picturephone call at the fair—or more precisely, if you wanted to talk with the Picturephone users at other booths—you simply pressed a button marked "V" for video; after that you could either talk through the handset or through a speakerphone on the picture unit.

Without question, the Picturephones were diverting. In several obvious respects, the device was less a radical innovation than an elegant melding of the established technologies of television and telephone. But it wasn't entirely clear whether the Picturephone actually solved a problem. Some Bell Labs engineers worried about this. As far back as the mid-1950s, John Pierce was exchanging memos with colleagues wondering about the utility of the new device: "The need for acceptability of such a service," Pierce wrote of the Picturephone, "has not been adequately evaluated, and the [phones] themselves were not at the point at which they could be put into commercial use."[3]

To be sure, the Picturephone technology had come a long way since then. By 1964, the video image was crisper than what Pierce had critiqued in the late 1950s; also, the entire device—which was to say millions of the devices—now seemed to have the potential to be integrated over the next few decades into the evolving telephone switching and transmission network. There were some concerns as to whether the phone network could handle the additional traffic. There were also concerns as to whether the nation was ready. But the response of visitors to the New York fair—usually a line of people were waiting to try the Picturephones—suggested a substantial degree of public curiosity, and perhaps even enthusiasm.

AT&T executives had in fact decided to use the fair as an opportunity to quietly commission a market research study. That the fairgoers who visited the Bell System pavilion might not represent a cross section of society was recognized as a shortcoming of the survey results. Nevertheless, researchers asked seven hundred users of the Picturephone for their reaction to its design and technology; they also asked whether they might want to use the Picturephone in the future. The overall reaction, summarized in a confidential report, was termed "generally favorable."

Users complained about the buttons and the size of the picture unit; a few found it difficult to stay on camera. But a majority said they perceived a need for Picturephones in their business, and a near majority said they perceived a need for Picturephones in their home.

Would they pay for it? Here, the results were less clear. For a price of between $40 and $60 a month, for instance, only 12 percent of the couples interviewed said they would want a Picturephone in their homes. Business customers, however, seemed more amenable. Even if the cost were substantially higher—$60 to $80 a month—29 percent said they would be interested in having the device at their place of business.

When the AT&T market researchers asked Picturephone users whether it was important to see the person they were speaking to during a conversation, a vast majority said it was either "very important" or "important." To phone company executives, this must have been deeply encouraging. Apparently the market researchers never asked users their opinion about whether it was important, or even pleasurable, that the person they were speaking with could see them, too.[4]

BY 1964, there were about 75 million telephones in the United States, meaning that Bell Labs had now created the means for about 2,500,000,000,000,000 possible interconnections between subscribers.[5] In light of this, the Labs' primary innovation of 1964, for all the attention the Picturephone received, was actually something nobody who used a telephone would see or even understand. It was known as ESS No. 1, a new electronic switching station, opened in a small modern building in the village of Succasunna, New Jersey.[6] The design for the switching station had taken two thousand "man-years" of work to create and used tens of thousands of transistors. Its complexity dwarfed that of other previous Bell Labs undertakings such as the transatlantic undersea cable. ESS, as Fred Kappel, the chairman of AT&T, pointed out at its opening in Succasunna, "was the largest single research and development project in Bell System history."[7] The costs were acknowledged by AT&T to be "more than" $100 million, which likely suggested they were much more,

Still, its expense was projected to drop dramatically as the ESS technology was deployed all over the country. By the year 2000, Bell executives maintained, all communications switching would be done through electronic means like that in Succasunna, and it would be "better and cheaper"—that golden combination—than the current system.

Why move in this direction? What kind of future did the men envision? One of the more intriguing attributes of the Bell System was that an apparent simplicity—just pick up the phone and dial—hid its increasingly fiendish interior complexity. What also seemed true, and even then looked to be a governing principle of the new information age, was that the more complex the system became in terms of capabilities, speed, and versatility, the simpler and sleeker it appeared. ESS was a case in point. Switching had always been immensely complex, both as a concept and an actual technology. Making untold connections between millions of subscribers, automatically routing a call from one central switching office to another, finding alternate routes (or alternate routes to the alternate routes) if the most direct path for a call was too busy—this was why those who understood the vast interlocking parts of the Bell System called it, as Pierce and Shannon often did, the largest and most complex machine ever built. A visit to a switching office in any city verified this observation. There, one could see the vast banks of crossbar switches— metal matrices, taller than a man, where calls came in and were routed out. To stand before the rectangular crossbars was to hear their constant clicking; to look behind them was to see the ropes of copper wiring, the logic of a programmed machine that encompassed bundles of thousands upon thousands of color-coded copper wires that snaked through and around the machinery so as to connect all to all. The "switching art," as it was known at Bell Labs, was suitably captured by a specialized technical jargon describing relays, registers, translators, markers, and so forth and a bevy of convoluted, mind-twisting flow charts. Those who had mastered the switching art were members of a technological priesthood.

ESS, which was meant to replace crossbars, looked simple on the outside—panels of glowing lights and data banks smaller than the crossbar matrices—but it was in fact far more sophisticated. In the moment during which a phone subscriber picked up the receiver and began dialing a friend—the moment during which the caller would receive a dial tone and begin pressing digits—the ESS followed thousands of separate, sequential instructions, a nearly instantaneous interplay of the system's logic and memory circuits, all carried out in microseconds. Bell engineers were asked by Bell management—ever wary of the federal government's order to stay out of the computer business—to describe the ESS only as "computer-like." (The internal memo warned, "Do not call ESS a computer," instead suggesting that ESS be described as "a large digital information processor.")[8] Of course, the great advance of ESS was that it *was* a computer, a highly sophisticated and programmable machine (unlike the crossbar) that could figure out phone connections in a few millionths of a second and could handle more traffic than any previous switching system. It could provide services that hadn't previously been available. And it could help integrate the Picturephone into the existing system.

Like the microwave towers that had recently been introduced to the long-distance transmission system—the ones that were a cheaper alternative to cables between cities—electronic switching was, from a user's point of view, an incremental improvement. If you weren't a phone engineer, it was hard to understand the excitement. After all, everyone could already talk with everyone else. Thus the phone company pushed various selling points. "A housewife will be able to turn on her oven while away from home by using her telephone . . . an office worker's call to a line that is busy will be completed automatically when the line is free," a *New York Times* article noted in a story about touch-tone phones and the new electronic switches.[9] And the ESS enabled other options, such as call waiting or conference calling, that seemed promising, too. The *New York Times* noted that ESS allowed for the possibility that "a couple out for an evening of bridge will be able to have their calls switched automatically to their host's home."

In other words, telephone customers could now move around—they could be *mobile*—and the system could still find them. Eventually, this would prove to be immensely valuable. Bell Labs' engineers had been encouraged by the public relations department to remark to journalists that ESS had the capacity "to provide services we haven't even thought of yet." It was a vague throwaway line that turned out to be exactly right.

THE FUTURE OF TECHNOLOGY is never particularly easy to discern. That was why John Pierce never ceased to point out that anyone in the business of making predictions was destined to make a humiliating false step. And yet if you worked at Bell Laboratories, and were therefore entrusted by the United States government with the future of telecommunications, you still had to have a plan. So how much should Pierce and his colleagues wager on one idea for the future or another? And how fast could—how fast should—the future happen?

Jim Fisk, the president of Bell Labs, presided over the ESS ribbon-cutting in Succasunna. Fisk was the physicist whom Mervin Kelly had hired right out of MIT and who had made his name designing magnetrons for radar sets during the war. As an executive, he had the same unflappable posture he'd had as a young researcher—an easygoing temperament, coupled with a strong scientific mind, that had prompted his friend Bill Shockley to remark to a colleague, as Fisk was visiting the Labs in the late 1930s, "If that man gets hired, we'll all be working for him in ten years." Now several decades had passed. To Fisk, looking out from the early 1960s toward the future, at least three things were apparent. The first was that the system would need to get faster—an imperative that would be solved in part through electronic switching systems and touch-tone phones. Second, the system would need to send more information digitally. Soon, Bell Labs engineers would put into place a system known as T-1 that used the pulse code modulation technique that Shannon and Pierce had long ago seen as the future. Instead of waves, transmission would consist of modulating voice signals transformed into on/off pulses that were effectively the same as the strings of 1s and 0s that guided the functions of computers. "The speech in each telephone channel," Fisk said of PCM, "is sampled at a rate of 8,000 times per second and coded signals describing each sample are transmitted to the receiving terminal where the original speech is reconstructed." It was in some ways the telecommunications equivalent of finely shredding a newspaper in one city, instantly sending those scraps of paper to another city, and then putting the sheets back together in such a way that no one could ever discern that the paper had once been sliced apart.

The third truth about the future was that the system would become more congested. Traffic— comprising voices, as phone subscriptions and calls continued to increase; data, as computers began conversing with one another over the phone wires; and video, as television and Picturephone devices became increasingly popular—would lead to overwhelming floods of information. How to accommodate it all? ESS would help. But ESS could only switch information and suggest a path, like a highly skilled traffic cop; you still needed to build pathways that were broad enough to *transmit* everything. A half-century-old trend suggested where those pathways might be found. "Historically," Fisk remarked, "the progress of radio and wire communications has required the use of ever higher and higher frequencies."[10]

To the engineers at Bell Labs, the implications of this statement were fairly clear. All forms of electronic communication use electromagnetic waves. And all electromagnetic waves have a place, classified by their length, on the electromagnetic spectrum. On one end are long waves— signals like the ones broadcast from huge antennas that project songs onto AM and FM radio stations. These undulating waves might measure several meters, or even hundreds of meters. Next come shorter waves whose lengths might only be measured in centimeters or millimeters. These wavelengths are commonly used for TV signals and radar. Generally speaking, the shorter the wavelength, the higher its frequency, and the more information it can carry.

By the early 1960s, Bell Labs executives had concluded that millimeter waves would serve as the communications medium of the future. The idea at Bell Labs was to send information through such waves not by wires or broadcast towers but by means of the circular waveguide, which had been developed down in Holmdel. "A specially designed hollow pipe," as Fisk defined it, the waveguide was just a few inches in diameter, and lined inside with a special material that would allow it to carry very high-frequency millimeter radio wave signals. The waveguide pipe would in effect do the same thing as an intercity phone cable—known as a trunk line—but with far more capacity. Indeed, each pipe would likely carry hundreds of thousands of calls at a time.

Labs engineers had looked beyond the current waveguide and the millimeter waves it carried to even shorter infrared and visible light waves. These waves are so tiny that they must be measured in arcane units known as angstroms. In a single millimeter, there are 10,000,000 angstroms. By 1960, the Bell engineers believed that within a few decades it might be possible to send data over such wavelengths—in other words, to send data through light itself. If they could figure out how to do that, the system would be able to transmit an unimaginably huge amount of information.

Directly overseeing Bell Labs' vast plans for the future were two of Jim Fisk's deputies, Julius Molnar and William Baker. Molnar, the Picture-phone's primary champion, believed that by the year 2000, "Picturephone will be the primary mode by which people will be communicating with one another."[11] Molnar oversaw the numerous development projects at the Labs—the digital transmission lines, for instance, as well as the new electronic switching centers—while serving as Fisk's executive vice president. He was "running the show," as one colleague puts it, much in the way Mervin Kelly did during Oliver Buckley's presidential tenure. A tall man with a friendly face, sparse hair, and pronounced bushy eyebrows, Molnar had been trained as a physicist at MIT before moving up through the Bell Labs ranks. He was legendary for the precision of his thinking, but also for his confidence. "He was a powerhouse," recalls Chuck Elmendorf, John Pierce's old Caltech friend, who was a friend of Molnar's as well. "Julius was extremely bright, extremely competent. But the thing about him was that he *exuded* power."

Indeed, those who knew Molnar believed that he knew more about the phone network and systems engineering than any person alive. "I can say, in front of any Bell Labs executive, without hurting anyone's feelings, that Julius was the greatest executive at Bell Labs," recalls Bill Fleckenstein, who worked under Molnar and eventually became the head of the Labs' switching development division. "He knew more about what was going on at the Labs than any of several people put together. I liked Fisk very much. But the combination of Fisk, who didn't know a lot about what was going on in the bowels of the place, and Julius, who knew *everything* about what was going on in the bowels of the place, was a good combination."[12]

It was hard to say whether Bill Baker, the head of Bell Labs' research division, knew what was going on in the bowels of the place. He did not, as a matter of course, tell people what he knew. He had nevertheless gathered a vast storehouse of information about Bell Labs' operations. Every day at lunch he would sit down with the first person he spotted in the cafeteria, whether he was a glassblower from the vacuum tube shop or a metallurgist from the semiconductor lab—"Is it okay if I join you?" he would ask politely, never to be refused—and would gently interview the employee about his work and personal life and ideas. "At the end of any conversation," Baker's friend and colleague Mike Noll recalls, "you would then realize that he would know everything about you but you would know absolutely nothing about him." His memory was as remarkable as his opacity. When several oral historians sat down with Baker in the mid-1980s, they asked him about his graduate school work of nearly five decades before; he spent perhaps close to an hour recalling in great detail his teachers, textbooks, and lectures. Then he recalled how each of his classmates from the late 1930s had spent their careers, and whom they had spent them with. Then he recalled the ideas and research they had produced, and why some of it mattered and some did not. Colleagues often stood amazed that Baker could recall by name someone he had met only once, twenty or thirty years before. His mind wasn't merely photographic, though; it worked in some ways like a switching apparatus: He tied everyone he ever met, and every conversation he ever had, into a complex and interrelated narrative of science and technology and society that he constantly updated, with apparent ease.

John Pierce—"so smart that he frightened people," as Pierce's friend Ed David says—was himself only frightened by his boss Bill Baker.[13] Baker was neither imposing nor aggressive; he was six feet tall and about 150 pounds, wore oversized glasses, and carefully combed over his thinning hair. His broad and frequent smile contrasted with a formal and patrician manner. He favored high-collared shirts, his tie always neatly knotted, yet his sense of fashion bowed to an extreme practicality. It was as if the Great Depression had made on him a permanent mark. He wore suits until they frayed and drove an ancient Buick, much to the surprise of some of his fellow executives who parked, in designated spaces, alongside him. When the Buick broke down, as it often did, he rode the bus.

He was thrifty with everything but words. Verbosity had long been Baker's defining characteristic. "A speech is a different format than writing," Bob Lucky explains, "but not with Baker. His speech was perfect grammatical sentences. He talked like a writer, and normal people don't do that. His cadence, his prosody, he was an amazing speaker—but always you had no idea what he said, even as you were mesmerized by the way he said it."

When Baker didn't want to answer a question, he would *appear* to answer, talking in circles, always using five words where one might do, and fogging a room with a nearly impenetrable rhetorical confusion. "Every month there was a Bill Baker meeting," recalls Henry Pollak, the director of the math department. All of Bell Labs' directors and executive directors of research—between fifteen and twenty people—would converge at a Murray Hill conference room to talk about their most interesting and most current research results. The Bill Baker meeting would begin in the morning and sometimes last through the afternoon if necessary. Baker's views on these results helped determine what kind of research Bell Labs would continue to fund. Invariably, after Baker had left the conference room and wandered off to some other pressing matter he would never explain or reveal, the men would regroup and try to decipher their boss's rhetoric.

These were among the smartest men in the world, but they could never cut through Baker's verbiage. "What the hell did Baker say?" they would ask one another. "What the hell did Baker *mean?*"[14]

Eventually they realized that when Baker showed modest enthusiasm—if something sounded *very* good to him—he didn't particularly like it. "If he really liked something," his colleague Irwin Dorros recalls, "then he would use about ten adjectives: *that is a terrifically outstanding and superb contribution that has exceeded all expectations,* or something like that."[15] As Ian Ross, who later became Baker's deputy, and ultimately Bell Labs' president, recalls, "The story Baker used to tell—not about himself, but it fitted him—was that there are two men sitting in a meeting where a man is making a presentation. And when the man finishes, one guy in the back turns to the other and says, 'What was he talking about?' And the other says, 'I don't know, he didn't say.' And that was Baker. He could speak for ten minutes and you hadn't the vaguest idea of what he said. It was habitual. And I think it was *willful.* He wanted to obfuscate."[16]

There could be exceptions. "He could be very blunt, and he could be very clear when he wanted to be clear," recalls Bill Keefauver, who headed Bell Labs' legal department during Baker's tenure. At those rare moments, Baker's equanimity would ebb away and reveal a kind of merciless, probing intelligence. At one of his monthly meetings, Henry Pollak recalls, the director of chemical research was taking a turn to give a presentation on some recent experiments. "He used an innocent sentence," Pollak recalls, "something like, 'and this particular aspect is completely understood.' And Baker didn't say anything, he just started asking him questions. He started with one thing, and then he asked a question about his answer, and then he asked questions about his answer to that, and so on—until he just demolished the guy. It was that statement—*this particular thing is completely understood.* He was trying to show him that it wasn't understood at all. And he didn't say, 'Oh, you don't want to say things like that.' He just cut him down, six inches at a time."

To Pollak, this was a demonstration not of Bill Baker's cruelty but of his acumen—in this case to push his deep belief that science rests on a foundation of inquiry rather than certainty. Also, it revealed how nimble

Baker's mind really was. "A very small number of times in my life I've been in the presence of somebody who didn't necessarily answer the question I asked. They answered the question I should have asked," Pollak says. "And Bill Baker was one of those people. And there are other people who just build a mystique and give the impression of a mystique around them. And Bill had that, too."

His mystique, perhaps even more than his intelligence, separated Baker from his colleagues. "Nobody knows what I do," he would sometimes say to his son. This was correct. And nobody really knew who he was.

"IN THE SPRING OF 1913," Bill Baker's mother, Helen, wrote about her family, "my husband and I yielded to that mysterious back-to-the-land urge and bought a farm." In fact they had left New York City and bought four hundred acres, on the Eastern Shore of Maryland on the Chester River where it empties into the Chesapeake. The area, known as Quaker Neck, was located just south of the small town of Chestertown; it was so rural and remote that it was hard to find on road maps. The Baker house was an old brick colonial, built in 1768 by a Dutch immigrant named Edward Cornelius Comegys. Bill, an only child, attended a one-room schoolhouse. It was Helen's idea that the family farm would raise turkeys, a potentially lucrative endeavor with which her son would help nearly every hour that he wasn't at school or studying. (Mother and son were inseparable until her death, writing each other sometimes twice a day when they were apart.) In the early 1920s, turkeys were actually rare at the American dinner table. "Turkey prices were so high," Helen Baker explained, "that few other than the rich were able to have more than the two proverbial holiday turkeys, at Thanksgiving and Christmas."[17] A bright woman with an untrained but scientific cast of mind, Helen Baker decided she would be an innovator. She would make turkeys plentiful in America as well as cheap. Though he almost never spoke of his childhood, Baker would later remark, "My mother in particular laid the basis for the broiler industry."[18] Baker boasted that her methods led to the techniques later used at industrial turkey producers like Perdue and Butterball.

From the start, raising the birds was difficult. Turkey production had suffered in the United States because the flocks were frequently stricken with parasitic diseases.[19] Helen Baker began working with pathologists at Harvard and at Merck, the New Jersey–based drug company, to fashion a medical treatment to prevent the illnesses. Before he became a teenager, Bill Baker was expert in inserting a three-foot-long catheter containing a parasite-destroying iodine treatment down the throat of a writhing bird. During the immunization season, he later recalled, "it had to be done bird by bird for thousands of them."[20] The farm was an endless workday—challenges of incubation, breeding, housing, feedings—which only yielded to the Bakers' tireless and deliberate work after a decade. By the early 1930s, the flock on Quaker Neck numbered about a thousand, and Helen, who had now renamed her breed Baker's Bronze Beauties, had spread her theories on turkey domestication through informal gatherings all over Maryland.

Other Bell Labs scientists would attribute their laboratory aptitude to their youthful efforts to take apart car engines or rebuild radios. But in helping Helen Baker pursue the perfect turkey feed, her son had found a crude but effective introduction to the precision of chemistry. They had spent years working out the correct recipe: "40 pounds ground whole wheat, 20 pounds ground whole oats, 15 pounds whole alfalfa, 25 pounds meat scraps." Then they added three pounds of fine charcoal, one pound ground-up shells, and one pound salt.[21]

Baker attended college near his home in Maryland, at tiny Washington College in Chestertown, before going on to graduate school at Princeton, where he finished a PhD in chemistry in the spring of 1939. By that fall, he had begun his job at Bell Labs. His parents sold the turkey farm and moved to New Jersey to be near him. Even as a young chemist, he retained a farm boy's habits. Rising early, he would make notes in his journal of the weather and any unusual birds he had spotted the day before. And then, after working hard all day at the Labs, he would return home to do his chores—mow the lawn, sweep the cellar, repair the screen door.

Jim Fisk and Baker had met each other at the very start of their careers. In early October 1939, just after they arrived at the West Street offices, they'd been seated next to each other in the orientation photo for new employees, an incoming class that at that time numbered only a few dozen. Mervin Kelly had hired both of them. But from the beginning of his career, Baker viewed the world somewhat differently than Fisk. Baker, after all, was not a physicist but a chemist—someone who perceived that progress, the means of moving science and technology forward, was really the struggle to understand the composition of materials and fashion new and better ones whenever possible. Materials, he would later say, represented "the grand alliance of engineering and science."[22] To Baker, chemistry was the discipline that made a global communications network feasible. He would often cite examples. By substituting the lead sheathing on telephone cables with a synthetic plastic created by Bell chemists, the Bell System saved "more than the total research budget of Bell Labs for the decade in which the innovation was worked out."[23] At one point Baker commissioned a study on the switch to plastic sheathing to satisfy his curiosity. The study concluded that the changeover saved the company about $2.5 billion. It also determined that if phone engineers had continued to sheathe telephone cable with lead, "it would require 80 percent of the total lead produced in the U.S."[24]

Baker's arrival at the Labs in the late 1930s roughly coincided with the arrival of its future intellectual stars—Shockley, Shannon, Pierce, and Fisk, among others—the men who were sometimes referred to by their colleagues as the Young Turks. As Baker later told the *New Yorker,* the Young Turks "came to Bell with an interest in attacking the hard, fundamental questions of science—something that not many people thought could be done in a place like this." In those days, Baker explained, it was assumed that such studies were done at the world's great universities. But Shockley and Pierce used Bell Labs' resources to create "a new kind of science—one that was 'deep' but at the same time closely coupled with human affairs." In Baker's view, the Young Turks succeeded for the first time in bridging the gap between the best science of the academy and the important applications that a modern society needed.[25]

Baker, too, was a Young Turk. Part of his early fame went back to Bell

Labs' involvement in the war effort, just after Pearl Harbor, when the supply lines of natural rubber to the United States were blocked. "The Germans were supposed to have had synthetic rubber," Baker recalled some years later, "which was a frightening hypothesis if true because we didn't." Without rubber, it would be difficult to carry on mechanized warfare—no tanks, no jeeps, no aircraft. "You couldn't have a lot of those things without rubber," he said, "and our domestic economy would have collapsed as well." Government officials asked Baker and some other Bell Labs chemists to travel to Akron, Ohio, in December 1942 and meet at the city's Mayflower Hotel with a number of other industrial scientists. The men in Akron—then the center of the U.S. rubber industry and the headquarters of Goodyear and Firestone—agreed to use various promising rubberlike compounds and in two years created a vast new industry that could produce hundreds of thousands of tons of a durable, synthetic rubber.[26] Baker succeeded spectacularly at his job of overseeing scientific planning and quality control: The scientists met all their production goals. He also discovered a synthetic polymer known as "micro-gel" that proved crucial to improving the manufacturing process for the synthetic rubber. In the years after the war, his rise at Bell Labs was meteoric. In 1953, when he was thirty-eight years old, *Fortune* magazine put Baker on its list of the "Ten Top Young Scientists in U.S. Industry." Another man on the list was Baker's colleague Claude Shannon, also thirty-eight at the time. Two years later, Kelly called Baker into his office and promoted him to vice president of Bell Labs in charge of research.

As they rose to influence, the web of relationships among the Young Turks was hard to discern. There was little doubt that Baker and Fisk, the most accomplished administrators in the group, admired each other; their lengthy private memos to each other from the 1950s attest to a mutual respect and deep trust. Baker and his research deputy John Pierce were even closer, though their temperaments differed greatly: Pierce was antic and impatient, whereas Baker was poised and diplomatic. The two nevertheless discovered that they were companionable. "Both were raised as only children," Mike Noll, who worked with them, points out. "And both had doting, intelligent mothers." Pierce's private memos to Baker, technical notes scribbled on small sheets of paper during the 1950s and 1960s, along with Baker's letters to Pierce in the 1970s, suggest a mutual dependency.[27] Pierce, in proposing why some research ideas could prove important, helped Baker perceive the future; Baker, in turn, ever enthusiastic about new knowledge and seeing Pierce as a tireless instigator of ideas, was willing to push for Pierce-driven efforts such as the Echo satellite or computer music. In Baker's yearly reviews of Pierce—always suggesting the maximum raise possible—he characterized Pierce as a vital asset to Bell Labs, a scientist of international renown.[28]

Baker's view of Shannon, meanwhile, approached awe. He regarded Shannon as the single individual who had laid down the theoretical basis for the information age. "On the tenth anniversary of the transistor," Baker recalled to an interviewer in the mid-1980s, "we had a big event here in 1958 or so, and we were all speaking in the [Murray Hill] auditorium. They were saying how [the transistor] would endure in history forever, and I said, 'Yes, and after it's forgotten in a few thousand years, communication theory will still be with us.'" The comment was telling. The public Bill Baker smiled often and doused his audience in a downpour of oratory and flattery. The private Bill Baker was not always so diplomatic. When Bill Shockley won the Nobel Prize, for instance, Baker sent to him in California a congratulatory telegram more verbose and effusive than any other Shockley received—wishing him "warmest felicitations" on his distinction. "We think with pride and honor of the part your gifted mind has played in making science at large, and communications science particularly, stand where they do."[29] But in May 1957, when Mervin Kelly sent Baker an advance copy of Shockley's actual Nobel speech, Baker took Shockley's grab for credit as an affront. "It is the expected egocentric and rambling discourse that unfortunately signifies Bill's preoccupations these days," Baker told Kelly. In a private letter to Jim Fisk a few years later, Baker seemed to enjoy passing on the gossip about the collapse of Shockley's transistor business on the West Coast. "You may know the Shockley saga has come full cycle in which he has been appointed the Alexander M. Poniatoff professor of engineering science at Stanford," Baker wrote. The new owners of Shockley's transistor shop had now "snatched *all* production and development operations away from the Shockley Laboratory in order to consolidate them at Waltham, Massachusetts. Shockley is identified as remaining a consultant to the research activities in the Shockley Laboratory, but, of course, I do not know what will come further in connection with this money loser."[30]

DESPITE BAKER'S APPARENT SOCIABILITY, no one at the Labs ever socialized with him. Though he served on many corporate and academic boards, his colleagues in those rarefied circles were no closer to him. "I talked to him on the phone many times," remarked a fellow Rockefeller Institute board member, Fred Seitz—the same Fred Seitz who had once shared a rollicking drive across the country from California with Bill Shockley as the two headed to graduate school in the East. But Seitz never went to Baker's house. "He kept his life very private," Seitz added.[31] More than three decades after a career during which he spent years working closely with Baker as his second in command at Bell Labs, Ian Ross would still wonder: How did Baker actually fill his days? "I never really knew what he did," Ross remarks.

Often Baker went to Washington, D.C. He was first drawn toward government work by Oliver Buckley, the Bell Labs president who preceded Mervin Kelly. After Kelly had refused the post of science advisor to Truman, it was offered to Buckley, who accepted. "I was invited to accompany [Buckley] to Washington," Baker later explained, "but felt that the impending emergence of the solid-state electronics-transistor era was so preemptive that I should stick with it here. The transistor's discovery in 1948 rather confirmed that view."[32] Nevertheless, when he became vice president of research at Bell Labs a few years later, Baker began forging contacts with some of Washington's wise men of science— Lee DuBridge, for instance, who succeeded Buckley as Truman's science advisor, and James Killian, the president of MIT.

By that point, it seemed a given that presidents and vice presidents of Bell Labs would contribute their time and opinions to the government's cold war intelligence endeavors. Jim Fisk, following closely in the footsteps of Jewett, Kelly, and Buckley, had already been pulled into Washington affairs through his earlier work on the Atomic Energy Commission. In the mid-1950s, Fisk was working with a technology advisory group to President Dwight D. Eisenhower. In 1956, Fisk responded to Eisenhower's request to set up a separate commission to figure out how to gather better information about the Soviet Union by suggesting Baker for the assignment. "There was the presumption that the Soviets had become undecipherable, that we would not have enough warning to respond defensively to their threats," Baker recalled.[33] The result was the Ad Hoc Task Force for the Application of Communications Analysis for National and International Security, otherwise known as the Baker Committee. The committee's conclusions would be directed to the then five-year-old National Security Agency, a new unit within the Department of Defense charged with securing the country's information networks and deciphering foreign intelligence. NSA's very existence was then considered a national secret. So Baker was organizing a committee that did not officially exist to write a top secret report about how to improve an organization that didn't officially exist either.

His group was not lacking in brainpower. Baker pulled in John Pierce and John Tukey from Bell Labs—"the country's keenest thinkers," both of whom now had top secret clearances—along with several other scientists, including the future Nobel physicist Luis Alvarez.[34] The goal, as stated in Baker's 1957 description of his committee, was "to search for new concepts of interconversion of information and intelligence." In other words, his group would consider all the ways that technology now allowed information to be hidden and transmitted—through encoded signals, and even through chemical and biological patterns—and then recommend how America's intelligence agencies might respond. More specifically, the Baker Committee sought insights into how the United States might develop the ability to crack any imaginable Soviet code. "Our history sustains the belief that for both the national security and the universal freedom of man," Baker wrote, in a passage that held its own with the most artful cold war rhetoric, "the applications of all science to foreign deciphering political and military communications (basically revealing attitudes of nations toward each other) is a suitable and worthy intent."[35] Much of the Baker Committee's work was done at Arlington Hall Station, an Arlington, Virginia, facility used by the NSA, in the spring of 1957. In mid-July, the committee delivered what came to be known as the Baker Report, still mostly classified five decades later, which urged the government to move strenuously toward using solid-state computers and digital technologies for cryptanalysis. In his history of the NSA, James Bamford described the Baker Report as recommending "a Manhattan Project–like effort to push the USA well ahead of the Soviet Union and all other nations" through the application of information-age tools. Bamford also noted that one of the committee's enduring legacies was its recommendation that the U.S. intelligence networks establish "a close yet secret alliance with America's academic and industrial communities."[36]

Few outsiders were more closely, or more secretly, allied than Baker. In a couple of years' time, John Pierce would come to despise Washington and avoid government work entirely. Claude Shannon, who was also involved with the NSA and CIA on several projects, would drift off to more eccentric pursuits at home in Massachusetts. Baker remained a consultant to the NSA for more than a decade, and began to branch out in Washington. Beginning with Eisenhower, and continuing through the Kennedy, Johnson, and Nixon administrations, and later again under Ford and Reagan—always leaning toward the G.O.P., a reflection of the fact that he was, deep down, a conservative, "the eminence grise of Republican science," as the British magazine *New Scientist* described him[37]—Baker served as a member of the President's Foreign Intelligence Advisory Board (PFIAB), a group that examined the operations of the CIA and other intelligence agencies. As part of his PFIAB work, he helped found in 1960 the National Reconnaissance Office, a government organization charged with planning, building, launching, and maintaining America's spy satellites. The NRO remained secret for its entire first decade.

Baker shuttled between New Jersey and Washington, among his five offices—at Bell Labs' Murray Hill and Holmdel locations, the White House, the Pentagon, and the State Department. His duties in Washington were often broader than his titles suggested. Soon after his inauguration, President Kennedy asked Baker to explain some of the protocols for ballistic missile attacks, a technology that relied on the early warning systems that Bell Labs had built for the military. The men chatted in the Oval Office. A crucial telephone with a direct line to the Defense Department happened to be missing from Kennedy's desk. "Kennedy and [his science advisor Jerome] Wiesner and I got down on our hands and knees and we got under the desk and found somebody had put it in a drawer," Baker later remarked. "And then we explained the whole technology." Kennedy often invited Baker into his private quarters for discussions; he likewise called him at his Bell Labs office, at least once trying to find out if Jim Fisk might take a job in the administration.[38] During the Cuban Missile Crisis—a crisis brought on by the interpretation of information, in this case aerial photographs—Baker also became a fixture in the cabinet room.[39]

In the early 1960s, Baker's closest associate in Washington was Clark Clifford, an advisor to Kennedy and the chairman of the Foreign Intelligence Advisory Board. At the PFIAB, Baker and Edwin Land, the inventor of the Polaroid camera, drove the agenda. "These two men were our teachers," Clifford would later recall, "turning all of us on the committee into missionaries for the view that the United States should vastly increase its commitment to the finest state-of-the-art technologies in the field of electronic, photographic, and satellite espionage."[40]

Baker was the stealthiest of cold warriors, who preferred shadows to the open air. In 1961, just after Kennedy was inaugurated, Baker trudged through a snowstorm for an appointment with Vice President Lyndon Johnson, whom Kennedy had asked to offer Baker the job as the head of NASA. Baker diplomatically refused, and instead suggested that Johnson offer it to James Webb, which he eventually did. As Baker related the story, he told Johnson, "You've got to have somebody who can reflect the public, political and popular attitudes in this business much better than I." Whether he believed this, or whether it was simply a politic excuse for maintaining his responsibilities at Bell Labs and his secret responsibilities in government, was a matter of debate. What was undeniable was that Baker wanted little to do with a highly visible public office and a large degree of public accountability. Over time, a half dozen U.S. presidents offered him the post of science advisor. He turned it down every time.

New titles might not have increased his influence. By the start of the 1960s Baker was engaged in a willfully obscure second career, much like the one Mervin Kelly had formerly conducted, a career that ran not sequentially like some men's—a stint in government following a stint in business, or vice versa—but in parallel, so that Baker's various jobs in Washington and his job at Bell Labs intersected in quiet and complex and multifarious ways. Baker could bring innovations in communications to the government's attention almost instantly. "Many thanks for the very interesting and profitable tour you arranged for me at the laboratory," Louis Tordella, the director of the National Security Agency, wrote Baker in 1959. "I particularly enjoyed the brief glimpses into the future shape of things to come. Even of more immediate value, however, were the talks I had, particularly with you and Dr. Pierce."[41] In understanding the leading research for coding and transporting information, Baker also had a unique ability to tell others how it might be intercepted. He therefore did more than connect government officials with technological hardware. He explained to Washington how information works, and how it flows.

Fifteen

MISTAKES

B ill Baker's view of the future was grand, and with good reason: In the late 1940s, as one of the Young Turks, he had observed firsthand the transistor's invention and early development, and he had since come to understand its possibilities both for society at large and the military in particular. "So often," says Ian Ross, who worked in Jack Morton's department at Bell Labs doing transistor development in the 1950s, "the original concept of what an innovation will do"—the replacement of the vacuum tube, in this case—"frequently turns out not to be the major impact."[1] The transistor's greatest value was not as a replacement for the old but as an exponent for the new—for computers, switches, and a host of novel electronic technologies.

Years before, when Claude Shannon had written of the transistor to his old schoolteacher, "I consider it very likely the most important invention of the last fifty years," he'd seen this. It helped that Shannon had spent most of his career dwelling on the possibilities of digital information. The transistor was the ideal digital tool. With tiny bursts of electricity, it could be switched on or off—in essence, turned into a yes or no, or a 1 or 0—at speeds measured in billionths of a second. Thus in addition to being an amplifier, a clump of transistors could be linked together to enable a logical decision (and thereby *process* information). Or a clump could be linked together to help represent bits of information (and thereby *remember* information). To put hundreds, or thousands, or tens of thousands of the devices alongside one another (the notion that billions would one day fit together was still unimaginable) might allow for extraordinary possibilities. It was a "wondrous coincidence," as Bill Baker described it, "that all of human knowledge and experience can be completely and accurately expressed in binary digital terms."[2] As usual, Shannon was ahead of his colleagues. But in only a few years, by the late 1950s, Baker, too, viewed the future of digital computing and that of human society as wholly interrelated.[3]

A STEADY STREAM of semiconductor inventions emerged from Bell Labs between 1950 and 1960. Some had arisen from Baker's research department and others from the much larger development department. As a result, there were now a multitude of new transistor types and important new methods of manufacturing, such as the technique—resembling art etchings done on a minuscule scale—known as photolithography. In keeping with AT&T's agreement with the federal government, the patents for these inventions and processes were licensed to a number of other companies, not only large industrial shops like General Electric and RCA, but also two fledgling semiconductor companies known as Texas Instruments and Fairchild Semiconductor. Fairchild had been established after its principal engineers, Robert Noyce and Gordon Moore, had fled Bill Shockley's unhappy company along with several other colleagues to start a company of their own. In that "precompetitive era," as Ian Ross terms it, information was freely exchanged. Ross and Bell Labs researchers such as Morry Tanenbaum, who had invented the silicon transistor, would often speak with the men at Fairchild in Palo Alto and Texas Instruments in Dallas; employees of those start-up companies would also swing by Bell Labs on their trips east to see what was new. Jack Kilby, a young engineer who eventually joined Texas Instruments, first made a pilgrimage to Bell Labs in 1952 to attend one of the earliest seminars on how to build transistors.[4] Bell Labs was where most of the world's knowledge about semiconductors resided. "Remember, we were the only game in town when it started," Ian Ross says. "If you wanted to know about semiconductor devices, you went to Murray Hill and Building 2."

The good thing about the transistor was that by the late 1950s it was becoming smaller and smaller as well as more and more reliable. The bad thing was that an electrical circuit containing thousands of tiny transistors, along with other elements such as resistors and capacitors, had to be interconnected with thousands of tiny wires. As Ian Ross describes it, "as you built more and more complicated devices, like switching systems, like computers, you got into millions of devices and millions of interconnections. So what should you do?" At Bell Labs, Jack Morton, the vice president of device development, had coined a name for the dilemma: "the tyranny of numbers." Morton believed that one way to tackle the tyranny of numbers was simply to reduce the number of components (transistors, resistors, capacitors, and so forth) in a circuit. Fewer components meant fewer interconnections. One way to do this, Morton thought, was to harness the physical properties of special semiconductors so that they might be made to perform multiple electronic tasks—turning them into a kind of electronic Swiss Army knife. Therefore "a simple thing" within a circuit could replace multiple components. Morton called these "functional devices," but they were proving exceedingly difficult to engineer. What's more, Morton, fiercely intelligent and widely feared, was largely unwilling to hear criticism or change course. As Bill Baker wrote in a wry note to Jim Fisk, Jack was someone with "an intense distaste for anyone opposing him."[5]

Jack Kilby at Texas Instruments and Robert Noyce at Fairchild had different, better ideas. Both men, nearly simultaneously, came up with the idea of constructing all of the components in a circuit out of silicon, so that a complete circuit could exist within one piece—one *chip*—of semiconductor material. By eliminating the tyranny of interconnections, the method seemed to suggest substantial advantages in manufacturing and operational speed. Their innovation could, in short, be better and cheaper. Kilby had the idea in the summer of 1958, probably a few months earlier than Noyce. But Noyce's design was arguably more elegant and more useful.[6] In the early days, the product that Kilby and Noyce designed was known as a "solid circuit." In 1960—the year that *Business Week* crowned solid-state electronics as "the world's fastest growing industry"— Kilby and Noyce's invention seemed promising but still unproven. Not long after, the new idea became known as the integrated circuit.

To the engineers and scientists at the Labs, the integrated circuit was not a complete surprise. "We knew we could make multiple transistors within a piece of silicon, we knew we could make resistors, we knew we could make capacitors," Ian Ross recalls. But it was the received wisdom under Jack Morton, Ross adds, that such devices could *never* be reliable. Even though the quality of manufactured transistors was improving, there was still a significant failure rate. And on a chip with hundreds or thousands of components? Some of those components would inevitably fail, thus rendering the entire device useless. Kilby and Noyce opted to believe, correctly, that the manufacturing challenges could be worked out later. Morry Tanenbaum happened to visit Noyce's shop during a visit to California just as Noyce was developing the chip idea. "One of the things he showed me," Tanenbaum recalls, "was one of the first integrated circuits. And I said, 'Boy, that's really neat.' And when I went back I told some people about it. It certainly looked important to me, but I can't say I had the imagination to see how far it would go."

Kilby or Noyce could not have understood the full implications either—personal computers, portable phones, deep space exploration, all of the things that defined the kind of future that John Pierce would write about in his memos and essays. But in just a few years' time, the integrated circuit would represent something new for Bell Labs: a moment when a hugely important advance in solid-state engineering, though built upon the scientific discoveries at the Labs, had occurred elsewhere. Such a development perhaps suggested that the landscape of competitiveness in American electronics, something that Mervin Kelly had written about in the closing days of World War II, was now very much a reality. At the very least, it proved that even the great technical minds at Bell Labs, Jack Morton especially, could misjudge the future. "We had all the elements to make an integrated circuit," Tanenbaum adds. "And all the processes—diffusion, photolithography—were developed at Bell Labs. But nobody had the foresight except Noyce and Kilby."

IN THE MID-1950s, Bell Labs hired back Charles Townes—Charlie to everyone at Bell Labs—the inventor of the maser. Townes was still a professor at Columbia, but he now visited Murray Hill regularly as a paid consultant. He had first come to Bell Labs in 1939, on that sinuous detour that took him as a young man through the American Southwest and Mexico and then back to the West Street labs in New York City; in the orientation photo for new employees from that year, he is standing two rows behind a young Bill Baker and Jim Fisk. The connections to the Labs went even deeper, though: Townes's brother-in-law, Arthur Schawlow, had joined Bell Labs and was now working as a researcher at Murray Hill, and the two men, whose wives were sisters, were quite close. One day in the fall of 1957, Townes and Schawlow had lunch together at Murray Hill and began talking about the maser, the device that had been used to amplify faint satellite signals in John Pierce's Echo experiment. The maser worked by bombarding, or "pumping," a crystal or gas with electromagnetic energy; a resulting reaction ensued (what was called a "stimulated emission") and a tightly focused beam of tiny, millimeter-length electromagnetic waves was released by the crystal or gas.

To understand where this might lead required understanding communication signals—the wavelengths at which they travel and how that related to the transmission of information. The same line of reasoning still applied:

The shorter the wavelength, the higher the frequency.

The shorter the wavelength and the higher the frequency, the greater the capacity to hold information.

When Townes and Schawlow talked about other possible ideas that day in 1957, they wondered if they could alter the output of the maser to produce even shorter waves than those in the millimeter range. Schawlow later recalled that he had been thinking of ways to make infrared masers, though he hadn't gotten very far. But then Townes told Schawlow he had been thinking along the same lines.[7] If they could produce infrared light, it was conceivable, too, that they might be able to build something that could produce even smaller waves than infrared—that is, light waves that existed in what is called the visible spectrum. The result of the collaboration between Townes and Schawlow was a paper, written in the summer of 1958, outlining the principles of the laser. The brothers-in-law received a patent, issued in 1960, for their invention.[8]

As scientists, Townes and Schawlow were probing into the unknown for new knowledge. But as employees of the Bell System, even at that early stage, they were aware of the practical reasons for pursuing a laser at Bell Labs. As John Pierce later explained, "The laser is to ordinary light as a broadcast signal is to static." Ordinary light radiates in a chaotic and scattershot manner. The laser does not. From the perspective of a communications engineer, it is *coherent*—meaning it is intense and ordered and nearly all one frequency, which are important qualities for carrying information. "In principle it makes it possible to do everything with light that one does with radio waves," Pierce added. What's more, the great advantage is that the "bandwidth" of such light—which is related to its capacity—"is hundreds or thousands of times greater than we now have." The very title of the Townes and Schawlow patent suggested a clear direction.[9] Bell Labs' claim for the laser was that it was a new method for communication.

The first working laser—the name came from a man named Gordon Gould, a former associate of Townes, who also made a successful legal claim to the invention—was not built at Bell Labs. Nor was it built by Schawlow and Townes. Rather, it was developed at Hughes Aircraft, in Malibu, California, by an engineer named Ted Maiman. In Maiman's design, which began functioning in mid-May 1960, a flash of bright light stimulated a small pink ruby that emitted a short and powerful pulse of focused light. The Maiman laser didn't prove that the laser was guaranteed to have any practical value—that question was unresolved and still far off in the distance. But it did prove that the laser could actually exist beyond the theory outlined by Townes and Schawlow. Indeed, almost as soon as the optical researchers at Bell Labs heard of Maiman's work, they built a near-exact copy of Maiman's device to replicate and verify his results.[10]

At around the same time, another group of researchers at Murray Hill—Ali Javan, Donald Herriott, and William Bennett—were trying to build a different kind of device. On December 13, 1960, the men succeeded in operating a laser that used a gas medium of helium and neon, rather than a ruby. Aided by a series of mirrors that could focus the energy emitted by the stimulated gases, the men succeeded in creating a narrow beam of light. Most important, it wasn't a pulse; it was a steady and continuous beam.[11] The day after Javan's team got their laser working, a team of Labs engineers used the focused beam of light to carry a telephone call.[12] That sort of thing made AT&T executives actually sit up and pay attention.

THE LASER WAS NOT so much a single invention. Rather, it was the result of a storm of inventions during the 1960s. Noteworthy improvements (like a new design) or variations (like getting a new material to "lase") followed one after another in rapid succession. Sometimes only a few days separated the announcements of new developments. "The whole business of [lasers] is one of these things you just can't afford to let go," John Pierce told an interviewer around that time. "You can't clearly see that it will be of any use in communication. I mean, we certainly can't *guarantee* that. But it has that potentiality." Pierce added that "when something as closely related to signaling and communication as this comes along, and something is new and little understood, and you have the people who can do something about it, you'd just better do it, and worry later just about the details of why you went into it."

Rudi Kompfner, Pierce's deputy and close friend, shared that sense of urgency. And so Kompfner, with Pierce's and Bill Baker's endorsement, started making a list. "He went around the world at that time, in 1960, trying to find good people—that's all he wanted, good people," recalls Herwig Kogelnik. "And then he would try to persuade them to switch their disciplines to take on what he called 'laser and optical communications research.'" Even within the scientific community, the terms were new and strange. Kogelnik was finishing up his PhD at Oxford when Kompfner showed up one day during his worldwide recruiting tour. Kogelnik was a plasma physicist, which meant he studied the physics of certain gases. Rudi Kompfner didn't care. His simple argument to Kogelnik on the day they met was that the young man should think of the high frequency of light and what that could mean in terms of its capacity for information. Colleagues all recall the charm as well as the passion that animated Kompfner. To Kogelnik, he just said, "Think of all this bandwidth!"—a line that inspired Kogelnik to switch from plasma to lasers. "I had invested many, many years in plasma physics. And he persuaded people like me to totally throw away their past and start in a new field."[13]

Simply put, if the Bell scientists could figure out how to use light's vast capacity to transmit phone calls, data, and TV, they could avoid future worries about congestion.[14] What was also attractive were the *economics* of optical transmission. For decades, the Bell System had realized that it was far more cost-efficient to mix together many hundreds of conversations on an intercity copper cable—by a complex technical means, the signals could be sent together at a higher frequency and then teased apart at the receiving ends. Sending more information and sending it more economically were often the same thing.

But as the 1960s wore on, this possibility seemed increasingly remote. Several years after Townes and Schawlow outlined the laser in their paper, a variety of ticklish problems, as one Bell Labs research director described them, remained unsolved.[15] For starters, there were a host of technical challenges in finding the best laser materials with the most useful frequencies for communications. At the same time, Bell engineers were still working on a technology that could modulate voice and data signals and then "impress" those signals upon the laser beam. Above all, there were nagging questions about transmission. Without question, light could carry voice and data—but how would it be sent around the country? Through the air? In a hollow pipe? In September 1960, just a few months after the ruby laser was invented, several scientists at Bell Labs succeeded in sending a pulse of laser light twenty-five miles through the air from Crawford Hill in Holmdel to the Murray Hill labs.[16] The laser group then began to realize that the earth's atmosphere creates all types of interference. "Rain, snow and fog can cause heavy power losses," a Bell Labs director, Stewart Miller, explained in *Scientific American*. That left another possibility—the waveguide, those hollow pipes, beset with problems and still in development, that were ultimately supposed to relieve the future traffic on the Bell System by carrying millimeter radio waves long distances. They could conceivably carry light waves, too. But sending light through a waveguide would be immensely difficult: The beam would have to go around turns and up hills and down hills and remain perfectly focused within the walls of the hollow pipe.[17] An inter-city waveguide might have to do this for hundreds of miles. Nobody at Bell Labs—not Fisk, not Baker, not Pierce, not Kompfner—seemed to think this was imminent. Light waveguides were, at best, a next-generation technology.[18]

There was, perhaps, another option. It was the idea of a scientist well outside the elite ranks of Bell Labs. In 1966, an engineer named Charles Kao, employed in England by International Telephone and Telegraph, visited the Labs to talk about a technology he was researching in Europe. Kao had recently delivered a paper at an engineering conference in London suggesting that transparent glass fibers, carrying waves of light, might solve the transmission problem. Some of the scientists at Bell Labs had already toyed with this possibility. It had long been clear to engineers that thin glass strands could transport light for short distances; such fibers, in fact, were already being used in medicine, where they were proving useful for gastrointestinal examinations. But those working in the communications field doubted that glass fibers could transport signals the much greater distances the phone system required. The glass just wasn't clear enough. A signal would be lost—technically speaking, it would scatter and attenuate—after only a few dozen meters. Fibers were, as the engineers put it, too "lossy."

Kao was more optimistic. He had spent the past few years in Europe doing deep exploratory research, and he had determined there was no fundamental reason why strands of exceedingly pure glass couldn't carry signals substantial distances. His paper was partly theoretical. Glass of the sort Kao was talking about didn't really exist; even the clearest glass on earth would effectively absorb or scatter light and thus kill a signal in a few dozen meters. But Kao hadn't said that pure glass would replace wires or waveguides immediately. He had only concluded it was *possible*.

He was also liberated to some extent from the pressures that shaped the views of the Bell Labs scientists. Labs upper management had bet the future on waveguides, but Kao had not. The fiber optic historian Jeff Hecht would later point out that Kao (unlike the accountants at AT&T) had no incentive to make years of investment, in both time and effort, pay off. What's more, there was a pull on Kao in a different direction. Innovations are to a great extent a response to need. Phone engineers in Europe—Kao included—weren't looking for a complex new technology, such as the waveguide, for intercity communication. They needed *intra-city* communication. Generally speaking, Europe's metropolitan areas were both denser than America's and closer to other metro areas. The British telecom planners, Jeff Hecht notes, "wanted better technology to send signals between local switching centers that typically were a few miles apart. They wanted something easy and inexpensive to install in heavily developed areas, not high priced, huge capacity systems to span vast distances."[19]

When Kao visited Bell Labs and urged the people there to follow up on this line of research, the optical team under Kompfner heeded him. "I remember very well that day," says Tingye Li, who would eventually help lead Bell Labs' optical research efforts and would become a close friend of Kao's. "We were having a picnic on top of Crawford Hill and he joined us. I had not met him before."[20] Kao would later say that he received a skeptical welcome at Bell Labs; but Li and several researchers who were there at the time recall it differently. "I don't think anyone pooh-poohed it," recalls Ira Jacobs, who would later become deeply involved in the Labs' fiber optics development. "But I think there were a lot of people who thought it would move slowly."[21] As the 1960s wore on, Kompfner's group became increasingly interested in the possibilities of fiber, and by 1969 there was a fiber research program up and running.[22]

Yet the Bell Labs brain trust had also concluded that their present strategy of hollow pipes remained more feasible. Apparently, Jim Fisk and Bill Baker agreed. "A telecommunications network is emerging in our nation which will match the historic upsurge of our society's demand for the transmission and distribution of human knowledge and personal experience," Baker said in a speech about networks for the future at a Chicago symposium in 1968. The "new science which will provide the massive telecommunications capabilities of the future," he explained, still depended on waveguide pipes, which would carry all the traffic and easily handle the demand for Picturephone circuits. The future, in other words, still looked the same to Baker as it had to Fisk at the beginning of the decade: the Picturephone, waveguides, electronic switching. These were the Bell System's bets, and they were sticking with them.

ANY SCIENTIST WHO WORKED at Bell Labs—especially anyone in Bill Baker's research department, whose job was probing the unknown—understood that failure was a large part of the job. Experiments sometimes literally exploded; results often disappointed; gut feelings frequently turned out to be indigestion. Moreover, new innovations that portended a grand future—the germanium point-contact transistor, for instance—could quickly be rendered irrelevant by a new iteration of a similar idea, such as the silicon transistor or (later still) the integrated circuit. In retrospect, of course, the evolution of technology looks like an ever-ascending staircase, with one novel development set atop another, leading incrementally and inevitably to all the benefits of modern life. Bill Baker was expert at crafting precisely this public narrative. "Within my lifetime," he testified to the Federal Communications Commission in 1966, "the United States has progressed from a nation held together by post and telegraph, in which the ability to span the country by telephone had barely been demonstrated, to one in which the complex telephone network is an indispensable component, intimately linked with the growth and operation of communities, private organizations, government, public safety and national security."[23] In truth, Baker, having spent years at the bench working on chemistry experiments, knew that science and technology weren't a matter of assured upward progress. The waveguide, for example—those hollow pipes that had entailed years of research and millions of dollars effectively died as a possible future in the fall of 1970. That September, the Corning glass company announced that it had succeeded in creating thin glass fibers so pure that they could transmit light with very low losses for thousands of feet. Corning did not yet have a manufacturable product, but they had demonstrated that Charles Kao's research and intuition were correct: There was a future for lightwave communications, and it was almost certainly in fiber optics rather than in waveguide pipes. By the end of November 1970, Rudi Kompfner agreed: "Optical fibers will be at the center of the stage in the future," he wrote to a colleague.[24]

No one was keeping an actual ledger during the 1960s on how well Bell Labs was doing in planning for the next century, but had they been, they might have put electronic switching and the laser work in the column of spectacular and prescient successes. During those same years, there were other achievements at Bell Labs that would, in time, alter the world. One occurred when several computer scientists at Murray Hill got together to write a revolutionary computer operating system they called Unix, which would serve as the basis for a host of other computer languages.[25] Another breakthrough—the charge-coupled device, or CCD— was invented during work on new forms of computer memory. The CCD was a light-sensitive electronic sensor that used the varying responses of electrons to different amounts of light to create photographs and images of extraordinary detail.[26] The CCD would become the foundation of digital photography as we now know it, but Bill Baker perceived its value to national security long before its commercial potential was understood. "I took it out to the NRO immediately," Baker would later recall, knowing that the National Reconnaissance Office, the spy satellite agency he had helped found, could use it for espionage.[27]

Meanwhile, the other side of the ledger was filling up, too. The integrated circuit of Kilby and Noyce, which built on the engineering and materials-science achievements of Bell Labs, seemed more a missed opportunity than a misstep.[28] But nobody could offer such a mitigating rationale to explain the waveguide and Picturephone, two interrelated and fabulously expensive follies.[29] It seems worth considering not only how those endeavors failed, but what those failures represented. Innovators make different kinds of mistakes. The waveguide, for instance, might be considered a mistake of perception. It was an instance where a technology of legitimate promise is eclipsed by a breakthrough elsewhere—in another corporate department, at another company, at a university, wherever—that solves a particular problem better. It was perhaps understandable, moreover, that a breakthrough in the creation of pure glass fibers wouldn't come from an organization such as Bell Labs, where materials scientists were experts on the behaviors of metals, polymers, and semiconductor crystals. Rather, it would come from a company like Corning, with over a century of expertise in glass and ceramics.

Mistakes of perception are not the same as mistakes of judgment, though. In the latter, an idea that developers think will satisfy a need or want does not. It may prove useless because of its functional shortcomings, or because it's too expensive in relation to its modest appeal, or because it arrives in the marketplace too early or too late. Or because of all those reasons combined. The Picturephone was a mistake in judgment.

The Picturephone began on a high note of optimism. "We have now received a clear go-ahead from AT&T on the Picturephone program we proposed," Julius Molnar, Bell Labs' executive vice president, informed the staff in late summer of 1966.[30] The actual Picturephone technology was being upgraded and redesigned; instead of the egg-shaped futuristic device that had made a splash at the World's Fair, Molnar told the staff, the set would be a "Model 2," or Mod 2, as it was called at Bell Labs, a squarish device, designed by a renowned industrial designer, that was both more elegant and more functional. Molnar's goal was to field test the device in 1968 and begin a rollout of "commercial face-to-face picturephone service" soon after.

A small exploratory marketing study of the Picturephone, comprising ninety-nine employees from major corporations and nonprofit institutions, was compiled at the end of 1967. Meant to investigate the views of "a cross-section of business customers," the study's conclusions sounded no alarms. The market potential for Picturephone service appeared to be strong among the survey participants, the study concluded: "Many would be willing to pay more than $50 monthly for a Picturephone service designed to meet their needs."[31] By the spring of 1968, there was no turning back. "The trials of Picturephone have now progressed to a point," Bell Labs president Jim Fisk said in a speech at the time, "where any skepticism as to its interest and utility is only a replay of the skeptical response which greeted [Alexander Graham] Bell when he tried to promote the telephone over 90 years ago."[32]

OFFICIALLY, the Picturephone rollout began with a trial at the Westinghouse Corporation in Pittsburgh, starting in February 1969; during the summer of that year Bell Labs devoted an entire issue of its magazine, the *Bell Laboratories Record*, to explain the science and engineering of the new launch. The possible impacts of the Picturephone, Julius Molnar suggested in an introductory note, could well be seismic: By lessening the need for shopping trips or for conducting in-person business, "there will be less need for dense population centers," as well as reduced traffic. "Picturephone is therefore much more than just another means of communication," Molnar wrote. "It may in fact help solve many social problems." AT&T began offering Picturephone service in Pittsburgh and Chicago at the end of 1970. Other electronics companies—RCA and GTE, for instance—readied similar technologies. If video telephones were the future, they saw no reason to let the Bell System reap all the rewards.[33]

But within about twelve months, Bell executives saw that the anticipated demand for the Picturephone service was not materializing. In a speech to Bell Labs' department heads in March 1972, Julius Molnar went through the results: "Most of you probably know that attempts to introduce it in Pittsburgh and Chicago have hardly been howling successes," he began. "In Pittsburgh after a year and a half there are only eight paying customers with 30 sets in service." The monthly price in that city for Picturephone service was $160. But in Chicago, where the price was set at a cut-rate figure of $75 per month, the results were nearly as worrisome. "After a vigorous sales campaign," Molnar acknowledged about the Chicago business, "they have 46 customers with 166 sets in service, with another 128 sets on order. Much better than in Pittsburgh, but still not quite encouraging."[34]

According to Irwin Dorros, one of the Bell Labs executives involved in the launch, the team working on the Picturephone had never doubted its eventual success. "Groupthink," as Dorros puts it, had infiltrated the endeavor. Yet as the Picturephone's demise became more evident, even its most ardent proponents began to ask why it was failing and why they hadn't anticipated that outcome.

Most began their list of reasons with one fundamental realization: People just didn't like the idea of Picturephones that much, and the market research indicating otherwise had been flawed and inadequate. Customers simply liked the impersonal aspects of a regular phone call—or they at least felt that video added little additional satisfaction to electronic communication.[35] Then came other realizations. The Picturephone equipment and service were all far more expensive than the cost of regular phone service. Picturephone calls required such tremendous (and costly) bandwidth that long-distance service was out of the question. For all these reasons, the technology couldn't attract enough users to attract even more users. "To start up a service, you have to think about: I have one, you don't have one—so I can't talk to you," Irwin Dorros says. "So I can only talk to you if you have one. So how do you get a critical mass of people that have them?" Many years later, a computer engineer named Robert Metcalfe would surmise that the value of a networked device increases dramatically as the number of people using the network grows. The larger the network, in other words, the higher the value of a device on that network to each user.[36] This formulation—sometimes known as Metcalfe's law—can help explain the immense appeal of the telephone system and Internet. However, the smaller the network, the lower the value of a device to each user. Picturephone's network was minuscule.

Price cuts didn't seem to be working. And so its value was vanishingly small, with little prospect of any increase.

For years, Bell Labs executives in Murray Hill, Holmdel, and Whippany had communicated with one another via Picturephones in their offices. "I used it all the time," Dorros recalls. "I used it every day. I thought it was terrific." Not all the directors and vice presidents, however, liked the service. Rudi Kompfner, for instance, positioned a still photograph of himself in front of his set—in John Pierce's admiring recollection, the image showed Kompfner to be remarkably attentive and invariably interested in whatever was being said—so that he could move about his office during a chat.[37] Eventually, as the Picturephone initiative died out and the reality of its technological and economic failure set in, the Bell Labs bosses stopped using Picturephone service or had it disconnected. Forty years later, some would still defend the device, seeing it as far ahead of its time and too ambitious for the technical capabilities of that era's network. The newfound popularity of video chats over the Internet might seem to validate this view. But to an innovator, being early is not necessarily different from being wrong. And in any event the Picturephone's rejection in the marketplace was swift and decisive. As he looked around his Holmdel office just a few years after the Chicago rollout and his eyes rested on his Picturephone, Bob Lucky recalls, "I thought I had the last one in the world." And on that day it occurred to him, he adds, that there was no one else to call.

Sixteen

COMPETITION

Jim Fisk announced his retirement as president of Bell Labs in December 1972. It was actually the second time Fisk had announced he was leaving the Laboratories. The first had been twenty-five years earlier, just after his successful work on World War II radar sets, when he had accepted a post in Washington as the chief of research for the Atomic Energy Commission. To mark that 1947 departure—a temporary one that lasted only a few years—Mervin Kelly had held a dinner party for Fisk. It was a classy affair, as well as a jovial one: Bell Labs' top researchers (dozens of men attired in ties and three-piece suits, and not one woman) came out to toast Fisk, first sharing martinis and oysters and then dinner and cigars. A professional photographer had been hired to document the festivities, and the record of that evening—an album of black-and-white photographs, preserved among the valued possessions of Bill Shockley—documents a group of men, none of them yet famous, on the cusp of an era that would soon be transformed by their work.

Before dinner, Julius Molnar, Walter Brattain, and John Bardeen mingled about in the crowd. Mervin Kelly shared a few confidential moments, his foot on a barstool, with his research chief Ralph Bown. Bill Baker was present, too. In the photographic portfolio from the evening, snapshots of Baker captured him surrounded by a small group, saying little, drinking nothing, but listening with polite enthusiasm, as was his habit, while Mervin Kelly held court. As dinner was served, the men were seated, and Fisk faced the audience at a long table on an elevated stage. Kelly and Shockley sat beside him on the dais. So did Harvey Fletcher, the head of physical research who so many years before had worked with Kelly counting oil drops at the University of Chicago laboratory run by their mentor, Robert Millikan. John Pierce was onstage, too, in a frenetic and lively mood, riling up the crowd with comic drawings he held up to the audience. Of all the elite Bell scientists and engineers gathered together on that evening in 1947—long before anyone had heard of a transistor or contemplated the idea of a laser—only Claude Shannon appeared to be absent.[1]

By the early 1970s, the crowd was gone. At Bell Labs, Bill Baker was the last of the Young Turks. John Pierce, feeling himself in a rut, had retired a few years before his sixty-fifth birthday, accepting a teaching position at his alma mater, the California Institute of Technology, and relocating to the West Coast, not far from where he had attended high school and flown glider planes. "John Pierce is the model of contemporary man," Baker said at Pierce's 1971 retirement ceremony, "because he is always ahead of his time by decades and hence can't avoid eventually being contemporary." It wasn't only men such as Albert Einstein or Niels Bohr, Baker added with his usual enthusiasm, who should be heroes of twentieth-century science; those men, brilliant as they were, could never have done what Pierce did, which was to "inject realism into every element of physical science pursued at the Bell Laboratories." In Baker's view, Pierce had helped create "usable and visible" technology. It wasn't theory; it was work that changed people's lives.

As he departed, Pierce asked if he could consult for Bell Labs. Bill Shockley—now teaching at Stanford—was consulting already, making some extra money by working for Jack Morton in the device department. But in the 1970s, the contributions of both Pierce and Shockley amounted to visits from California to New Jersey for a few weeks a year. They were at most an occasional presence, their influence minimal. Julius Molnar, meanwhile, the czar of the development organization and Fisk's heir apparent, had taken ill during the latter part of 1972. When he died of colon cancer the following January, he was fifty-six years old. As Molnar's condition grew dire, Fisk had called Bill Baker at his home one evening and asked him to come over. "I bet he's going to ask you to be president," Baker's wife said to her husband after he hung up the phone. When the prediction turned out to be true, Baker hesitated before accepting. Being director of research at Bell Labs had been the only job Baker ever wanted. Over the years he had turned down a number of government positions and university presidencies so he could remain. Only when he decided that as president he could still keep a hand in managing basic research did he accept.[2]

Assuming the presidency meant Baker would be addressing the same kinds of questions that Frank Jewett had faced when he established Bell Labs nearly fifty years before. What problems did AT&T and the local phone companies need Bell Labs to solve for them? And what new communications technologies might exist—or should exist—ten or twenty years hence? But Baker would also have to address some challenges that Jewett never faced. Over the past decade or so, some of the phone company's problems had become problems of business rather than technology: Several new companies, most of them using innovations developed by Bell Labs, were trying to compete in both telephone equipment sales and long-distance services. These competitors were helped in the late 1960s and early 1970s by a burgeoning philosophy, now finding adherents among politicians and lawyers in Washington, D.C., that American consumers would be better served through competition rather than a tight federal control of industry. The U.S. Congress was already getting poised to loosen the rules that had long governed the airline, railroad, and securities businesses. In the argot of the day, these industries would soon be "deregulated." Should telecommunications be next? The dilemma was whether it remained in Americans' best interests to have a regulated phone monopoly such as AT&T—a monopoly that had "an end-to-end responsibility" for telephone service—or whether the phone giant should be dismantled in the expectation of more competition, lower costs, and perhaps an even greater rate of innovation.

Such questions had little to do with science and engineering. And thus they went far beyond the purview of Bill Baker, and farther still beyond the problem-solving capabilities of even the brightest minds at Bell Labs.

FOR MORE THAN FIFTY YEARS, AT&T had remained intact at the pleasure of the United States government—always, as the economist Peter Temin points out, "operating at the limit of what the antitrust laws would allow."[3] The company's unprecedented size and reach and power—ownership of Bell Laboratories and the equipment manufacturer Western Electric; ownership of the long-distance company known as AT&T Long Lines; and shares in or outright ownership of twenty-two local phone companies— had repeatedly been the cause for legal action on the part of the U.S. government. Yet each time, AT&T had found a way to dodge or duck; the company would appease regulators and politicians and continue on its usual business. Each time, moreover, the legal arguments subsided into a renewed understanding that telecommunications were "a natural monopoly" and were therefore best carried out by a single entity.

That perceived natural monopoly wasn't only justified by the phone system's technological complexity and interdependence; it was also—in an argument that telephone executives made over and over again—a matter of economics. With one company in effect serving the country's phone customers, some parts of the phone business that were highly profitable, such as long-distance service, could subsidize other aspects that were less profitable, such as local calling. Profits from high-paying corporate customers, moreover, could subsidize service to residential customers. Profits from dense urban areas could subsidize expansion into sparse rural areas. All in all, this kind of "averaging," as it was sometimes called, helped make telephone service available and affordable for most Americans. At the same time, thanks to technological innovations, the quality of the service had steadily improved over the years, even as many of the costs had steadily decreased.

The history of relations between the Bell System and the U.S. government seemed to follow a pattern, with truces occurring roughly every other decade. In 1913, in response to an antitrust suit by the U.S. Department of Justice, AT&T had agreed to stop buying up local phone companies and to offer the ones that continued to do business access to its long-distance phone network. In the late 1930s, the newly minted Federal Communications Commission had embarked on an investigation of the phone company that had resulted in a scathing evaluation of its business practices. In the end regulators had homed in on questionable billing practices AT&T had used in providing equipment to customers through its manufacturing arm, Western Electric. Yet in 1939, the government's line of attack stalled: World War II intervened, and Western Electric and Bell Labs demonstrated in the work on radar and other devices that they were essential gears in the nation's military machinery. It wasn't until 1949 that the Justice Department, picking up on existing government complaints, filed a new antitrust suit against the Bell System. What followed were years of hearings and a series of cozy negotiations between government officials and AT&T lawyers, until finally the 1949 suit was settled through an agreement known as the 1956 consent decree. AT&T could continue its regulated monopoly on phone service. But the decree forced the company out of the computer business and insisted that it make its older patents freely available and its new patents available for a reasonable charge.

In the wake of the 1956 agreement, AT&T appeared to be indestructible. It now had the U.S. government's blessing. It was easily the largest company in the world by assets and by workforce. And its Bell Laboratories, as *Fortune* magazine had declared, was indisputably "the world's greatest industrial laboratory."

And yet even in the 1960s and 1970s, as Bill Baker's former deputy Ian Ross recalls, the "long, long history of worry about losing our monopoly status persisted." To a certain extent, Bill Baker and Mervin Kelly believed their involvement in government affairs could lessen these worries. In the view of Ross and others, such efforts probably helped delay a variety of antitrust actions. Ross recalls, "Kelly set up Sandia Labs, which was run by AT&T, managed by us, and whenever I asked, 'Why do we stay with this damn thing, it's not our line of business,' the answer was, 'It helps us if we get into an antitrust suit.' And Bell Labs did work on military programs. Why? Not really to make money. It was part of being invaluable."

But being invaluable to the Department of Defense was not necessarily the same as being invaluable to the Department of Justice or the Federal Communications Commission. And the Department of Justice still employed lawyers who harbored a deep mistrust of the Bell System from the manner in which the 1949 antitrust suit had been swept aside by the 1956 consent decree. "Throughout the 1960s," the journalist Steve Coll noted in his account of the 1970s-era battles between AT&T and the federal government, "the [antitrust] division maintained files about AT&T's activities, waiting for the right moment to go after Western Electric again."[4]

Relations between Ma Bell and various regulatory commissions were also strained. A number of service problems in the late 1960s, especially in New York City, had undercut the phone company's image of quality and perfection. At the same time, a variety of independent manufacturers had won the right, granted by the FCC, to connect their equipment, such as office switchboards, into Ma Bell's precious network.[5] Most important, however, was that certain lawyers at the FCC were increasingly sympathetic to a new long-distance company, run out of a modest Washington office by a brilliant, bare-knuckled businessman—a smoker and a drinker, and judging by the risks he took on his telecom business, a gambler, too—named Bill McGowan. The company was known as Microwave Communications Inc., but was better known as MCI.[6]

IN THE LATE 1960s, McGowan had teamed up with an entrepreneur named Jack Goeken to start building a nationwide network of microwave towers, just as Bell Labs had done in the early 1950s. Their company offered long-distance service, first to businesses, then to ordinary residents. MCI's service wasn't any better than the Bell System's—indeed, the quality of calls and the network connections were sometimes poor, and the technology was in certain respects inferior. But the service was significantly cheaper than AT&T's. Unlike Ma Bell, MCI didn't have to build things to last forty years. It likewise didn't have to subsidize its local phone service by charging high long-distance rates. MCI didn't *own* any local phone companies. The executives at AT&T therefore maintained that MCI was "cream-skimming." What they meant was that MCI was homing in on the most profitable part of the phone business and taking none of the responsibility for a nationwide "end-to-end" network, as AT&T was obliged, by law and tradition, to do. The FCC had nevertheless concluded that MCI was a worthy competitor that offered businesses and individuals a useful service. In 1971 the FCC commissioners maintained that AT&T had to connect MCI's long-distance microwave network into its local switching centers.[7]

To Bell Laboratories' top managers, the decision was both astounding and deeply threatening. The Bell System had long coexisted with a modicum of competition: Smaller local phone companies still operated in rural areas, relics of an agreement struck between AT&T's Theodore Vail and the government early in the century, when Vail had agreed to stop buying up his competition. But MCI's pursuit of the profitable long-distance monopoly was different. It left some Bell Labs executives to wonder if they were seeing the start of a potentially mortal threat. Indeed, MCI was finding sympathy from the FCC as well as from various U.S. congressmen, who were now warming to the idea of greater competition and a less monolithic AT&T. A breakup, once deemed unthinkable, had moved into the realm of the possible.

In testimony before the U.S. Senate's Subcommittee on Antitrust and Monopoly, Bill Baker asserted that "the notion that Bell Laboratories could endure and function away from AT&T, Western Electric, and the operating integrated Bell System would be laughable were it not so sinister and so ominous." It was an argument like the one a gifted child might make in favor of preserving his parents' marriage. The strength of Bell Labs, Baker declared, was in its links with other parts of the monopoly. It was what allowed the Labs' scientists and engineers to "think of new digital networks, or new telephone instruments, of new modes of distribution like satellites and fiber optics." It was, Baker added with a typical flourish, what allowed "human creativity [to be] converted to human benefits." The arrangement must continue.

Four months later, on November 20, 1974, the Department of Justice filed a sweeping antitrust suit against AT&T, naming Bell Labs and Western Electric as codefendants. The government alleged that AT&T and Co. had engaged in "an unlawful conspiracy" to monopolize communications service. Among other things, the phone giant had made it difficult for equipment providers to compete against Western Electric and had obstructed service providers such as MCI from connecting to its network. The defendants, the government said, "are continuing and will continue these violations" unless something was done. The Department of Justice not only implored the courts to break off Western Electric from AT&T; it urged that some or all of the local phone companies be split off as well.[8]

John Pierce had never been enthusiastic about AT&T's stuffy corporate management. In his Bell Labs office, Pierce had placed a large sign on his desk that asked, in reference to AT&T's headquarters, "195 Broadway: Is there intelligent life there?" But the threat of breakup stirred in Pierce, now an academic in California, a fervid willingness to defend his old employer. "It is certainly within the law for the Department of Justice to institute a suit against the Bell System," Pierce wrote in an essay he submitted to the *New York Times* op-ed page, "but I think that it goes beyond all reason when they assert that this could possibly result in cheaper or better telephone service for the public. We have too many examples in other countries of the high costs and poor service created by the very sort of situation that the Justice Department is trying to create." John deButts, the powerful chairman of AT&T, read one of Pierce's defenses of the phone company in a financial magazine and sent Pierce a warm note on his personal stationery. "Not many men have so broad a grasp of how crucial organization is to innovation, and we are fortunate to have so authoritative a voice as yours to point out what mischief the Justice Department suit would do should it succeed. Which it won't."[9]

John deButts considered AT&T's vast communications network to be unique in all the world. No one else could replicate it; no one else could run it. Its construction and maintenance, done over the course of a century, had been Herculean. Its electronic architecture was the product of genius and hard work. He was correct in all these respects. He did not seem to grasp, however, how quickly technology could now be replicated, in part thanks to Bell Labs' widely available patents. Mervin Kelly had predicted during World War II that the telecommunications industry would eventually begin to compete against the larger electronics industry, where radio and television makers constantly battled over products and prices. "We have been a conservative and non-competitive organization," Kelly had pointed out to his colleagues in 1943. "We engineer for high quality service, with long life, low maintenance costs, high factor of reliability, as basic elements in our philosophy of design and manufacture. But our basic technology is becoming increasingly similar to that of a high volume, annual model, highly competitive, young, vigorous and growing industry."[10]

MCI was proof of that striving and competitive future. Bob Lucky recalls a day in the early 1970s when several AT&T executives were discussing with Bell Labs executives the prospect of upstart companies offering long-distance service. "You don't have to worry about this," the AT&T executive assured them, "because we have the network. No one else has the network." For a short while, at least, that was true. They didn't realize at the time that anyone could build a network.

As LEGAL MANEUVERS BEGAN in late 1974—the trial, most observers agreed, would take years, perhaps a decade, to resolve—one of the points of discussion had to do with whether telecommunications was now a "mature" industry. A mature industry shouldn't need the protective hug of federal regulations. A mature industry was ready for jolts of competition. More than half a century before the Justice Department filed its latest suit, AT&T president Theodore Vail had set forth a goal for his phone company: "One policy, one system, universal service." Had that goal been achieved? There were now cables and microwave links and electronic switching stations that connected all Americans; there were likewise satellite links and undersea cables that connected the country to the rest of the world. At Bell Labs, Bill Baker, though loath to concede that a mature industry like his should be deregulated, agreed with this assessment. "I think that the era of having to demonstrate new technical communications functions is largely behind us," he said. "It's no longer necessary to show, for example, that all of the people of the world can be connected together, or that voice and video can be transmitted with fidelity and reliability. Now we have to concentrate on maximizing the efficiency, performance and economy of all these facilities."[11]

Baker chose his words carefully; they were meant to be strategic. Improving efficiency and lowering costs could help the Bell System maintain an advantage in what was shaping up to be a new era of competition. A look back on the network of the early 1970s, however—especially with a knowledge of what it would become in the decades following— erodes any belief that it was near completion. The businesses and citizens of the world had only begun to consider how they might send, and how they might use, information. What altered their understanding were two complex and expensive projects, both undertaken at Bell Labs amid the efforts to break its parent company apart. These projects—the first in exploring how to manufacture and install glass fiber to carry light pulses, the second in mobile telephones—actually transformed the network into something else. Those efforts made global communications into something thoroughly modern.

ALMOST FROM THE START—from the time that Rudi Kompfner in 1960 went around the world to recruit scientists to work on what he called "optical communications"—Bell Labs' managers understood that moving the world's communications systems from electrical waves to light waves would require several inventions. The first was a laser. It wouldn't be the kind of laser that existed already in the 1960s, however. It would be a new kind with certain desirable characteristics. First, it would have to be durable, so that it could emit a beam for years without dying out. Second, it would have to work at wavelengths that were appropriate for communications. Third, it would have to work at room temperature, meaning it wouldn't need to be supercooled, which many lasers required to prevent overheating. Finally, it wouldn't be the size of a table top. It would be small, like a transistor, and made from some combination of solid materials, most likely semiconductors. "At this moment we have a pretty good notion what an eventual system would look like, and what some of the essential ingredients are going to be," Kompfner wrote to John Pierce in April 1970. One of them, he noted, was a room-temperature laser, which his group was pursuing aggressively. "That does not exist," he noted, but "that does not cause us to sit around waiting."[12] If there was going to be a functioning optical communication system in ten years, Kompfner added, "much of it will be based on what is being done here now."

The laser—or rather the lasers—that Kompfner and Pierce were looking for arrived fairly quickly. In the early 1970s Bell Labs scientists came up with several types of revolutionary devices known as room-temperature, continuous-wave, heterostructure semiconductor injection lasers.[13] The fearfully long name belied their size. The lasers were tiny, no larger than a grain of sand. Essentially, they consisted of a rectangular slab, polished on one face to a mirrorlike finish, made from a sandwich of several specially prepared semiconducting materials. (Usually they relied on a combination of gallium and arsenic.) On the top and bottom of this tiny crystal were two metal contacts. When these were connected to a power source, such as a common battery, the holes and electrons inside the semiconductor crystal would combine to emit a steady beam, which would emerge from the slab through a "stripe" on its polished face.[14] Significantly, that steady beam could be "modulated" to transmit a multitude of digital signals.[15] In other words, a conversation in the form of an electrical signal, moving through a copper cable, could be transformed into the on/off pulses of a laser. Tens of millions of these digital pulses could be sent each second.[16]

Kompfner sensed an urgency to the work. There was now competition wherever he looked. In every part of the developed world, scientists and engineers were seeking new kinds of semiconductor lasers. At the same time, a number of companies were aggressively pursuing the second essential invention: a medium that could carry the laser pulses long distances. By October 1971, about a year had elapsed since Corning announced it had created the first glass fiber practical for optical communications. Kompfner now seemed convinced that fiber was the answer to the future. As he saw it, though, Corning was still ahead of Bell Labs. "The progress made with optical fibers" at the Labs was outstanding, he noted, but it was not in the "breakthrough" class like Corning's.[17] Kompfner's main advantage was that Bell Labs could throw almost unlimited amounts of talent and money at a problem, especially when solving it had such tantalizing potential. Optical fiber was exactly that kind of problem.

From the beginning, a basic architecture for fiber was generally accepted. Not much thicker than a human hair, the glass strand was composed of two kinds of exceedingly clear glass. On the inside was a solid "core" of glass; on the outside was a "cladding" of glass. There was an important difference between the two: The cladding must have what's known as a smaller *refractive index* than the core glass. That difference would allow pulses of light, sent through the core, to be confined as they traveled along. Essentially, the light waves were reflected off the walls of the core glass as they moved forward. Sometimes, optical scientists compared the reflective effect inside the fiber core to looking up at the sky while underwater in a swimming pool. The difference in the refractive index of water and air makes the surface appear like a mirror.

The fundamental goal in making transistor materials is purity; the fundamental goal in making fiber materials is clarity. Only then can light pass through unimpeded; or as optical engineers say, only then can "losses" of light in the fiber be kept to an acceptable minimum. Two problems stand in the way of this objective, and both plagued early fiber makers.

When a fiber shows too much "absorption," it means that too much light is being lost thanks to traces of impurities—metals such as nickel and iron—within the glass. The other problem is called "scattering." A more complicated phenomenon, scattering often arises from imperfections—infinitesimal bubbles or cracks, for instance—in the glass crystal itself.

In the early 1970s, Corning and Bell Labs struck an agreement to share their patents on fiber production. Over the next few years, both companies came up with sophisticated ways to reduce absorption and scattering. Making fiber had by then become a profoundly involved process. It was done in several steps, and at high temperatures, and involved a bath of chemical vapors and exacting mechanisms to "draw" the fiber out from a molten glass rod and stretch it to an ethereal thinness.[18] At Bell Labs, the fiber became clearer and clearer. Indeed, it reached astonishing levels of clarity. Within a few years, if you were to somehow look through a kilometer-thick block of the kind of glass being made by Corning or Bell Labs, it would be roughly as clear as looking through several panes of window glass.[19] The fiber also became more flexible, which would prove essential in any system where fiber, bundled together into cables, would have to snake through underground ducts and up into buildings.

The process was rapid enough that a group of Bell Labs engineers began meeting in Holmdel to conceive of a test project for the new material.[20] By around 1975, not all the kinks were worked out. Fibers were still hard to make. Lasers were not always durable—they could burn out after a few thousand hours. The whole installation process looked to be far more expensive than copper cable. Also, there were a variety of challenges involved in splicing fiber cables together and building repeaters to amplify the telephone signals when they weakened after a certain distance. John Pierce, coming through the Holmdel area on one of his consulting trips from California, nevertheless wrote an exultant letter to Bill Baker not long after about the progress he encountered. The only problems, as Pierce saw it, were to demonstrate that Americans "will *want* and *use* all the communication that optical fibers can provide, and that light-wave communication is bound to get cheaper."[21] It was now obvious that the Picturephone wasn't going to soak up all that extra bandwidth. Perhaps something else would come along.

AT THE PRECISE MOMENT that optical systems were ready to be field-tested, a group of Bell Labs engineers were putting the final touches on a test system for mobile phones. The two technologies were not in a race. One had moved fast and the other slow. Whereas lasers and optical fiber represented the culmination of fifteen years of rapid innovation, mobile phones had undergone a longer, stop-and-start evolution. In fact, to locate their origins you had to go back several decades, through a technical and political history so convoluted that many managers at Bell Labs didn't even know the particulars.

The beginnings of the wireless radio business dated back to ship-to-shore calls that were first made in the early decades of the twentieth century. Beginning in 1929, a small number of ocean liners were equipped with radio-telephone service—a business that catered to an affluent clientele stuck at sea for a week or more.[22] A call from a ship's passenger would be sent over radio waves to the shore, where it would be received by massive antenna systems on the New Jersey coastline and then connected into the phone system. At about the same time, land-based mobile service began. The first mobile radios were introduced by the Detroit Police Department in 1921, and the concept quickly caught on.[23] The early devices for fire and police vehicles were used only for work. A policeman could call a colleague in another patrol car, for instance, but to call a colleague at his home would require getting patched into the larger phone system through the police station switchboard. At the time, a caller couldn't dial directly from a mobile telephone into what was known at AT&T as "the switched network."

During World War II, radio communications took a great leap. Bell Labs, at the military's behest, had worked on compact and sophisticated communications systems for tanks and airplanes. Meanwhile, Motorola, a small company out of Chicago, built a rugged "handie-talkie" for soldiers. In Europe, Asia, and North Africa, these portable devices changed the nature of battlefield communications and military improvisation. To electronics executives back home, it was also apparent that there would be applications for these technologies in postwar society. As early as 1945, Mervin Kelly was reviewing plans for Bell Labs and AT&T to create a business selling mobile telephone service to car owners.[24] In select cities, the phone company would mount a powerful antenna on the tallest hill or building to send and receive calls from drivers nearby. The equipment in the cars was heavy—the early sets, often made by Motorola, were mounted in the trunk, and were connected to a handset near the driver (later the sets were shrunk to about the size of an attaché case and mounted under the seat). The mobile phones worked—you could now call an operator who would connect you to any number—but they were pricey: Costs in the late 1940s ran about $15 a month (about $145 in 2010 dollars) and fifteen cents for every minute you used the service. The main shortcoming, however, was that the Federal Communications Commission had made only a narrow portion of the radio spectrum—a portion just above the frequency of FM radio—available for mobile telephone service. The narrow spectrum meant there were only a few channels available to make calls. In all of Manhattan, fewer than a dozen people could use their car phones at any one time.

One of the keys to unlocking the mobile business was better technology. But another key was dangling on a chain in Washington, D.C., at the FCC. The radio frequency spectrum, as Bell Labs' Bill Jakes once put it, is a natural resource. The FCC, as the custodian of this resource, is given the authority to decide how to "balance the sometimes conflicting needs of the different services required by society."[25] Should more of the spectrum be given to TV broadcasters? Or should it go to radio? What was clear to Bell Labs' researchers after the war was that if regulators made more of the radio spectrum available, their mobile phone system's capacity would increase dramatically, and car telephones would presumably become instantly more popular. Already there were enormous waiting lists in some cities for the devices. In 1947 AT&T began to petition the FCC for more spectrum in what was known as the ultra-high-frequency, or UHF, range. These were frequencies just slightly higher than what was already available to mobile phone users. As part of the company's petition, engineers presented a plan that was extracted from a long technical memo, completed in early December 1947, by a Bell Labs employee named Doug Ring, who was assisted by a colleague named Rae Young. The two men worked out of the small rural radio lab at Holmdel. Their memo happened to be written the same month that John Bardeen and Walter Brattain were perfecting their transistor at Murray Hill, thirty miles to the north, and that Claude Shannon was finishing up his paper on information theory.

No one would have put Ring and Young in such rarefied company. Their memorandum—"Mobile Telephony: Wide Area Coverage"—merely outlined how an "adequate mobile radio system should provide service to any equipped vehicle at any point in the whole country." Nevertheless, it entertained an intriguing idea: Rather than continue with the idea of placing a single high-powered antenna in a city center, there might be an advantage in spreading a multitude of low-powered antennas over a wider area to service mobile phones. More strikingly, Ring and Rae suggested considering an area of mobile coverage not as big circles with an antenna in the center but as a honeycomb of small hexagons, perhaps with antennas on the corners. Each hexagon could represent a distinct region for mobile reception, perhaps a few miles in diameter. Each hexagon, moreover, could have its own range of frequencies.

This approach would allow for a very efficient use (and reuse) of the limited radio spectrum—that valuable natural resource. You could connect far more calls with many small coverage areas than a single large coverage area. If the FCC allowed a block of frequencies to be used for mobile radio channels, Bell Labs could cut that block into, say, five slices. It could then assign a different slice to each of five hexagons in the honeycomb. This would help minimize interference and increase capacity, since the hexagon next door to the first hexagon would have a different slice of frequencies—and when you drove from one hexagon to another your phone would automatically switch frequencies. The hexagon next to that would have still a different range—and your phone would switch frequencies again. Eventually, once you got far enough away from the first hexagon, you could drive through another clump of five hexagons with the same frequencies as the first clump. That was feasible since the distance now precluded any interference. Once the pattern was repeated, it could be repeated again in the neighboring area of hexagons. And the pattern could effectively go on forever. The capacity for mobile calling would be far larger than what presently existed. Mobile radio didn't have to be local. It could be national.[26]

Ring and Young hadn't used the word "cellular" in their presentation. Nevertheless what they outlined—in the honeycomb of hexagons and repeating frequencies—was exactly that. Those hexagons were cells. After Ring's memo was forwarded to the FCC it was filed away in New Jersey with the thousands of other technical memos that were churned out every year by Bell Labs scientists. It was not published in the prestigious *Bell System Technical Journal*. It was not published anywhere, in fact. And there is nothing to suggest it was considered a landmark of engineering. More likely it was thought to be merely helpful in the Bell System's lobbying efforts for more radio spectrum. In June 1950, Oliver Buckley, the president of Bell Labs, testified before the FCC and alluded to the 1947 plan. "We are ready to proceed with the development as soon as we have assurance of frequency space," Buckley told the commission. "I am convinced that the benefits to the public at large of having such an integrated mobile telephone system will be very great indeed."[27]

The FCC had other ideas. In the late 1950s, the commissioners awarded a large block of radio frequencies to television broadcasters. The broadcasters were to create eighty new channels in the UHF range.[28] Had it been given to cellular service instead, which requires less bandwidth than TV, the same block of spectrum could have created as many as eight hundred or a thousand new phone channels. (Each channel, in turn, could serve many mobile phone users.) It was a decision that maddened John Pierce, who was a fierce advocate for mobile radio and believed that wireless phones would someday be small and portable, like a transistor radio. Pierce's notion seemed utopian to many radio engineers at Bell Labs. Most considered mobile phones as necessarily bulky and limited to cars, due to the power required to transmit signals from the phone to a nearby antenna. Pierce, in any event, wryly observed that "the FCC has decided pretty clearly that what the American people want is mass communication rather than individual communication." In choosing television over telephony, Pierce thought the FCC had picked a communications technology that suppresses individual expression rather than encouraging it.[29]

He resolved to keep trying. For the next few decades Pierce and a variety of other Bell executives lobbied to change the FCC's decision. And in 1964 and 1965, several AT&T executives began meeting in New York to discuss reviving the company's petition for mobile radio spectrum.[30] "The consensus of the group," an AT&T memo from that era explained, "was that a substantial market does exist."[31] Those business planners brought in several engineers from Bell Labs to consider how a national "high capacity" mobile system might be installed. The Bell Labs engineers imagined such a system might take ten years to plan, develop, and deploy.

Soon after, AT&T was quietly informed that the FCC might be willing to reconsider the mobile radio proposals. Apparently the commission was disappointed with the lackluster content and low popularity of UHF television. And so the mobile radio initiative fell, quietly and momentarily, into the lap of Chuck Elmendorf, John Pierce's old Caltech friend and New York City roommate. Elmendorf was now an assistant vice president at AT&T, working out of a big office in the gilded headquarters at 195 Broadway. In March 1967 he composed a letter to a director at the Bell Labs office in Holmdel: At the FCC, Elmendorf wrote, "it appears that some conditions may have changed." The commission was now "giving serious consideration" to mobile radio, which in turn meant that "the Bell System should be prepared with some concrete proposals."[32] The suggestion wasn't subtle. The FCC was about to make a move, and Bell Labs should get ready.

Seventeen

APART

A few years before the phone company executives began to try build-ing a national mobile telephone system, a mechanical engineer named Dick Frenkiel set out for his first day of work in Holmdel, New Jersey. It was July 1963. The isolated Holmdel radio labs that had stood on the site for decades—the vast green fields where children would throw boomerangs on weekends, the gracious woodframe building around which engineers would test radio transmissions—were gone. Those labs had been razed. In their stead, on the center of 460 acres of former farm-land, Bell Labs had commissioned an enormous modern building to ac-commodate its growing ranks. The employees at Bell Labs now numbered around thirteen thousand. The new Holmdel lab housed about twenty-six hundred people, but it was eventually intended to hold about five thousand, which would make it even larger than Murray Hill. For obvious reasons, the building was soon nicknamed the Black Box. It was a steel-and-glass six-story structure, serious and austere, designed by the Finnish American architect Eero Saarinen.[1] It was also a monument to architec-tural presumption. Saarinen, who died before his design was actually built, saw his creation as having the same kind of flexibility as Murray Hill—offices could easily be moved and partitioned, for instance—but with a crucial difference. He placed the building's long connecting hall-ways on its glassy perimeter, with the windowless offices and labs in the interior. "Gone completely are the old claustrophobic, dreary, prison-like corridors," Saarinen remarked with pride. Thanks to the floor-to-ceiling windows, the members of the technical staff would be liberated by un-obstructed views of the countryside rather than chance encounters in the hallways.[2]

On his first day, Dick Frenkiel turned into the long driveway and drove toward the new building, now barely a year old. "A wide, dark rect-angle centered on flat, empty grassland," he later wrote of his impres-sions. "It was still a half-mile distant as one entered the property and started down the long esplanade road. Size and space conspired to create a strange optical illusion, in which the road turned out to be much longer, and the building much larger, than one at first perceived."[3] The fields, he noticed, were brown that summer from a drought.

Frenkiel accepted a dollar bill on his first day in exchange for his fu-ture patent rights. It was the same ritual that new Bell Labs employees had enacted for decades. But by Frenkiel's own account, he soon came to realize that he had joined an organization that differed from its myth. The Black Box represented one aspect of this evolution. More to the point, the thrust of the work at Bell Labs seemed to have shifted decisively to big projects involving hundreds of people. Frenkiel's Bell Labs didn't seem to have anything to do with heroic research on a new amplifier, done by a few men in a hushed lab. It was about large teams attacking knotty problems for years on end. Jim Fisk's 1947 dinner party—the Bell Labs of Kelly and Pierce and Shockley, "an ancient place of grainy and fading photographs," as Frenkiel later viewed it, "where giants of sci-ence produced their individual monuments and left behind their personal legends"—was a ghost story. Frenkiel heard about but never bumped into any of those legends. They seemed to have vanished along with Holm-del's woodframe laboratory building.

In his first few years at Holmdel, Frenkiel floundered about. He was assigned to work on heavy-duty machines that produced prerecorded phone messages, such as ones that give callers the exact date and time.

In January 1966, however, his boss came to him with a different project. Apparently the rumors about mobile radio were getting around. As Frenkiel recalls, "He was saying that someone told Bell Labs that the FCC was willing to consider mobile again, because UHF television was a disappointment. It was not in the news, but he said, 'Somebody whispered in our ear that we should start thinking about a proposal again.' "[4] Among other things, Frenkiel was handed Doug Ring and Rae Young's old 1947 memo.[5] He was also teamed up with an engineer named Phil Porter and asked to write a report laying out some ideas for a mobile phone system. On January 17, 1966, Frenkiel and Porter met to talk about their ideas for the first time. Frenkiel scribbled some notes that day on a sheet of looseleaf paper. "A system could be developed to locate mobile at all times," he noted. Also, there should be "hexagonal cellular array of areas." Neither Frenkiel nor Porter knew precisely how this would be achieved. "It was just two of us," Frenkiel says. "Nothing important."

They spent most of 1966 working on the problem—or rather, the problems. The two men covered the walls of their offices with maps and climbed on ladders in various parts of the country to count hills. There were thousands of questions they would need to answer eventually. Many of these were extremely technical, regarding reception and transmission. They talked about signal strength and interference and channel width. They knew every cell would need to be served by what they called "base station" antennas. These antennas would (1) transmit and receive the signals from the mobile phones and (2) feed those signals, by cable, into a switching center that was connected to the nationwide Bell System. Still, several big conceptual problems stood out.

The first was, *How large should a hexagonal cell be?* Base station antennas would be expensive. How few could they install and still have a high-functioning system?

The second was, *How could you "split" a cell?* The system would almost certainly start with just a few users—meaning big cells. But as the number of users grew, those cells would subdivide to accommodate the traffic. And more, smaller cells would require more base stations. What was the best and cheapest way?

The third was, *How would you "hand off" a call from one cell to another?*
It had never been done. But it would be the system's essential characteristic. As a mobile telephone user moved around, how could you switch
the call from one antenna to another—from cell to cell, in other words—
without causing great distraction to the caller?

Frenkiel and Porter began working out some approximate answers to
cell size and cell splitting and handoffs. "Those kinds of things were in
our memo," he recalls. "But the FCC hadn't done anything by the end of
that year so we went on to other work." Their plan sat on the shelf, and
they were drawn into a different Bell Labs project, paid for by the U.S.
Department of Transportation, that involved putting pay phones on the
Metroliner, a new express train that would run between Union Station
in Washington, D.C., and Penn Station in New York City. Frenkiel and
Porter's presence made sense: This was a simplified application of the
cellular idea. The Metroliner route was divided into cells of different
frequencies. Markers were put on the tracks—coils, actually—that could
be tripped by the train as it passed; these signaled that a call had to be
handed off from one cell, and one frequency, to another. "It was not great
technology," Frenkiel recalls, "but it was the first cellular system."

AROUND THAT TIME, Frenkiel and Porter met a systems engineer named
Joel Engel, who had just joined the Holmdel office. Engel—bright, opin-
ionated, and energetic—soon noticed that even though Frenkiel and Porter were working on the Metroliner system, they were fixated on the
questions from the previous year, when they had put forward the basic
ideas for a larger mobile phone service. It was Engel's understanding that
to get ahead at Bell Labs, "you were supposed to work on more than you
were asked to work on."[6] It was necessary, in other words, not only to do
your assigned work but to devote 20 or 30 percent of your time to another project. At the time Engel was assigned to work on Bell Labs' pag-
ing systems; AT&T was now selling a bulky, bricklike box, known as "Bell
Boy," that doctors or other busy professional people could use to alert
them with a buzz that someone had called them. That buzzing urged a

Bell Boy user to get to a pay phone and call their office in case there was an emergency. To a systems engineer, the Bell Boy was not terribly interesting. Engel, as a result, was soon drawn into Frenkiel and Porter's spare-time obsession. "We would meet," he recalls, "the three of us, and we would grab a conference room and stand around a blackboard and draw hexagons."

The three men standing by the blackboard in Holmdel, New Jersey, in early 1968 did not have grandiose plans. "We were not visionaries," Engel says of the early cellular meetings. "We were techies. If there was a vision it was primarily as a business service. Real estate agents. Doctors who made house calls." The men believed that if they could get the system to work, the economics of cellular service could be compelling: A trucking service, for instance, could boost its efficiency with a fleet of phones in its vehicles. Increased productivity would justify the cost of the phones. The men around the blackboard were thinking car phones. Perhaps small handheld phones would emerge at some point, but not for decades yet.

In the summer of 1968, the FCC officially informed the Bell System that it might be interested in hearing what it might do if some of the channels being used by UHF television were reapportioned. Though Bell Labs engineers had already been warned by Chuck Elmendorf at AT&T, in Engel's recollection the reaction to the FCC's invitation was tinged with panic. Some of the old-timers at Bell Labs doubted, from looking at the early plans of Frenkiel, Porter, and Engel, that such a cellular system would work. "Microwaves don't travel in hexagons," Engel recalls hearing. There was further concern that the system wouldn't be able to "find" a mobile phone subscriber and connect a call to them. Nevertheless, the FCC's invitation represented an extraordinary opportunity for AT&T. The company had waited twenty years for this. Porter, Frenkiel, and Engel estimated that it would take about three years to deliver a cellular plan, which turned out to be correct.

"We got a lot less attention than you would think," Frenkiel recalls of the efforts to create the cellular system. "We were really just another project." Indeed, to stroll around inside the Black Box at that time, one would not have imagined that a mobile telephone system was an on-ramp to the future. The thing about Bell Labs, Frenkiel remarks, was that it could spend millions of dollars—or even $100 million, which was what AT&T would spend on cellular before it went to market[7]—on a technology that offered little guarantee it would succeed technologically or economically. Indeed, a marketing study commissioned by AT&T in the fall of 1971 informed its team that "there was no market for mobile phones at any price." Neither man agreed with that assessment. Though Engel didn't perceive it at the time, he later came to believe that marketing studies could only tell you something about the demand for products that actually exist. Cellular phones were a product that people had to imagine might exist.

"You have to understand," Joel Engel says of the entire effort, "we were all very young, we were unscarred by failure. So we always knew it was going to work." Not all of the AT&T executives were as optimistic. But anyone worrying that the cellular project might face the same disastrous fate as the Picturephone might see that it had one advantage. A Picturephone was only valuable if everyone else had a Picturephone. But cellular users didn't only talk to other cellular users. They could talk to anyone in the national or global network. The only difference was that they could move.

ENGEL WAS PUT IN CHARGE of the group planning the cellular system design. He would later look back and see the early 1970s as a perfect example of what engineers sometimes call "steam engine time." This term refers to the Scottish engineer James Watt, the inventor of the first commercially popular steam engine, whose name is also memorialized in the term we use to measure power. In the late 1700s, Watt made startling improvements upon more basic ideas of how to use compressed steam to run heavy machinery. The knowledge needed to make such an engine had by then coalesced to the point that his innovation was, arguably, inevitable. By the 1970s, the mobile business was ready to happen, Engel was sure, even if the marketers had their doubts. The technology was there.

It was now just a matter of who was going to do it, and how fast they could make it work. "It was," he says, "steam engine time for cellular."

The FCC's decision to consider proposals for mobile radio had been the spark. But a number of other technologies made it steam engine time, too. To Engel's colleague Dick Frenkiel, it seems unlikely that the early cellular pioneers at Bell Labs could have actually implemented their designs in the 1950s. "Cellular is a computer technology," Frenkiel points out. "It's not a radio technology." In other words, engineering the transmission and reception from a mobile handset to the local antennas, while challenging, wasn't what made the idea innovative. It was the system's logic—locating a user moving through the cellular honeycomb, monitoring the signal strength of that call, and handing off a call to a new channel, and a new antenna tower, as a caller moves along. One necessary piece of hardware for this logic was integrated circuits, those silicon chips on which a tiny circuit and thousands of transistors could be etched. They had only been developed a few years before Frenkiel's mobile work at the Labs. And then, as the cellular team at Bell Labs began working on its FCC proposal, a Santa Clara, California, semiconductor company named Intel—formed by Robert Noyce and Gordon Moore, both refugees from Bill Shockley's first semiconductor company—began producing a revolutionary integrated circuit called the 4004 microprocessor. Measuring only one-eighth by one-sixteenth of an inch, and containing 2,300 transistors, the 4004 was essentially a tiny, powerful computer. It was the first generation of devices that, when inserted into a mobile phone unit, could do a host of essential and highly complex calculations.

What also made cellular possible were the phone network's new electronic switching stations, or ESSs. In 1964, when Bell Labs had opened its first ESS center in Succasunna, New Jersey, the public relations department had urged Bell engineers to explain that the ESS had the ability "to provide services we haven't even thought of yet." Six years later, Frenkiel and Engel and the rest of the Bell Labs cellular team were envisioning what those services might be. A cellular phone would need to send a digital signal every few seconds to the nearest base station antenna. The base station would in turn send that information to a mobile switching center.

But really that was just the start. A vast amount of data needed to go back and forth, almost constantly, between the mobile phones, the base stations, and the mobile switching center. The switching system would have to communicate with the base stations so as to keep track of who was where. In addition, Frenkiel explains, "every few seconds, signal strength data would have to be taken at surrounding base stations to find out if a better one now existed." If another base station had a stronger signal, the computerized system would hand the call off to the next cell. The catch was that before a call could be handed off, the switching system had to identify a channel in the new cell, set up that call, and send a message to the mobile so it could switch frequencies. As Frenkiel remarks, the ESS was invented to be a central office switch, meaning it was created to simply direct calls between landline phones.[8] "But it's sitting there and it's got the ability to be *programmed* for something no one ever expected it to do—all those instantaneous decisions that were never necessary. And now we come along to say we need to do locating and handoffs. And also, by the way, we need to keep track of the health of every base station in the system, we need trouble reports, we need to gather data for traffic." The ESS—a switching system with a powerful computer embedded within it—could do all these things.

Frenkiel, Engel, and Porter could identify these challenges, but they couldn't solve them. As systems engineers, they were looking at a big project in a comprehensive but somewhat general way. Systems engineers consider all the standards and technologies and economics necessary to make a project work. They worry, moreover, about how to integrate a complex new technology with the rest of the system. Cellular phones were an ideal case, since the new technology had to (1) work, (2) work affordably, and (3) work seamlessly with the rest of the existing phone network. What the systems people couldn't do was actually make those projects function the way they dreamed.

Help came from several different directions. Bill Jakes, the lead engineer on John Pierce's Echo experiment, had already been asked by Pierce to look deeper into the science of microwaves.[9] He was in the research department; he was therefore looking for new knowledge. When

Jakes set out to work on the cellular research, he bought a van with Bell Labs money and hired a driver. Then he and some colleagues piled in the back with recording equipment and headphones, and the van drove thousands of miles along the highways and byways of New York and New Jersey. "We went everywhere," Jakes recalls, "through tunnels, up on top of hills, in woods, near lakes. All day long, day after day after day." Their goal, in part, was to study the effect of obstructions on transmission and reception. Over the course of months, they puzzled over why microwaves behaved one way in a particular situation, and another way in a different situation. The radio transmission problems involving a moving vehicle and a cellular system, Jakes later wrote, were "so difficult they challenge the imagination."[10]

THE WORLD IS FULL OF NOISE. It blocks our attempts to talk with one another. Claude Shannon had philosophized about this, but to radio engineers in the field, noise is a slightly different phenomenon. You think about noise not so much as an idea that interrupts a message containing information. You think about noise as clicks and static and fadeouts—a physical or electrical problem that must be overcome by engineering or by savvy. Urban noise is often more intrusive than rural noise. Car engines and trains and legions of electronic devices create bursts and blizzards of interference. Also, there are big buildings, throwing shadows over reception areas or bouncing signals unpredictably to and fro. Even in the countryside, though, noise can interfere with radio transmissions. There are highway overpasses and high-tension wires; there are mountains and trees.

"The guys who made cellular real," as Dick Frenkiel says, were recruited from the Bell Labs Whippany office, where most of the work was military-related. These men were from the development department, the place at Bell Labs that actually made ideas and new knowledge into innovations. For the cellular project, their job was to build the components—low-power antennas, experimental phone sets, and so forth—that the systems team had described. They would spend most of the 1970s worrying about noise and interference of one type or another. A small but typical concern: How high do you actually need to place an antenna to make it work? One of the early engineers in the cellular development group was named Gerry DiPiazza. He was not the sort of Ivy League type that clogged the hallways of the Murray Hill research labs. Rather, he was an engineer's engineer, who had gone to a local New York City school and had attended "Kelly College," the Labs' continuing education program. He had essentially lifted himself up through the Bell Labs ranks by his bootstraps.

By 1968, DiPiazza's career with the Labs had already involved a surreal tour of duty in the South Pacific. In the early 1960s, Bell Labs had established a tiny laboratory outpost, totaling about forty people, on a remote coral atoll, only about 820 acres in size, known as Kwajalein.[11] It was part of the Marshall Islands and had been the site of a significant battle during World War II. A largely secret installation, where engineers now worked on radar communications and antiballistic systems known as Nike-Zeus missiles, Kwajalein, or "Kwaj," as the Bell Labs employees called it, was several thousand miles southwest of Hawaii. In 1965, a Bell Labs vice president had recommended that DiPiazza go there as a good career move, and DiPiazza agreed. He and his wife packed their belongings and their young children. They had never ventured far from New Jersey. They boarded a jet for San Francisco, then a propeller plane to Hawaii, then another propeller plane to Kwaj. They landed on a small airfield in the tropics amid stifling humidity. "We were greeted by a family with a truck, and we were driven to a household," DiPiazza recalls. "We were given dinner. And then we were taken to a concrete duplex home. All the furniture was rattan. It was unkempt and hot. It needed to be rehabbed. There were rat droppings in the crib. My wife cried all night, and I did, too. She looked at me and asked, 'What did you get us into?'"[12] His family was allowed to fly home for a two-week vacation, but only after DiPiazza had worked on Kwajalein for 510 days.

Kwaj was not only isolated. It was strange. A local Marshallese population traveled to the island each morning on U.S. military launches to work as merchants and maids. The island's king and landlord carried around with him his rent money—purportedly a million dollars, paid by the U.S. government—in a bag. Bones of Japanese soldiers, now dead for at least twenty years, turned up regularly on the beaches. On the main street of town was a grocery store where everything was sold frozen, even the milk and meat. There were few cars; instead, DiPiazza and his wife rode their bikes everywhere. And in time, they grew to like it. A tight camaraderie existed among the families there, and the leisure pursuits—sailing, swimming, and free movies—helped pass the time. Most important, DiPiazza's work on highly sophisticated radar systems held his interest.[13] Bell Labs and Western Electric were testing the capabilities of the radar systems against target-practice missiles shot toward Kwajalein from Vandenberg Air Force Base in California. At the time, military strategists were concerned that foreign missiles would be built to scatter a cloud of foil decoys—they called it chaff—before detonating. The foil confused radar defenses and obscured the warhead inside the chaff cloud. DiPiazza's job was to develop techniques that allowed U.S. radar to discriminate between the decoys and the actual explosive, so that a U.S. missile could take out the enemy warhead.

The work put DiPiazza on the vanguard of radio engineering. And two years later, when he finished his tour of duty on Kwaj and returned to Bell Labs' Whippany office, he was asked to continue with military radar. Not long after, though, Bell Labs decided to get out of missile work when it became apparent that the United States would sign an anti–ballistic missile treaty with the Soviet Union. "Bell Labs wasn't going to fire us," recalls DiPiazza. "They were going to tell us to find a job within Bell Labs." Around that time, he ran into a good friend of his at the Labs. "My friend said, 'You know, you guys should be doing radio telephone.'" And so DiPiazza and his boss asked around and soon found themselves in a meeting with Dick Frenkiel and Phil Porter in Holmdel, listening to them talk about cellular telephony. They pitched their own talents in the hope of being brought in.

There may have been only a few people in the world who knew as much about building high-frequency military communications systems as DiPiazza. And none of them were drafted, serendipitously, into a cellular telephone project. As DiPiazza saw it, the systems engineers like

Engel and Frenkiel were standing around blackboards or sitting at their desks with sharpened pencils. His team of development engineers could actually build what they needed—a complex mobile radio that could automatically change frequency, for instance, as a driver moved from cell to cell. For the next few years, DiPiazza, like Bill Jakes, spent most of his time in the field, driving around New Jersey and Philadelphia with his colleagues in an effort to construct a real-world cellular experiment. The men had purchased a small trailer home, then they ripped out the bathrooms and kitchens and hauled in small computers and electronic detection equipment. They called the trailer their mobile technology unit. Often the team would stay up all night, driving through various Philadelphia neighborhoods, testing signal strengths in their trailer home and tinkering with their hardware.

Mainly, they needed to know what radio signals did in urban and suburban settings. As DiPiazza says, "You had to find out, What is the noise level in a suburban environment? How far would a signal go if the antenna was at ten feet, twenty feet, fifty feet? Would it go one mile, two miles, four miles? How many antennas do you need? How do you build an antenna? What are you going to put the antenna *on*?" Nobody had ever built those things before. And if a cellular system was going to someday expand across the country, the core challenge was coming up with standards. They had little time to contemplate their legacy, but some of the decisions made by DiPiazza's team proved indelible—for instance, the height of a cellular tower and the close arrangement of its antennas. Meanwhile, Phil Porter, who had worked with Frenkiel on the original system, came up with a permanent answer to an interesting question. Should a cellular phone have a dial tone? Porter made a radical suggestion that it shouldn't. A caller should dial a number and then push "send." That way, the mobile caller would be less rushed; also, the call would be connected for a shorter time, thus putting less strain on the network. That this idea—dial, then send—would later prove crucial to texting technology was not even considered.

In December 1971, AT&T submitted a long and detailed cellular proposal to the FCC, and the regulators began their deliberations. Motorola, whose radio business could conceivably be jeopardized by the Bell Labs plan, made a variety of seemingly contradictory arguments against it. The company objected to the proposed cellular system on technical grounds, indicating that they believed it simply wouldn't work; at the same time, Motorola claimed that AT&T would enter the cellular market and use its monopoly power to crush any competition.[14] Not long after AT&T submitted its proposal, the company announced that it would only seek permission to build and operate cellular networks. "The company felt it had to make some concession, so they said they would not make handsets," Dick Frenkiel recalls. Thus the handset business would be opened up to companies like Motorola or Japanese vendors. Such moves were meant to appease regulators concerned about competition. As the phone company executives now realized, the pervasive feeling in Washington was that the current monopoly was big enough. Perhaps it was too much. The cellular business, if it caught on, was going to be different.

IN 1975, the Bell Labs optical team decided to test their new product. Executives at AT&T and Bell Labs had concluded that fiber, at least at first, would be most useful for high-traffic areas within cities. Even without the Picturephone, copper lines were becoming congested with the steady increase in phone calls and computer data.[15] The first fiber was fabricated at a new manufacturing facility in Atlanta. But Bell Labs engineers decided the Atlanta factory would also serve as an experiment—the first building where fiber would be installed.[16] The Labs engineering team planned to string a glass cable—actually, 144 fibers arranged on a flat ribbon, and then bound within a protective cover—through the factory's underground ducts. Afterward, they would spend several months testing the fiber, lasers, repeaters, splicing, transmission quality, and a host of other technical matters. Though it was only the thickness of a person's thumb, the cable could carry forty-six thousand two-way conversations.

It went even more smoothly than anticipated.[17] Encouraged by the success, the team began a more practical test a year later in Chicago, where fiber would be installed in the field to transmit voice, video, and data to customers and between switching offices. Chicago, like Atlanta, proceeded without a serious hitch. This was cause for elation within a company that was now engaged in a legal struggle for survival. "I have taken very seriously the principles of innovation that you and I followed so long, and that you emphasized about lightguide systems years ago," Bill Baker wrote at the time to John Pierce in California. Pierce should take pride in the Chicago effort, Baker suggested. "Your spirit is embodied in it."[18]

Chicago was also chosen as the test site for the new cell phone technology. The FCC seemed to be moving cautiously toward approval. But it would first review how the Bell Labs system worked in the real world. In the summer of 1978, Bell set up a working cellular service in Chicago, limited to two thousand paid subscribers, over a two-thousand-square-mile area of the region. For the most part, the test was meant to satisfy the FCC's desire to better understand the market—not only how cellular subscribers judged the service, but how equipment from other manufacturers, like handsets built by Motorola, meshed with Bell's new network.[19]

The Chicago tests of fiber and cellular were largely completed by the late 1970s. Bell Labs executives then began planning for a major installation of fiber along the northeast corridor. As for mobile phones, AT&T planned a rollout in several cities—pending a green light from the FCC. By 1980, the success of both technologies seemed assured.

THE FUTURE OF THE BELL SYSTEM was far less clear. Between 1978 and 1980, the legal landscape for Ma Bell had grown increasingly grim. The government's lawsuit asking for a breakup of AT&T had proceeded steadily forward. A new lawyer at the Justice Department's antitrust division was now overseeing the suit—a Stanford academic, William Baxter, who promised to litigate it "to the eyeballs."[20] Also, there was a new judge hearing the case, a brilliant and mercilessly efficient former government lawyer named Harold Greene. Finally, the powerful and pugnacious John deButts had retired as chairman of AT&T. The new AT&T chairman was a mild-mannered electrical engineer who had begun his career at AT&T as a ditch digger and had worked his way up through the company via twenty-three different jobs over several decades.[21] His name was Charles Brown. Everybody called him Charlie.

One could see AT&T's fundamental problem as arrogance. The phone company's broad powers and influence, Judge Greene would later say, gave it "both the ability and the incentive to prevent competitors from gaining a substantial foothold."[22] This in turn had had a negative impact on consumers. During the trial, the judge would eventually hear a mountain of persuasive evidence put forward by government litigators; one after another, competitors would testify how Ma Bell had thwarted their efforts to enter the telecommunications business. At the same time, one could view the trial as a great war over ideas. To William Baxter, the driving force at Justice, AT&T's fundamental problem was that it was both vertically and horizontally integrated. Vertical integration meant that the company controlled its research, development, manufacturing, and deployment. One way to visualize the verticality was to see Ma Bell as a series of boxes, stacked one atop another, each representing essential units of the company. The bottom box, where ideas and innovations began, was Bell Laboratories. Above that was Western Electric, where those innovations were in turn mass-produced. Above that was the top box, AT&T, which deployed the new technologies in the long-distance and local markets.[23]

Baxter did not object to the idea of vertical integration. It made sense that companies should fund research and bring the fruits of those investments to the market. It was the horizontal integration he considered unacceptable. Horizontal integration could be seen as a series of boxes, too, but these boxes stretched from side to side, or really from coast to coast, instead of from top to bottom. These boxes comprised AT&T's long-distance group as well as the local Bell operating companies—New York Telephone, New England Telephone, Southern Bell, Northwestern Bell, and so forth. The government believed that AT&T's control of almost all the local networks created a bottleneck preventing long-distance competition, such as MCI, from thriving. It also created a barrier to companies that wanted to build their own telecommunications equipment. As it was, Western Electric, a close corporate partner to phone companies around the country, had a near hammerlock on the telecom manufacturing business.

The lawsuit was not, therefore, about Bell Labs. Executives at AT&T had nonetheless realized from the start that whatever fate befell the larger phone system would befall their laboratory, too. By 1980, Morry Tanenbaum, who invented the silicon transistor at Bell Labs years before, had become involved in the decisions over the company's destiny. Tanenbaum had risen to become AT&T's executive vice president, effectively making him one of Charlie Brown's top deputies. "I was the only one of the senior officers who had spent time at Bell Labs, and Charlie was extremely concerned about Bell Labs," he recalls. Brown worried that if he spun off Western Electric—the middle box on the vertical stack— then AT&T wouldn't be able to take new discoveries from Bell Labs and move them up into the telecommunications network. Spinning off the local operating companies presented a similarly unappealing prospect: It would drastically reduce the funding available to Bell Labs, since those local companies represented a large portion of AT&T's annual revenue.

Brown asked Tanenbaum to advise him on what each choice could mean to Bell Labs' future. By mid-1981, as the trial was in full swing, it was clear to both men that spinning off Western Electric would not satisfy the government. So there were really only two options: AT&T could agree to spin off the local operating companies, or it could continue with the trial and see what happened. Charlie Brown leaned toward a settlement. The problem, as both Tanenbaum and Brown saw it, was Judge Greene. His preliminary comments from the bench suggested he was not sympathetic to the phone company's arguments in favor of the status quo. In mid-trial, Greene had gone so far as to say that from what he had heard, the government's evidence "demonstrate[s] the Bell System has violated the antitrust laws in a number of ways over a lengthy period of time."[24]

Some of the most farsighted thinkers at Bell Labs had long believed that the phone monopoly might not endure. Mervin Kelly, for one, constantly had that possibility on his mind, from the mid-1940s onward. Their reasoning was neither legal nor philosophical. Popular technologies spread quickly through society; inevitably, they are duplicated and improved by outsiders. As that happens, the original innovator becomes less and less crucial to the technology itself. "I think they knew that," says John Mayo, a legendary engineer who began working at Bell Labs in the 1950s and rose to become its president some years after the litigation had ended. An intriguing question, at least to Mayo, is why the leadership that preceded him at Bell Labs—Jewett, Kelly, Fisk, Pierce, Baker, and the rest—nonetheless decided to invest so heavily and so consistently in research and in exploring what he calls "the unknown." They were not forced to; other government-run phone companies around the world did not. Arguably, Bell Labs could have existed as a highly competent development organization without doing much in the way of basic or applied research. In Mayo's view, "it's not clear what possessed them to do such a unique thing, because in the long term it clearly was not something that assured their future."

Morry Tanenbaum puts it somewhat differently. "Technology would have destroyed the monopoly anyway," he says. Tanenbaum notes that Bell Labs' most significant research and development efforts—transistors, microwave towers, digital transmission, optical fiber, cellular telephone systems—all fit a pattern. They took years to be developed and deployed, and soon became essential parts of the network. Yet many of the essential patents were given away or licensed for a pittance. And those technologies that weren't shared were duplicated or improved upon by outsiders anyway. And eventually, the results were always the same. All the innovations returned, ferociously, in the form of competition.

BACK IN 1974, just after the Department of Justice filed its suit, Bill Baker was asked what it would mean to Bell Labs if the government won. "Well, I think that Bell Laboratories as we know it now would just disappear," he replied. As he saw it, parts of Bell Labs would have to be chopped away. "As Western Electric and Long Lines or the operating companies go off," he noted, "parts of the Labs would doubtless spin off with them, and Bell Laboratories would cease to exist." Baker also remarked that his beloved research division would ultimately face extinction, too. "There'd be no reason for it," he said. It could not be properly funded without the operating companies, which comprised two-thirds of the phone company's assets.[25]

Baker remained president of Bell Labs until he retired in 1979. Though he still kept an office at Murray Hill—and kept his secret schedule, too, with frequent trips to Washington—he handed the reins to his deputy Ian Ross. On Friday morning, January 8, 1982, news leaked out at Murray Hill that AT&T's president, Charlie Brown, and the Justice Department's William Baxter had struck an agreement. On its face, the pact was simple: AT&T would agree to divest its local phone companies, which would all become separate corporations in their own right. At the same time, AT&T would be released from the old consent decree, made in 1956, that prevented it from entering into other industries. "AT&T would be free to enter such previously prohibited fields as data processing, communications between computers, and the sale of telephone and computer terminal equipment," the *New York Times* reported. Also, it would retain its long-distance service.

The agreement didn't immediately realize Baker's worst fears. Some of the engineers at Bell Labs would have to leave to assist the local phone companies, but most could stay. Western Electric and Bell Labs would officially remain part of AT&T. So for the moment at least, the company's vertical structure was intact. But the divestiture itself—figuring out how to split up the world's biggest company—would take two years in itself. Surely the residual effects wouldn't be understood for years, or even decades, after that. It was almost impossible to see what would happen in the long run.

Media reports at the time focused mainly on whether phone customers would be better off with more competition or would pay higher rates to their new local phone companies. Americans would now be allowed to buy their own phones and pick their long-distance provider—a prospect that some politicians worried could lead to mass confusion. Still, in the aftermath of the announcement, most business forecasts for the new arrangement were upbeat. "What they've really done," one communications analyst told the *Times* about AT&T, "is taken the most capital intensive, politically intensive, labor intensive part of their business and given it to someone else."[26] A former FCC chairman, Richard Wiley, simply noted, "The settlement was a brilliant master stroke on the company's part."[27]

The fate of Bell Labs was more complicated. Looking back on the decision with thirty years of hindsight, Morry Tanenbaum would regret that AT&T didn't let the case go to the end of the trial—not for the sake of AT&T, but for the sake of Bell Labs. "Because things couldn't have turned out worse than they did." At the time, however, he notes that Bell Labs' management perceived a silver lining in the breakup. The Labs would no longer be forced to give away its technologies—as it had done, for instance, with the transistor. Now it could push its innovations to get into whatever businesses it wanted. "The agreement should unleash us," Sol Buchsbaum, a Bell Labs vice president, said at the time.[28] Judge Greene seemed to echo these sentiments. In 1984, just after the divestiture was complete, Greene noted, "Bell Laboratories, that greatly acclaimed resource for both basic research and applied technology, will now be able to use the talents of its inventors, scientists, and engineers in the information industry in all of its varied aspects."[29]

The business philosopher Peter Drucker saw a murkier picture. "Does Bell Labs have a future?" he asked as he sorted through the implications of the decision just after the breakup. Like John Mayo and Morry Tanenbaum, Drucker believed that Bell Labs' technical contributions over the course of fifty years had essentially made its continued existence untenable. "Bell Laboratories' discoveries and inventions," he wrote, "have largely created modern electronics." As those discoveries and inventions had spread around the world, however, they had made telephone technology indistinct. To Drucker, telecommunications was now just a part of the immense field of information and electronic technology. There were many competitors and many competing ideas in this field. And therefore, going forward, no single lab could on its own provide the new technology for the entire electronics and information industry.

At the same time, he noted, the reverse was true: The scientists and engineers at Bell Labs had been producing too many ideas over the past half century for a single company to handle:

> In a wide array of areas, from the transistor to fiber optics, and from switching theory to computer logic, the Bell System has been no more adequate as a conduit for Bell Labs' scientific contributions than an eye dropper would be to channel a mountain freshet. The main users have been others—that is, non-telephone industries—with Bell Labs getting little out of its contributions other than an occasional footnote in a scientific paper.[30]

Drucker saw two possible roads ahead. On the one hand, Bell Labs could become a standard industrial lab, much like the ones that supplied technology to General Electric or RCA. Or the Labs could take a "far bolder, but also far riskier course" by going into business for itself, making money from its patents and products. It could become a kind of unique and monolithic brain trust, one that did research for AT&T but also for any company or part of the government that was willing to pay for access to its people and resources. "Nothing like this has ever been done," Drucker noted. "And no one knows whether it could succeed."

It was a tantalizing idea: Bell Labs would remain intact as a citadel for problem-solving. And it would be a citadel of capitalism, too. But perhaps this was too tantalizing. Drucker wondered if the notion was simply too experimental and too radical, and that it therefore could not actually come to pass. A conventional future, he concluded, seemed far more likely.

Eighteen

AFTERLIVES

Mervin Kelly did not live to see the breakup of his beloved Bell System.

A celebration of Kelly's retirement as president of Bell Laboratories was held at the Savoy Hilton, a grand hotel overlooking New York's Central Park, in January 1959. That evening, Kelly's former Bell Labs deputies—Bill Baker, Jim Fisk, and John Pierce, among others—presented him with a silver tray inscribed with their names. Their former boss promised he would use it for his evening cocktails. He meant it in all seriousness. Kelly and his wife enjoyed a drink before dinner each night in their wood-paneled library in Short Hills. "It will thus be a frequent reminder of all of you," Kelly wrote in a note to Baker.[1] The retiring boss meanwhile gave Baker the impression that he was embarking on a new life of leisure. He would spend his days on his two main passions, gardening and golf. Of course, that had never been the way he lived. "Relaxation did not come easy to him," his son-in-law, Robert Von Mehren, remarks. Indeed, Kelly took a long European vacation with his wife. Then he got back to work.

A retiring executive of Bell Labs had entrée to a vast world of genteel opportunity. Kelly soon accepted seats on the corporate boards of several large technology companies; he also joined a variety of committees that advised the science and engineering departments at MIT, Harvard, and the University of Rochester.[2] Kelly's contacts within the U.S. government and military, meanwhile, still reached to the highest levels. If he'd wanted to, he could have moved in a more political direction, too. Yet he decided against it, just as he had decided against becoming the president's science advisor a decade before. Other than a part-time advisory post in Washington—as a special advisor to James Webb, NASA's top administrator—Kelly chose to focus on the one thing he knew better than anyone: industrial research.

Repeatedly during his career, Kelly had fended off job offers—offers from big companies that would pay him more money than AT&T could. In the end, he had always decided against leaving Ma Bell. "He wouldn't have left that laboratory for anything, no matter what anybody offered him," his wife recalled. "It was just his whole life."[3] But now Bell Labs was no longer his whole life. And this time, when an offer arrived directly from IBM's chairman, Thomas Watson Jr., Kelly accepted. He agreed to work as a consultant, reporting directly to Watson, for several months a year. "He was trying to take the pulse of what we were doing and then relate his opinions to Mr. Watson," Robert Gunther-Mohr, who worked with Kelly at IBM, recalls. Kelly's job was to tour IBM's research offices around the country and in Europe. He would visit the labs, interview promising scientists, and then evaluate the company's technical pursuits and staff. The job played to his strengths—a lightning-fast grasp of technology and an instantaneous decisiveness. What's more, IBM's business now revolved around silicon integrated circuits, the latest iteration of the transistor revolution Kelly had begun decades before. Kelly would sit with the IBM scientists at labs around the world and listen to their presentations with his eyes closed. He still smoked compulsively. His fingers were now yellowed from the habit.

Kelly could perceive the obvious differences between IBM and Bell Labs. IBM was a computer company, first and foremost, and not a communications company. "We were moving faster than Bell Labs would," Gunther-Mohr says, noting that Bell Labs had a thirty-year schedule for applying its inventions to the phone network. IBM was attempting to best its competitors as quickly as possible. At the same time, though, IBM had the same concerns as any other technology company. Which lines of research were worth funding? Which were worth discarding? And in which direction would (or should) the future turn? Kelly would never show his hand during his lab visits. And he would leave for posterity no records of his evaluations. Instead, he would take his observations to IBM's headquarters in Armonk, New York, and share them in private conversations with Watson, the IBM chairman. It was Kelly's habit to single out researchers whose work or manner impressed him. "His evaluation and identification of people had a profound effect on their careers," Emmanuel Piore, IBM's chief scientist, once remarked. Yet it seems likely these men and women never knew it. An unseen hand, Kelly's own, had plucked them out as IBM's future scientific leaders.

In the mid-1960s, Kelly's work for IBM tapered off. And by the late 1960s he was becoming increasingly unsteady. The man who would speak at a mile-a-minute clip, and whose walking pace was nearer to a gallop, could now do neither. "He had to stop driving, which he didn't like to do, which bothered him quite a bit," Von Mehren says. He was diagnosed with Parkinson's disease. In 1970, when Kelly was asked to speak at a conference about the past and future of communications—an event being held in his honor at his old college, the University of Missouri at Rolla—he asked Jim Fisk to go in his stead. Kelly died not long after, in March 1971, at a country club in Port St. Lucie, Florida, where he owned a second home. It was not the Parkinson's; he choked on a bite of steak. His ashes were sprinkled into the Gulf of Mexico.

The National Academy of Sciences has long had a tradition of publishing biographical tributes to mark the passing of members. Usually these tributes are concise and informational. Occasionally they are concise and picaresque. Upon Kelly's death, John Pierce was asked to compose a brief biography. His tribute was lengthy and exhaustive, an examination of Kelly's upbringing in impoverished circumstances in Missouri and his congenital impatience and character quirks, as well as a discourse on his theories of management and innovation. Pierce researched

Kelly's life with care. The effort led him deep into his boss's secret military work. "I learned a lot on my own," he said of his attempts to dig into Kelly's career, "and then I had to unlearn a lot of that when it was reviewed by other people."[4] The finished tribute, in any event, was appreciative and lively. "During his life," Pierce began, "I regarded Mervin Kelly as an almost supernatural force. While I saw him many times in the course of my work at Bell Laboratories, usually with others, and a few times in his home, I did not seek him out for fear of being struck by lightning."

Pierce later remarked that one thing about Kelly impressed him above all else: It had to do with how his former boss would advise members of Bell Labs' technical staff when they were asked to work on something new. Whether it was a radar technology for the military or solid-state research for the phone company, Kelly did not want to begin a project by focusing on what was known. He would want to begin by focusing on what was not known. As Pierce explained, the approach was both difficult and counterintuitive. It was more common practice, at least in the military, to proceed with what technology would allow and fill in the gaps afterward. Kelly's tack was akin to saying: Locate the missing puzzle piece first. Then do the puzzle.

"It, of course, has the elegance of a Pierce essay," Bill Baker told Pierce, now living in California and teaching at Caltech, after he had read his published tribute to Kelly.[5] Jim Fisk wrote to Pierce to say that the Kelly biography had captured the man they both knew so well. "I'm glad," Fisk noted, "that you did it rather than anyone else."[6] Baker was then assuming the Labs' presidency. Fisk was retiring. Pierce filed both letters away with his personal papers.

AFTER HE TENDERED his Bell Labs resignation to Kelly in 1955, Bill Shockley's life descended, slowly and inexorably, into paranoia and disgrace. But not right away. In the very beginning, as he departed from New Jersey in the mid-1950s, there was almost unimaginable promise. He arrived back in Palo Alto, the town of his childhood, with an assured belief in the golden possibilities of California. Fifteen years before, the Stanford graduates Bill Hewlett and Dave Packard had founded a successful technology business in the area. Their company, which the men had named Hewlett-Packard, was focused mostly on building precision laboratory equipment. Shockley's would be the first business to design and engineer transistors. It would thus bring the silicon into what would eventually be called Silicon Valley. Arnold Beckman, a wealthy California entrepreneur, bankrolled Shockley's efforts and agreed to pay him $30,000 a year to run the lab. Beckman also gave Shockley a sizable chunk of stock options in his own company.[7] Shockley assumed he was on his way to earning millions. He was almost certainly the first to understand the enormous financial potential of his invention; he would later recall that two years after the transistor's invention, as he was writing a textbook on solid-state physics, it had dawned on him that "we had opened up a field" as big as the electronics field that already existed at the time.[8] Still, it was the men Shockley hired for his California venture—proof of his uncanny ability to spot talent—who actually went on to create the Valley's extraordinary wealth. In the process, several became billionaires.

One of those men was Gordon Moore. Eventually, Moore would become famous for cofounding the Intel Corporation and for making an observation about the rapid rate of progress in semiconductor engineering—namely, that the number of transistors that can fit on a silicon chip tends to double about every two years. This was not a law in any scientific sense, but after having been proven true for several decades it nevertheless became known as Moore's law.[9] In his recollection, working for Shockley was largely an exercise in frustration. In a *Time* magazine story he authored some years after the fact, Moore recalled that Shockley "extended his competitive nature even to his working relationships with the young physicists he supervised. Beyond that, he developed traits that we came to view as paranoid." Shockley suspected his employees were purposely trying to undermine the semiconductor work; in response, he refused to let some of the scientists and engineers have access to the lab. "He viewed several trivial events as malicious and assigned blame. He felt it necessary to check new results with his previous colleagues at Bell Labs, and he generally made it difficult for us to work together." The young men under Shockley now understood what Bell Labs' research staffers had long known: Shockley was an exceedingly poor manager, or perhaps something worse. In 1957, Moore and seven other colleagues, later nicknamed the "traitorous eight," decided to leave Shockley's company to form their own. Shockley felt that someone within the office was sabotaging the firm's work. "The final straw," Moore noted, had been when Shockley asked his entire staff to take polygraph tests.[10]

"One of Shockley's characteristics," his former Bell Labs colleague Ian Ross recalls, "was that he didn't like to be ignored. He didn't like people to reject his ideas. And that's really what got him started on this unfortunate path that he ended up on." The unfortunate path that Ross describes was not Shockley's struggling transistor business, which was later sold to another company and was folded in the late 1960s. Rather, the path was Shockley's near-total immersion, during the final decades of his life, into what he termed "dysgenics." At some point between 1963 and 1964, Shockley came to the conclusion that the long-term health of the human race was imperiled by the reproductive tendencies of society's least intelligent members. "The unfit," he noted, "may increase faster in our population than ever was true in the past." He believed that too many children hailed from these "inferior strains."[11] At first, he didn't connect this perceived inferiority with race. But that would soon change.

There are varying explanations as to why Shockley became fixated on genetics and race. Some former colleagues speculate that a serious car accident in the early 1960s may have led indirectly to changes in cognition. It's worth asking, however, whether Shockley's fixation was merely an accentuation of beliefs he had long harbored. Fred Seitz, who accompanied Shockley on a cross-country car trip in 1932 from California to the East Coast, remarked that even as a young man, Shockley was warm to the idea that intellectual superiority should be rewarded with authority and influence. "He was inclined to believe that society should be governed by what one might regard as an intellectually elite group . . . rather than by majority decisions as in a democratic society."[12] Shockley maintained a heightened interest in IQ scores all through his life. He would become notorious for screening some of the applicants at Shockley Semiconductor, in 1956 and 1957, through psychological and intelligence tests. But he had been interested in applying such tests to people he hired at Bell Labs well before that, perhaps as early as 1950. "It was at least a year, maybe two, after I started working for Shockley that he came up with the test idea," recalls Walter Brown. "It was given by a psychologist in New Jersey, but I don't actually remember where. It was essentially an IQ test. It did not have mathematical- or scientific-type problem-solving questions, but I believe there were logical-thinking questions. Shockley didn't require me to take it, but posed it as an interesting effort he was making to try to quantify potential for creativity, or something like that, with the idea that a test of this kind might be of value in future hiring." Several Bell Labs veterans recall that the difference between Shockley's own scores and other Bell Labs employees was not significant. It must have been a blow to Shockley's self-esteem.

In fact it was only a temporary setback. In Shockley's case, the leap from his early question (How does intelligence translate into scientific creativity?) to his later question (How do genes influence intelligence?) was not especially controversial. But then came a more precarious leap: How does race determine intelligence? It was likely the case, as some of Shockley's former colleagues contend, that he didn't begin his effort out of any deep-seated bigotry. But it is almost certainly true that Shockley's sensitivity to rejection led him in that direction.[13] He would often say that his views on genetics were clarified when he read a news account in 1963 of a deli owner in San Francisco who was blinded by a teenager hired to throw acid in the owner's face; the perpetrator, according to Shockley, was one of seventeen children borne by a woman with an IQ of 55. Whatever the precipitating event, as Shockley encountered increasing resistance to his views—they were first made public in a 1965 interview in the weekly newsmagazine *U.S. News & World Report*—it stoked the fires of his defiance. And after that, his opinions, though cloaked in scientific jargon and the rhetoric of statistical analysis, took a turn toward the grotesque.

All during 1966 and 1967, Shockley urged the National Academy of Sciences, the organization of America's most distinguished scientists, to focus more deeply on the question of how heredity affects intelligence. In April 1968, at a meeting of the academy, Shockley charged that the country's leading thinkers were showing a "lack of responsibility and courage" by not examining correlations of race and intelligence.[14] He delivered a paper at the meeting as well. "An objective examination of relevant data," he declared, "leads me inescapably to the opinion that the major deficit in Negro intellectual performance must be primarily of hereditary origin and thus relatively irremediable by practical improvements in environment."[15] He sent copies of the speech to Bill Baker, Jim Fisk, Mervin Kelly, and John Pierce.

By his own choice, Shockley then began transforming himself from the most esteemed solid-state physicist in the world to a fringe eugenicist. He was likewise starting to think that his work on genetics could become far more important than anything he had so far accomplished in his lifetime. Echoing the honorary language of the Nobel Prize, he told friends it was how he would now make "the greatest contribution to the benefit of man."[16]

DURING THIS INCENDIARY PERIOD, Bell Labs hired Shockley back as a consultant. At the time he held an endowed chair at Stanford's engineering school. He had already become the subject of student demonstrations. Shockley seemed to relish the attention. At the same time, it didn't seem to detract from his abilities as an instructor. Students who were willing to take his physics class—sometimes he gave it in the living room of his own home—were fortunate to encounter one of the great college physics teachers of the twentieth century. The problem with this arrangement, at least to Shockley, was that his Stanford salary was modest. And so at about the time Shockley first went public with his views on genetics and intelligence, he began talking with his old friend at Bell Labs, Jack Morton, about coming back to earn some extra money. Years before, Morton had spearheaded the development of the transistor. The two men had a great deal in common. Both were driven and competitive; both were also bullies, especially when confronted by an idea they disagreed with. During the early 1950s, Morton would often confide in Shockley, unloading his problems on the research physicist. Both men had reached a place in the Bell Labs hierarchy where it was clear that their personalities (and, for Morton, his drinking) precluded further advancement. But in the mid-1960s, as a Bell Labs vice president, Morton nevertheless wielded a certain amount of power. It likewise helped that Shockley's old MIT friend Jim Fisk was now president of the Labs. In late 1965, Shockley secured a consultant deal.

The plan was that Shockley would advise Morton, several weeks a year, on device development. He was no longer a research man. During his periodic visits to the East Coast, Shockley would take a room in Summit, New Jersey, in a comfortable Tudor-style hotel known as the Summit Suburban. Visitors to the Labs offices in Murray Hill were often directed to stay there, and Bell Labs would pick up the tab. One week in the mid-1960s, Chuck Elmendorf—John Pierce's old college friend— ran into Shockley at the Suburban. Elmendorf, too, was visiting the Murray Hill offices; his main job at the time was running a branch laboratory for Bell Labs in Massachusetts. The two men shared dinner in the hotel dining room. Years before the transistor was invented, Elmendorf had sat on Shockley's living room couch while Shockley patiently tutored him on the principles of solid-state physics. Several years after that, Shockley had again tutored Elmendorf, this time on the principles of radar. Now, Elmendorf recalls, his old friend who at one time ate, slept, and breathed physics—a teacher who was "decent, wonderful, pleasant"—wanted only to talk about race and genetics. Like other colleagues from long ago, Elmendorf found Shockley both single-minded and intolerable.

Almost certainly the dinner conversation exemplified how Shockley seized every moment that he was not at the Bell Labs offices as an opportunity to discuss eugenics. The agreement he had struck with Morton stipulated that he would refrain from mixing his racial theories, which he pursued most of the year out of his home in California, with his development work at Bell Labs. It soon became clear, however, that Shockley could barely abide by such an agreement. Indeed, after only a few years of consulting work at the Labs, he asked to scale back. Dysgenics was now his actual career; the main thrust of Shockley's life from 1970 on was to attract attention to the idea. He had always enjoyed publicity, but now he actively courted the press, making sure that any speech or controversy he initiated—a suggestion, for instance, that the United States might begin a voluntary sterilization plan for those with low IQs—would be mentioned in the next day's newspapers. He meanwhile began a systematic effort to record and catalog all his phone conversations and correspondence. Often he would ask journalists who sought to meet him to undergo a series of "pre-interview interviews"; sometimes, too, he would provide inquiring reporters with quizzes on science and statistical concepts and belittle their performance. His ultimate intention was to pass judgment on their abilities and intentions. He now carried a tape recorder everywhere he went. On occasion, he surreptitiously recorded conversations without the knowledge of the participants.

Shockley and his wife seemed well aware that there would be consequences to a life on the scientific fringe. "Someday," Emma Shockley told the writer Rae Goodell in the late 1970s, "we may actually be terribly alone."[17] To a large extent this was already true. Shockley had become mostly estranged from the children of his first marriage. His friends in Palo Alto had distanced themselves, too. Fellow academics, and especially members of the National Academy, feared to be seen with him. His old friends at Bell Labs, meanwhile, were slipping away. Pierce and Shannon avoided contact with him. Jack Morton, Shockley's sponsor at the Labs, died in December 1971, a tragedy that rattled Shockley. (Morton had been murdered following a disagreement at a New Jersey bar—his body was retrieved from his burning car, which had evidently been set alight by his attackers.) Shockley's consulting job at Bell Labs ended not long after. Of Bell Labs' Young Turks, perhaps his only remaining friend was Jim Fisk—and then Fisk died, too, after an illness and brief retirement, in 1980. In 1981, Shockley contacted Bill Baker, but not to reminisce. He was looking for funding—several hundred thousand dollars, according to Baker's notes—for an organization Shockley had started called FREED. The name stood for the Foundation for Research and Education on Eugenics and Dysgenics. Baker, who sat on the boards of numerous philanthropies and foundations, told him, "I can't think of any source of funds."[18]

The great tragedy of Bill Shockley's life, Ian Ross remarks, was that he did almost nothing of scientific worth after leaving Bell Labs. Arguably, his successes were as a teacher and as an impresario of talent—that is, in assembling the team at Shockley Semiconductor that ultimately drove the success of America's computer chip industry. "Had he stayed in that environment," Ross says of Bell Labs, "it would have been a very different story." With a larger enveloping group, in other words—with a slew of supervisors who understood how to compensate for his weaknesses, and a veritable army of technicians who could prepare semiconductor materials to exacting specifications—Shockley might not have been overcome by his poor management skills and defective judgment. As it was, his final years were a succession of serial embarrassments. He decided in his seventies to declare publicly, in the *Los Angeles Times*, that he had donated his sperm to a project that was attempting to create a sperm bank of Nobel Prize winners. When *Playboy* magazine asked him about it in a lengthy interview, he defended the idea; at least to him, a sperm bank of superachievers made perfect sense. When he was asked in the interview how his own children turned out, he replied, "In terms of my own capacities, my children represent a very significant regression. My first wife—their mother—had not as high an academic achievement standing as I had."[19]

In 1982, he ran for the Republican nomination for U.S. Senate in California. He acknowledged that it was unlikely he would win. Mostly, he did it to air his racial theories. "My conscience will be clear if I give it a good try and succeed in getting this idea around," he told the Associated Press.[20] He received less than 1 percent of the vote.

These were last-gasp efforts to garner attention. He had no more real research to do, and no more teaching. Mostly, Shockley organized his vast cache of telephone recordings and papers and did his best to court publicity for his eugenics foundation. All the while, he made notations in his daily diaries. He had always used multiple books for organization. During any given year, going back to the late 1930s, there had been pocket-sized diaries, booklet diaries, large notebook diaries, and so forth. Some were used to record his personal trials ("sore throat, temperature 98") and others his professional obligations. Almost compulsively, he made entries on gardening, doctors, business appointments, store purchases, gas mileage, lunches, and exercise regimens; there were notations, too, on colleagues, ideas, and dreams. It was as though there had always been too much occurring in his life, and too much occurring in his mind, to possibly hold it in one place. But now that was not quite the case. In the mid-1980s, Shockley began keeping a diary on a new computer. He seemed mainly interested in observing the behavior of the birds in his backyard, where he would sometimes sit for hours and note their habits and gestures. He fed the birds lovingly. When he was diagnosed with prostate cancer in 1987, it had already metastasized. Lumps soon began to form in his legs and neck. Shockley's biographer, Joel Shurkin, relates that in 1989, during the final months of life, Shockley suffered in extraordinary pain, made barely tolerable only by large doses of morphine.

His wife and hospice workers cared for him at home in Palo Alto during his final weeks. There were few visitors.

JOHN BARDEEN AND WALTER BRATTAIN, Shockley's colleagues on the transistor invention, fared better after leaving Bell Labs. The two men remained friendly until the end, and whatever bitterness they harbored toward Shockley during the late 1940s seemed to ebb as the years passed. When the men were forced to meet, as was sometimes the case, their relations were civil. In April 1972, for instance, on the occasion of the twenty-fifth anniversary of the transistor, they were interviewed together in Washington, D.C. Bardeen graciously praised Shockley's insights that led to the junction transistor, and Shockley in turn praised his two former colleagues for their pioneering work on the point-contact transistor. The trio were asked to identify the most important developments that had arisen from the transistor. Brattain said he cherished the fact that transistor radios could now bring information to people living in the most remote and impoverished corners of the world. Bardeen pointed to communications with the Apollo astronauts. He also noted the invention's tremendous financial impact in the fledgling computer industry, "which wouldn't have been possible without the transistor, and now runs in this country—I think—the order of $20 billion a year."

Shockley suggested that the greatest innovation arising from the transistor might be the compact tape recorder.[21]

Brattain, still garrulous and rough-edged, had long expressed a desire to someday return to the West. Like so many of his colleagues, he had begun his life with humble prospects and ended with epic achievements. It amazed him. He had grown up amid latter-day pioneers in Washington State, where he spent the days learning his way around heavy machinery and rifles; fifty years later he had ended up on a stage, bowing his head before the king of Sweden, so he could receive a Nobel Prize. All the time he had insisted he was lucky rather than some kind of genius. But there was one part of Brattain's past that he considered formative above all others, and often it nagged at him. In the 1920s, a professor named Benjamin Brown, at Whitman College in Walla Walla, Washington, had changed Brattain's life by introducing him to advanced physics. Brattain wanted to do the same thing for someone else. "I've gotten to the place where I will retire to my small alma mater out there and do a little teaching," he told the sociologist Harriet Zuckerman, who interviewed him at Murray Hill in the early 1960s. He left Bell Labs soon after for a second career in academia at Whitman College. Before he died in 1987, Brattain spent his last few years in a nursing home, suffering the effects of Alzheimer's. John Bardeen later wrote a memorial essay about his friend for the National Academy of Sciences, where he noted that their old office mate at Bell Labs, Gerald Pearson, an inventor of the solar cell and one of the men who signed Brattain's notebook testifying to the transistor demonstration on December 24, 1947, died just two weeks after Brattain. Pearson, an Oregon native, had also returned to the West after leaving Bell Labs. He had finished his career at Stanford.

Bardeen himself died a few years later, in 1991. He had worked at the University of Illinois–Urbana from the time he left Bell Labs, in 1951, until then. A good deal of his subsequent research at Illinois had focused on the phenomenon of superconductivity. In other words, he and several colleagues explained the reasons why some substances, at very cold temperatures, can conduct electricity without any resistance. Their ideas took shape in a formidable tract that merged theoretical physics with complex mathematics. In 1972, Bardeen was awarded a second Nobel Prize in Physics for the work. He is the world's only physicist to have such a distinction.

AFTER MOVING FROM BELL LABS to MIT in the late 1950s, Claude Shannon continued to publish important papers on communications. But his most productive years as a mathematician were behind him. Like Shockley, he had left the Bell cocoon; the difference, perhaps, was that Shannon understood the implications. "I believe that scientists get their best work done before they are fifty, or even earlier than that," he told an interviewer late in life. "I did most of my best work while I was young."[22] Indeed, a decade after arriving at MIT, his work on information began to trail off. "He was concerned that he had nothing left to say," Len Kleinrock, who was a PhD student under Shannon at MIT in the early 1960s, recalls. This may have been another instance of Shannon's typical modesty. Still, says Kleinrock, "He had his doubts about keeping up with his own field."

Shannon nonetheless remained interested in the implications of his work. His speeches from that era suggest a man quietly convinced that information—how it moved, how it was stored, how it was processed—would soon define global societies and economies. A few years after he entered academia, in 1959, he lectured to an audience of students and faculty at the University of Pennsylvania. "I think that this present century in a sense will see a great upsurge and development of this whole information business," Shannon remarked. The future, he predicted, would depend on "the business of collecting information and the business of transmitting it from one point to another, and perhaps most important of all, the business of processing it—using it to replace man at semi-rote operation[s] at a factory . . . even the replacement of man in the things that we almost think of as creative, things like doing mathematics or translating languages."

When he started at MIT, Shannon gave regular lectures on topics that held his interest. He also invited speakers to campus. In 1961, for instance, he arranged for John Pierce to visit and give a talk entitled "What Computers Can Do Better—and How." Pierce, though enthusiastic about the potential of computers, seemed slightly less optimistic than Shannon (who introduced his old friend with a brief speech of his own). By the late 1960s, though, Shannon's talks at MIT were becoming rare occurrences. What's more, his visits to the MIT campus were beginning to grow infrequent. Mostly he stayed home to tinker with his gadgets. Visitors to his large, elegant house in the Boston suburbs would get a tour of his juggling machines, his unicycles, his collection of pianos and musical instruments, and all his hand-built gadgets and automata. One week, he liked to show off a robotic lawn mower he'd constructed; another week, it was a flame-throwing trumpet. Much of it was deposited in what he called his "toy room." When he visited Shannon, John Pierce seemed less interested in the gadgets than some others; he sometimes told people Shannon built his automata to "show off." Mostly, Pierce wanted to chat with his friend about ideas and the future. When Pierce was at the Shannon house, Betty Shannon recalls, he and Claude would take a boat out on the lake that adjoined the property and paddle around, talking for hours.

At almost every other opportunity, Shannon seemed eager to project an air of frivolity. "I think he just loosened up," his wife recalls.[23] To some fellow academics, his lighthearted obsessions remained puzzling: Wasn't he wasting his time? But it was worth considering whether his interests were largely consistent with his habits, decades earlier, at Bell Labs. With information theory, Shannon had never had any intention of changing the world—it had just worked out that way. He had pursued the work not because he perceived it would be useful in squeezing more information into undersea ocean cables or deep space communications. He had pursued it because it intrigued him. In fact, Shannon had never been especially interested in the everyday value of his work. He once told an interviewer, "I think you impute a little more practical purpose to my thinking than actually exists. My mind wanders around, and I conceive of different things day and night. Like a science-fiction writer, I'm thinking, 'What if it were like this?' or, 'Is there an interesting problem of this type?' . . . It's usually just that I like to solve a problem, and I work on these all the time."[24]

The stock market was an interesting problem. In the late 1960s and early 1970s it became something of an obsession. Shannon did not invest because he needed money. He had his MIT salary and pension; he also earned a comfortable sum consulting at Bell Labs. (Though Shannon had long ceased being involved at Murray Hill, Bill Baker, the president of the Labs, kept him on the payroll anyway. Baker insisted it was an honorable thing to do "should the man who came up with information theory" suffer any kind of financial hardship.)[25] Shannon had become wealthy, too, through friends in the technology industry. He owned significant shares in Hewlett-Packard, where his friend Barney Oliver ran the research labs, and was deeply invested in Teledyne, a conglomerate started by another friend, Henry Singleton. Shannon sat on Teledyne's board of directors. The stock market was therefore just another puzzle, albeit one with a pleasant proof of success. He was convinced that the stock market was less efficient than some economists believed, and that a smart investor who took advantage of mispriced stocks could do quite well.[26]

Len Kleinrock, Shannon's former student, recalls that one day at MIT, Shannon mentioned that he was making a mathematical model of the stock market. "I said, 'Mr. Shannon, you're interested in making money?'" Kleinrock recalls. "He said, 'Why yes, aren't you?'"

AS HE TURNED AWAY from academic pursuits, Shannon also focused on juggling. To him, the sport had a number of inviting aspects: It was a game, a problem, a puzzle. It produced motions he considered beautiful. And it was something he simply could not master, making it all the more tantalizing. Shannon would often lament that he had small hands, and thus had great difficulty making the jump from four balls to five—a demarcation, some might argue, between a good juggler and a great juggler. Old friends—fellow jugglers from the Bell Labs days—wrote encouraging letters suggesting he was closer to five balls than he realized. It's likely Shannon never quite achieved that. Nevertheless, in the late 1970s he found himself consumed by the question of whether he could formulate a scientific theory of juggling to explain its unifying principles. Just as he had done years before—for his papers on cryptography, information, and computer chess—he delved into the history of juggling and took stock of its greatest practitioners.

He began to seek out data. "One day he came to the MIT juggling club, basically with a tape measure and a stopwatch," recalls Arthur Lewbel, an MIT graduate student who later became an economics professor at Boston College. "And he came up to some of us who were juggling. He didn't say who he was—none of us would have known him even if he said his name. And he said he just wanted to see if he could measure our juggling." Shannon came back a number of times, and eventually he became friendly with the students. He invited them for pizza at his house. In turn, when the juggling club decided to go to the Big Apple Circus together, they called Shannon, who was thrilled to be invited. He was now part of the gang.[27]

In December 1980, he asked various jugglers he knew to wear electromagnetic sensors—"a flexible copper mesh which was fitted over the first and third fingers of the juggler's hand"—and then had them juggle lacrosse balls covered in conducting foil. When a ball was caught, it closed the circuit between the first and third fingers, and thereby started a precision time clock; when the ball was released, microseconds later, a circuit opened and the clock stopped. Shannon took exact measurements, comprising hundreds of pages of data, about how long each ball stayed in the juggler's hand before release.[28] From this information, he put forward a theoretical equation—$(F + D)H = (V + D)N$—that governed juggling's physics. (As Shannon's juggling friend Arthur Lewbel explains, F is the time a ball spends in the air, D is the time a ball is in a juggler's hand, H is the number of hands, V is the time a hand is vacant, and N is the number of balls juggled.)[29]

This was only one aspect, however, of a longer treatise he was composing at the time. Shannon had traced the origins of juggling back to Egyptian murals from 1900 BC and then through to the ancient Greeks and the jesters and minstrels of the medieval era. "Jugglers," he had concluded, "are surely among the most vulnerable of all entertainers. Musicians and actors can usually cover their slips, but if a juggler makes a mistake, *it's a beaut!*" When he showed the juggling draft to some editors at *Scientific American*, they let him know they were interested in publishing it. But then Shannon balked. He didn't think the work was yet polished or insightful enough to merit publication. In an exchange of letters that continued for several years, the magazine would beseech him to let it print the manuscript. Shannon would pleasantly change the subject. Sometimes he would send his poetry, mostly rhyming doggerel, and suggest the magazine publish that instead. In response the editors would politely decline and change the subject back to publishing the juggling essay—would he please agree to let them?

He never did. And in truth Shannon had no compelling reason to publish anything anymore. His legacy was secure. And his reputation was not harmed by his reclusive tendencies.[30] On the contrary, his legend only appeared to grow. Each year seemed to bring a new honorary degree or prize that he and Betty collected by traveling the world. In one famous instance in 1985, Shannon showed up, unannounced, at an international conference on information theory in Brighton, England. He blended into the crowd—just another kindly, polite, slender, white-haired gentleman. He'd been absent from academic meetings for so long that apparently no one recognized him. Then a rumor spread that Shannon was there. As one of the attendees later told *Scientific American*, "it was as if Newton had showed up at a physics conference."[31] Later, when Shannon was asked to speak, he grew anxious, believing he had little of value to say, and took several balls out of his pocket. And then he juggled for the crowd. Afterward, the attendees, some of the leading mathematicians and engineers in the world, lined up to get his autograph.

It may have been the case that Shannon, like Shockley, deserved a Nobel Prize. But the Nobel is not awarded for mathematics or engineering. In the mid-1980s, however, an award was established in Japan known as the Kyoto Prize that was meant for outstanding contributions in the field of mathematics. Shannon was voted the first recipient. "I don't know how history is taught here in Japan," he told the audience when he traveled there in 1985 to give an acceptance speech, "but in the United States in my college days, most of the time was spent on the study of political leaders and wars—Caesars, Napoleons, and Hitlers. I think this is totally wrong. The important people and events of history are the thinkers and innovators, the Darwins, Newtons, Beethovens whose work continues to grow in influence in a positive fashion." Shannon also conveyed little doubt that machines would soon outpace humans in some respects. Forty years before, he had been one of the first to pursue a computerized chess program; now, in the 1986 Kyoto speech, he noted that chess programs had become so sophisticated that they could beat chess masters. He believed they would soon beat grandmasters. And after that he believed they would win a prize by dethroning a world champion. "If I were a betting man," he said, "I would bet that this prize will be won before the year 2001."

The joke, perhaps, was that Shannon wasn't much of a betting man. He made wagers when logic assured him he could win. Or he built games tilted in his favor. He won the chess bet in 1997, when Deep Blue, an IBM chess computer, beat the Russian grandmaster Garry Kasparov. The result, however, gave Shannon little satisfaction. At that point he was living in a Massachusetts nursing home, his mind lost to Alzheimer's. It had begun, some of his friends recall, in the late 1980s, when there had been lapses when he tried to answer questions. At first, small things were forgotten. Then it became larger things, to the point where efforts to recall events left him skidding through broad and frightening blank patches in his memory. His handwriting became shaky. He forgot about some of his Bell Labs work. He would get lost coming home from the store. Eventually he lost the ability to place names and faces.

It took Betty Shannon years after her husband's death, in 2001, to clear out their Massachusetts house. There were so many awards and robes from honorary graduations, and so many books and papers, many of which Shannon had forgotten or decided never to publish. And there were, finally, so many machine parts and tools, probably tens of thousands of dollars' worth. Betty donated some of the games and juggling toys he had built to the MIT Museum, which in a 2007 exhibit dubbed them *Claude Shannon's Ingenious Machines*. His juggling clowns were included, along with games and machines such as THROBAC, the useless but amusing hand-built computer that could calculate in Roman numerals. Theseus, a remnant from the Bell Labs days—the mouse-machine built by Shannon, mostly late at night, which could navigate any maze— was one of the featured museum pieces, too.

JOHN PIERCE OUTLIVED CLAUDE SHANNON by one year. At the very end of their lives, the two men were too ill to communicate. As Shannon suffered from Alzheimer's, Pierce suffered from Parkinson's. Long before either had become sick, however, they enjoyed a final hurrah.

In the spring of 1978 the men spent a semester together in England. The idea had been Rudi Kompfner's, Pierce's friend and collaborator on the traveling wave tube and the Echo satellite, who had retired from the Labs and was teaching at All Souls College in Oxford. He had arranged for Shannon, Pierce, and Barney Oliver (the former Bell Labs researcher who had gone on to run Hewlett-Packard's research labs) to come to Oxford as visiting fellows for a college term. The three men had something in common: Thirty years before, they had collaborated on a paper about pulse code modulation that had proven prescient. The men had put forward the notion that digital pulses, rather than waves, were certain to be the future of information transmission. At Oxford, they were to have only one obligation: to lecture about any aspect of their research work they deemed important. "To get something out of Claude may be the problem," Kompfner admitted in a note to Pierce. "Could you help?"[32] Pierce, too, worried that this could ruin the plan. When Shannon didn't want to do something, Pierce knew, Shannon didn't do it. "I don't know how to tackle Claude effectively," Pierce confessed to Kompfner. "Maybe Barney and I could *interview* him about information theory, the stock market, and other matters?"[33] As it turned out, the triumphant reunion of the soothsayers of the information age became infused with melancholy. It's unclear what, if anything, Shannon contributed while in England, with the exception of some notes he made for a mirrored contraption allowing American drivers to adjust more easily to the English custom of driving on the left side of the road. Pierce, for his part, delivered a lecture in memory of his friend Kompfner, who had died of a heart attack a few months before the reunion took place.

Pierce went back home to Caltech. For six years he had been doing research and advising graduate students, but he was finding the adjustment difficult. At Bell Labs he had spent his days doing whatever suited him. The brunt of his management work there had consisted of dropping in, unannounced, on colleagues in their labs to ask how work was progressing. But at Caltech he had to give lectures at a prearranged time and then had to spend hours explaining complex ideas to grad students. At Bell Labs, as he recalled it, the same conversation with his colleagues would usually take minutes. (Whether his colleagues actually understood his explanations, or whether he simply walked away before he could field their questions, was a matter for debate.) "I didn't adapt well to Cal Tech," he later admitted. "Not that there was anything wrong. For years and years I'd had it too easy. There were very few times when it mattered where I was. I had very few obligations to be at a particular place at a particular time to do a particular thing at Bell Labs." Pierce obviously seemed to favor the Bell Labs arrangement. As he saw it, the work at the Labs was vital; it was required to improve the network. "People cared about everything," he said of colleagues there. On the contrary, he noted, in the university "no one can tell a professor what to do, on the one hand. But in any deep sense, nobody cares what he's doing, either."[34]

Pierce may have complained about academia merely because he complained about a lot of things. He was caviling and argumentative by nature. It nevertheless remained the case that the good fortune he enjoyed throughout his life lasted long after his Bell Labs retirement. His advice to his students in California was that the key to a good life was to be lucky *and* smart. To be sure, he considered himself both. His complaints to the contrary, he often enjoyed working with students, and he did serious research during his spare time. He and his wife lived in Pasadena, in a home with a serene Japanese rock garden and a small bubbling waterfall. He had a vast home library where he could spend entire days reading. And like Shannon, Pierce now had the pleasure of collecting a trove of honorary degrees and prizes from around the world. Meanwhile, he maintained a large network of friends, some of whom continued to offer him work. When Pierce retired from Caltech in 1979, for instance, he spent a few years working as a technology advisor at the Jet Propulsion Lab in Pasadena. And a few years later, at an age when most of his friends had retired or died, he decided to devote the rest of his life to music. Stanford offered him a visiting professorship in its Center for Computer Research in Music and Acoustics. The school went by the abbreviation CCRMA, but it appealed to Pierce that it was commonly pronounced as "karma." And so in 1983 Pierce moved north, from Pasadena to Palo Alto, to start anew.

By his own admission, John Pierce could not carry a tune. And there was no way of avoiding the truth: He was not a good piano player. His limitations, though, had never diminished his interest in music or his conviction that it was possible to innovate within the art. In the 1950s, for instance, Pierce and Shannon—a devoted clarinetist, oboist, and jazz fan—made several attempts at Bell Labs to merge their interest in music and information. "John and Claude tried to figure out the information rate of music as well as sound," recalls Max Mathews, who worked on digital transmission and acoustics at Bell Labs. Neither was successful in these efforts, Mathews says, "and neither were they successful in creating programs to compose interesting music." But Mathews recalls that one evening in 1957, at a classical music concert that Pierce and Mathews attended, "Pierce said, 'You get sound out of a computer now, you get numbers out of sound, if you write a different program, maybe you can get computers to make music.' He said, 'Take a little time from your computer work and try that.'" Mathews went on to become one of the pioneers of computer-generated music. He and Pierce and Baker justified the research to AT&T management by explaining, truthfully, that it would yield insights into computer-synthesized speech, which was considered useful for the phone system.

By the time Pierce arrived at Stanford, computer-generated music had become a thriving field. There were now electronic tools to create and manipulate almost any sound. At Stanford, he became particularly interested in what he called "psychoacoustics." This was the relation, as Pierce described it, "between the acoustic stimulus and what we perceive internally—how it strikes us, what we can distinguish."[35] More enduringly, Pierce helped invent a scale, known as the Bohlen-Pierce scale, that was not built upon a standard octave but a different arrangement of thirteen ascending tones. It was a characteristically complex endeavor, and his efforts to explain it were laced with technical jargon about the scale's frequency ratios.[36] A composer and friend of Pierce's described its musical effect more directly, in words that almost seemed to describe Pierce himself. Pierce's scale, the friend said, has "ear-catching dissonances [and] warm and pure consonances." But music tends to resist easy description. The experience of listening to compositions written in the scale—easily done through an Internet search—can be Pierce-like, too: quirky, ethereal, intriguing. You are certain you're not listening to anything you've heard before.

On Pierce's eightieth birthday, a friend at Northwestern University decided to honor him with a concert. Several pieces of music were commissioned for the event. One of these, "I Know of No Geometry," was based on the Pierce scale; another was named "Echo," in honor of Pierce's satellite experiment.[37] Pierce traveled to Illinois for the ceremony. "My life," he said in a speech at the concert, "has been full of surprises." He dutifully recalled his work on satellites, digital communications, and telephone switching. "Ideas and plans are essential to innovation," he remarked, "but the time has to be right." He wondered if that moment had arrived for computer music. He also wondered if the composers and innovators gathered at his party were taking advantage of that fact. "While I am amazed at the wonderful sounds that have been generated and used musically," Pierce told his friends, "I am somewhat disappointed that the variety of sounds used effectively has been narrow compared with the range of natural sounds—bird sounds, the sly sounds of creatures hidden from sight, ferocious sounds, sea sounds, stormy gales, waves, the creaking of spars, and, too, the sounds of brooks, and the soughing of winds and rustling of branches. In part, such sounds have been organized into music. But somehow, I had dreamed of more." Apparently, he was still trying to instigate something.

BILL BAKER, like Mervin Kelly, had spent his entire working life at Bell Laboratories. He had risen from a lowly member of the technical staff to president. What should he do afterward? The answer was that he wouldn't leave. After Baker's term as president ended around 1980, the Labs' management agreed to provide him with a driver and an office and a secretary, and life went on much as before. The part of his workday that involved managing research projects and telecommunications was now over; in Ian Ross and John Mayo, the two men who followed Baker as president, the Labs had a succession of able and independent-minded executives. But Baker was still involved with a dizzying number of committees and corporate boards. Through the 1980s, moreover, he still had his work in Washington to occupy him. Some of his advice was given to presidents and intelligence agencies informally; most was solicited through his work on various government committees, some of which still met in secret. Even into the early 1990s, as Baker neared his eightieth birthday, his office at Murray Hill boasted a direct line to the White House and a highly secure STU-III phone to the Pentagon. As his secretary was instructed, if Baker happened to be out of the office and either phone rang, "do not answer it."

But the phones were ringing less and less. And—like Shannon, like Pierce—Baker spent a fair amount of time out of the office collecting awards and honorary degrees. He seemed especially willing to rehash old histories from the glory days. In his personal letters, in fact, Baker sometimes seemed as if he were writing for posterity, creating long and often simplistic narratives of what had been complex innovations. His rendering of the past seemed geared to redound to the greater glory of the Young Turks' Bell Labs, the Bell Labs that existed before the phone monopoly was broken up by lawsuits and regulators. "Happily, the science and technology that you remember in some of your writings is flourishing still," he wrote to his old friend Clark Clifford in 1991. "I am ever grateful that the new phenomena of lasers and photonics that came from our research in the early and mid-60s is finally transforming the world of telecommunications. We were not encouraged for the acceptance of this industrially or economically at first, when millimeter wave technology seemed so appealing. But we dug in our heels, and I must say that our corporation did give us opportunities, even though it was made very clear that the risk was our own."[38]

Meanwhile, old friends paid tribute to Baker even while sending along condolences on the passing of his old world. One high-ranking Washington insider, a cabinet member to several U.S. presidents, wrote to Baker on his eightieth birthday, "You, of course, have been in a position to observe the shortening of the time horizon (and the shrinkage of curiosity) at Bell Labs, since the demise of Ma Bell and the growth of unfettered competition." Baker was too much of a gentleman to agree. In a long and convoluted letter back—"equally necessary was a versatility and sharing of knowledge for a coherent policy formation in the aggregate," he wrote in a moment of reminiscence—he seemed mostly intent on revisiting his old intelligence work and recounting its triumphs. He came across as nostalgic for the cold war.

What pushed Baker from private regrets about the state of telecommunications to forthright disapproval was the Telecommunications Act of 1996. A huge and complex piece of federal legislation, the Telecom Act altered the structure of the communications business by allowing, among other things, the former regional telephone companies (now known as the Baby Bells) to compete nationally with AT&T and MCI.

In short order, the 1996 rules created a mad frenzy for telecom equipment and network infrastructure, resulting in absurd stock valuations for some of the companies involved, as well as fraud and malfeasance. Baker viewed the results with disgust. The country's telecom system, he told a journalist not long after, was "utterly disordered and needs some system of regulation that is publicly and politically acceptable." He had scorn for the Federal Communications Commission, too, which "has no centralized philosophy or objectives" and seemed to spend its time squabbling.[39] His clear message was that it had been a mistake to break the old system up in favor of a more chaotic marketplace. And in his view, as several more years passed, the situation only grew worse. By 2002, the institution Baker had helped build had become unrecognizable to him. "There isn't any institution," he dismissively told an interviewer when asked about his former employer. "Bell Labs does not exist as an institution."[40]

This was not precisely true. Yet to Baker, for a number of reasons, it recalled something he'd said years before. If the phone company were broken up, he remarked in 1974, its long-term research efforts would become untenable and "Bell Laboratories would cease to exist." In his view, the prediction was now as good as prophecy.

Nineteen

INHERITANCE

B aker died in 2005, after a long stay in a nursing home. He was secre-
tive to the end. His burial was limited to family members. None of
his Bell Labs colleagues attended. In his home—the home no one from
Bell Labs had ever visited—he left behind an immense cache of books
and papers from his past, hoarded and piled in disarray.

Whether or not one agreed that the institution of Bell Labs had disappeared, it was obvious, by the time of Baker's passing, that Bell Labs
hardly resembled his old home. To be sure, companies, or parts of companies, can collapse or vanish into liquidation with great rapidity. Bell
Labs was more an example of how an organization could endure through
a process of downsizing and adaptation. True to Peter Drucker's analysis,
over the course of twenty years following the breakup, Bell Labs became
a respectable industrial lab. The tragedy to Bill Baker was that it also,
slowly and steadily, ceased being essential to America's technology and
culture.

One way to think about the fate of Bell Labs is to think of the institution as something akin to a vast inheritance. While staggering as a
combined sum, it somehow becomes more modest once it is split, and
then split again, in various ways over time among various descendants.

On January 1, 1984, the Bell System breakup officially went into effect. AT&T and Western Electric—now one combined company—were severed from the local phone companies, such as New England Telephone and Southern Bell Corporation. Most Bell Labs employees stayed within AT&T. Yet a significant number (about 10 percent) went to a new research institution called Bellcore, which was established to serve the research and development needs of the new "Baby Bells."

In a number of ways, life at Bell Labs went on as before. The AT&T management continued to make a commitment to funding basic research; the big Murray Hill facility and the Black Box at Holmdel continued to bustle. Some Bell Labs discoveries in the 1980s were as noteworthy as what had come before. For instance, a young physicist named Steven Chu, who would later become the U.S. secretary of energy, figured out a way to "trap" and study atoms at freezing temperatures by means of laser beams. Another Bell Labs team discovered and explained a complex physical phenomenon known as the fractional quantum Hall effect. Both groups were ultimately awarded Nobel Prizes for their work.

In the meantime, engineering did not suffer perceptibly. The digital era had at long last arrived, along with staggering jumps in how much information could be sent over fiber optic cables, which were beginning to connect not only regions of the country but also, through undersea cables, the continents. Many of the innovations that emerged from the Labs during this period would never become familiar to Americans in the way that communications satellites or the transistor had been. Still, such advances—in technologies known as digital signal processing and wavelength multiplexing for fiber optic lines, for instance—laid the foundation for the present era. John Mayo, who was soon to become president of the Labs, observed presciently in *Science* magazine in the early 1980s that the employees of the Labs were aiding in "the merging of telecommunications and computer technologies." To Mayo, Bell Labs and the telecommunications industry appeared to be leading the country into a future "where most of the work force will create, process or disseminate information."[1]

Usually it is not difficult to locate a point in time at which a business, or a scientific organization, pivots in one direction or another. One can look back at the calendar and compare events with performance. A new boss takes the helm, for instance, or a canny new strategy is imposed, and the results are clear to see. It's conceivable that the old Bell Labs, the institution of Bill Baker, was transformed at a certain moment in time. The 1984 split is one possibility. "But still, after that, we felt that the world's still okay," Bob Lucky recalls. "But it was slipping slowly away from us and we didn't realize it. And it's hard to put my finger on just where it started to come apart."

In 1986, for instance, the challenges that lay ahead remained indistinct. John Pierce, watching the fate of his old employer from his perch in California, set down some thoughts at the time in a letter to a friend. As Pierce saw it, the great laboratories of the twentieth century had a clear purpose: "Someone depended on them for something, and was anxious to get it. They were really needed, and they rose to the need." For Bell Labs, Pierce noted, the need was modern communications. That future rested upon the institution and the researchers who worked there. Pierce was now watching "as an interested onlooker" to see if the new AT&T Bell Laboratories could figure out a new mission, a new purpose. He wasn't skeptical; he believed it was indeed possible. But he wasn't terribly optimistic, either. The old world was already gone, he explained, it was just that most people hadn't yet noticed. "It is just plain silly," he wrote, "to identify the new AT&T Bell Laboratories with the old Bell Telephone Laboratories just because the new Laboratories has inherited buildings, equipment and personnel from the old. The mission was absolutely essential to the research done at the old Laboratories, and that mission is gone and has not been replaced."[2]

IT'S ARGUABLE THAT NOSTALGIA was now coloring the opinions of people like Pierce. As part of the lost world, he had no place in the new one. The flaw in this argument is that Pierce was not a nostalgic person. Moreover, what he said had the sharp resolution of truth. What was the purpose of his old company and old laboratory now? One answer was that it would still supply the best long-distance service in the world, and that it would compete ferociously in a number of new markets, such as computers. So therefore it would become the premier computer and communications company in the world. At the time of the breakup, in fact, it was widely assumed in the business press that IBM and AT&T would now struggle for supremacy. What undermined such an assumption was the historical record: Everything Bell Labs had ever made for AT&T had been channeled into a monopoly business. "One immediate problem for which no amount of corporate bulk can compensate is the firm's lack of marketing expertise," one journalist, Christopher Byron of *Time*, noted. It was a wise point. Bell Labs and AT&T had "never really had to sell anything."[3] And when they had tried—as was the case with the Picturephone—they failed. Government regulation, as AT&T had learned, could be immensely difficult to manage and comply with. But markets, they would soon discover, were simply brutal. AT&T's leaders, such as CEO Charlie Brown, "had never had the experience or the training to compete," Irwin Dorros, a former Bell Labs and AT&T executive, points out. "They tried to apply the skills that they grew up with, and it didn't work."

In later years, the downsizing at Bell Labs, in terms of both purpose and people, would mostly be linked to this inability to compete. (The worse AT&T did, in other words, the tighter the constraints on Bell Labs, where funding depended on the health and revenue of its parent company.) What's more, a company that had always focused on building things to last three or four decades was now engaged in a business where products and ideas became dated after three or four years. AT&T's attempt to enter the computer business, by purchasing a company known as NCR, failed. And it struggled in the markets for telephones and equipment, too, as it now faced a number of low-priced electronics competitors from Asia. As the 1980s wore on and AT&T entered the 1990s, its mission became even more uncertain. Bell Labs, in turn, began to shed good people, who either left to go to Bellcore or departed for academia. Some, too, began to hear the call of California, where private companies were eager to pay far higher salaries than anyone at the old Bell Labs had ever imagined possible.

Perhaps the most fundamental difference between the old and new Bell Labs was that its focus had become more constrained. In 1995, a Bell Labs researcher named Andrew Odlyzko, who worked as a manager in the mathematics department, circulated a paper he had written that considered what was happening to American technology and, in effect, the world of Bell Labs. Odlyzko pointed out that while it was easy to blame the narrowing ambitions on shortsighted management that aimed to turn a buck more quickly, the actual forces involved were somewhat more complex. "Unfettered research," as Odlyzko termed it, was no longer a logical or necessary investment for a company. For one thing, it took far too long for an actual breakthrough to pay off as a commercial innovation—if it ever did. For another, the base of science was now so broad, thanks to work in academia as well as old industrial laboratories such as Bell Labs, that a company could profit merely by pursuing an incremental strategy rather than a game-changing discovery or invention.

Odlyzko quoted MIT's Jay Forrester, who, in 1948, thinking that the new transistor might be an ideal component for a computer, had written to Ralph Bown to request a sample. In 1995, Forrester remarked that "science and technology is now a production line. If you want a new idea, you hire some people, give them a budget, and have fairly good odds of getting what you asked for. It's like building refrigerators." Perhaps this was an exaggeration. But there was something to it, too. The number of transistors on a chip kept increasing; it was changing the nature of computers and transforming the business of the world. But it was mostly a result of deft and aggressive engineering rather than any scientific breakthrough. The Internet, meanwhile, was already becoming a powerful force for communications. When Odlyzko wrote his paper, a small company called Netscape had just gone public, with a valuation that astounded the business world. And yet Netscape's innovative product—a viewing browser for the World Wide Web—was largely the beneficiary of scientific and engineering advances that had been steadily accruing through academic, military, and government-funded work (on switching and networks, especially) over the past few decades.

In sum, it had become difficult, and perhaps unnecessary, for a company to capture the value of a big breakthrough. So why do it? To put it darkly, the future was a matter of short-term thinking rather than long-term thinking. In business, progress would not be won through a stupendous leap or advance; it would be won through a continuous series of short sprints, all run within a narrow track. "In American and European industry," Odlyzko concluded, "the prospects for a return to unfettered research in the near future are slim. The trend is towards concentration on narrow market segments."[4]

THE LARGE INHERITANCE of Bell Labs was divided again in 1996. AT&T executives concluded that their company needed to be more tightly focused around the long-distance telephone business and a new cellular business. In short order, the company sold its struggling computer company. More drastically, it sliced off its huge telecom equipment division— what had essentially been Western Electric—into a new company called Lucent. In the course of this split, most of the Bell Labs staff went to Lucent, which retained the Bell Labs name for its research and development department. Yet a number of the Labs' researchers, including many mathematicians, were whittled off from Murray Hill to go with AT&T. This group was relocated to a new AT&T facility, now known as the Shannon labs, named after Claude Shannon, in another part of New Jersey.[5]

The old Murray Hill complex, meanwhile, became Lucent's global headquarters. By a number of measures—patents and awards, for instance—the company still retained a first-rate industrial laboratory with a skilled staff. And from the start, the prospects for Lucent and Lucent's Bell Labs were considered promising. The company would design and build the next generation of wireless and wireline equipment. But things went even better than expected, and Lucent's first few years proved to be the kind of fairy tale that the business press and financial investors adore. As wireless phone services boomed, and as the Internet exploded in popularity, so, too, did the need for telecommunications equipment in the United States and abroad. A host of companies embarked on an extraordinary buildout of the country's telecommunications and data infrastructure; Lucent, in turn, began reaping enormous profits. Just two years after it split from AT&T, Lucent's stock valuation—$98.5 billion—was higher than its onetime parent. The next few years became known variously as the telecom boom and the dotcom (a nickname for new web-based companies) boom. The assumption, as one financial columnist described it, was "that the explosive proliferation of dotcoms would send endlessly expanding amounts of data, voice and video streaming across larger and larger networks."[6] At its peak, Lucent was valued at $270 billion. Bell Labs, in turn, enjoyed ample funding. It seemed like another golden age of communications research was on the wing.

Things fell apart quickly. By 2000, it was understood that the predicted demand for telecommunications switching and transmission equipment was a fantasy. To compound Lucent's problems, it was soon discovered that the company's profits had been inflated by a practice of helping outside companies finance purchases of its equipment. The subsequent fallout was devastating. Lucent's revenue plunged. Its stock price, which had peaked at about $84 a share, fell below $2. The company slashed tens of thousands of jobs, including thousands within Bell Labs. Some researchers and engineers were cast off when the company, desperate to alleviate its losses, divided the Bell Labs' inheritance into even more parts—to smaller companies that took the name Agere and Avaya, for instance. Others were unceremoniously laid off. In the New Jersey suburbs, workers found they were embarrassed to wear Lucent shirts or hats to the store. In previous years, as the company's stock price climbed, they would receive slaps on the back. Now they were greeted with angry reprisals of "What happened?" or "I lost a lot of money." In the end, the company reduced its workforce from a high of 150,000 to about 40,000. And in its omnibus efforts to cut costs and energy consumption, every other light inside the vast buildings at Murray Hill was turned off. The acres of lawns in front of the buildings were mowed less frequently. Meanwhile, the remaining employees—at the company whose engineers perfected the telephone—were asked to limit their calls at work.

Then things got worse. In 2002, a panel of experts concluded that a prominent Bell Labs researcher, J. Hendrik Schon, had published a series of papers relying on fraudulent data. One of Schon's claims was that he had created molecular-scale transistors. "I am convinced that they are real," Schon wrote of his results, "although I could not prove this to the investigation committee."[7] It was arguable that the scientific misconduct was Schon's alone, and not an indictment of Bell Labs. But it was difficult to believe such an incident could have occurred years before. "What does the Schon scandal mean?" an interviewer from the *New York Times* asked a young physicist named Paul Ginsparg. "The demise of Bell Labs by becoming corporate," Ginsparg replied.[8]

LUCENT'S EXECUTIVES PREDICTED that their company, as well as the rest of the global telecom industry, would rebound. But the next few years were equally difficult. In 2005, it was believed a merger with the French telecommunications company Alcatel might save Lucent, and perhaps restore the luster of Bell Labs, which was now about a third of its former size. The resulting company was named Alcatel-Lucent. And to some extent, in a volatile market for telecommunications equipment, an era of stability resumed. A bright, capable engineer named Jeong Kim was appointed Bell Labs president. Kim decided that to save Bell Labs he would need to change its structure, vision, and strategy. As he saw it, the Labs had to focus single-mindedly on how its innovations—products that were geared both for the near and medium term—could help Lucent's bottom line. Kim, the *Wall Street Journal* noted, "was perhaps the laboratory's best hope of maintaining its relevance."[9] Kim, however, didn't want Bell Labs to be a citadel of science and scholarship. He wanted it to be a hotbed of entrepreneurial thinking. He did not downplay the challenges of reviving the laboratory. "I did not take this job because it was easy," he said of his work to reinvent the Labs. "I took this job because it was difficult."

The Bell Labs Jeong Kim was running—the Bell Labs Kim felt he needed to run in the face of the telecom market's fierce competition—was barely recognizable as the shop of Kelly, Fisk, and Baker. But at that point these changes were almost certainly necessary. The scientific press nevertheless mourned the pragmatic turn the Labs had taken. When *Nature*, the esteemed British science magazine, discovered that only four researchers were now working in basic physics at the Labs, it ran an article entitled "Bell Labs Bottoms Out."[10] Meanwhile, grown men who had worked at Bell Labs during its golden age would sometimes confess to driving by the Murray Hill complex and experiencing an emotion close to bereavement. A few would weep. The Murray Hill parking lot, where scientists and engineers once had to fight to get a spot, was now half full on some days.

In 2006, Bell Labs' other main campus—the Black Box at Holmdel, Eero Saarinen's architectural tribute to Bell Labs' innovative capabilities—was shut down. What remained of the staff there, originally about fifty-five hundred employees, but now far fewer, were either brought to Murray Hill, reassigned elsewhere, or laid off. Alcatel-Lucent put the 1.9-million-square-foot building up for sale. At first a real estate developer expressed interest in the property, but a deal soon collapsed. Perhaps it could serve another company as an office center, some local officials speculated, or perhaps it could be transformed into a public park. The 472-acre site, once the expansive spread of mown grass where scientists experimented on antennas and microwave transmission, resumed its lonesome character. Except for a roaming security guard, no one remained on the premises. On the long driveway leading to the immense and empty glass building, weeds began poking through the tarmac. The property stayed on the market for months, and then years. There were no buyers.

Twenty

ECHOES

It seems likely that the Black Box at Holmdel will ultimately disappear. The building's almost two million square feet of space were designed for workers of a bygone age who were pursuing a mission of universal connectivity that now seems largely completed. Still, looking around any-place in America on any given day, one cannot miss enduring aspects of the empire designed by Kelly and his colleagues. There are many physical reminders—the wooden phone poles and wires, for starters, that support our current phone and data networks and were made almost absurdly durable thanks to years of research at Bell Labs. The local switching exchanges in most American towns, those sturdy, windowless buildings, also still stand, though switching has become so efficient and miniatur-ized through the use of integrated circuits that the buildings are mostly empty, and chronically dark. The clicking racks of crossbar switches, and the staffs of engineers needed to attend to the machines, are gone. As the engineering writer Brian Hayes points out, "a machine the size of a pizza box can handle a town's telephone system."[1] In the event of a problem, a computer switch can alert someone at network headquarters to come take a look in person.

Bell Labs' original building on West Street in Manhattan—the office where Kelly, Shockley, Pierce, Shannon, Fisk, and Baker first reported to work when they began their careers—was sold long ago. Its interior was cut up and converted to apartments. Other pieces of the phone company's broken empire, however, have been kept in service by Alcatel-Lucent, Bell Labs' current parent company. About forty-five miles southwest of the city, Crawford Hill, the flat-topped ridge where John Pierce and Bill Jakes tracked the progress of the Echo balloon during summer nights in 1960, is sometimes used for wireless research. Atop the hill, light breezes blow in from the Atlantic, and the old horn antenna, though rusted now, remains intact. A plaque nearby designates the site as a national historic landmark. And at the bottom of the hill, there is the small three-story Crawford Hill lab, built in the early 1960s to house a corps of researchers while the Black Box was under construction a few miles away. The scientists and engineers there, about a hundred in all, continue to work on optical fibers and lightwave communications.[2]

At Murray Hill, the sprawling complex twenty-five miles west of New York that Kelly imagined as an escape from the noise, dirt, and vibrations of the city, the buildings are largely unchanged from the 1980s. The computer labs stay busy. The same can't be said for the physical science and electronics laboratories. On workdays many of the labs are locked and dark. Walking down those extraordinarily long hallways can be a haunting experience. On the fourth floor of Murray Hill's old Building 1, a small plate affixed to a wall marks the approximate spot where the transistor was invented. Downstairs, in the lobby, where an exhibit marks the historic sweep of Bell Labs' innovations, the original transistor is on display, protected by a glass case. Not far away is a bronze bust of Claude Shannon, rubbing his chin and looking slightly amused.

According to the U.S. Space Objects Registry, as of late 2010, Telstar—the first active satellite—is no longer functional. Yet it still orbits the earth. An exquisite machine requiring a year of nonstop work by hundreds of engineers, it is now a rotating piece of space junk.

BELL LABS' most durable contributions were not things that could be touched or seen. Many of the ideas and innovations that came out of the

Labs have been subsumed into a global electronic network far larger and more wondrous than existed when Shannon and Pierce agreed, in the late 1940s, that the Bell System was the most complex machine ever created. At the same time, those innovations have been improved upon to the point that the original leap that made them possible has been mostly forgotten. To use the most obvious example, the transistor, countless engineers around the world have taken the original idea and made it better and better, as well as smaller and smaller. For instance, a recent Intel computer processor chip not much larger than a postage stamp contains two billion transistors. (Intel, moreover, manufactures about 10 billion transistors every second.)[3] Lasers and optical fibers and CCD chips—the charge-coupled devices, invented at Bell Labs, that allow for digital photography—have followed a similarly astounding trajectory. Each second, the newest lasers can send nearly unimaginable quantities of information through a single strand of glass fiber. "The commercially available rates are eight terabits per second per fiber," explains Herwig Kogelnik, the laser scientist who began his career under Rudi Kompfner in 1960. To put that number into context, a terabit contains one trillion bits, and a bit, as Shannon formulated long ago, is a unit of information represented by a 1 or 0. A fiber strand can therefore carry millions of voice channels or thousands of digital TV channels. "It's just enormous," Kogelnik adds. "Every ten years it increases by a factor of 100." He points out, too, that some researchers have created systems that can transmit 100 terabits per second. His point is that the future capacity of our networks, now being worked out by his younger colleagues, will dwarf what we have today.

Finding an aspect of modern life that doesn't incorporate some strand of Bell Labs' DNA would be difficult. The transistors, lasers, quality assurance methods, and information technologies have been incorporated into computers, communications, medical surgery tools, factory productivity methods, digital photography, defense weaponry, and a list of industries and devices and processes almost too long to name. Scores of Bell Labs veterans have meanwhile taken jobs in technology companies such as Google and Microsoft; even more have gone into academia, following Shannon's and Shockley's example, and passed along their ideas to the next generation.

Many information-age advances, a large percentage of which trace back to Bell Labs, are less clearly good news. The National Security Agency—Bill Baker's old stomping ground, which continues to monitor the transmission of information around the globe—recently revealed that every day it intercepts and stores two billion phone calls, email messages, and other data transmissions.[4] Therein lies evidence of how much information now moves around the globe. Therein also lies evidence of how the rapid, easy, and cheap exchange of information exposes individuals to intrusions of privacy while also exposing societies to vulnerabilities that could be exploited by hackers and cyberterrorists. And still, these concerns seem only one aspect of a much larger question: Has access to so much information not only expanded our lives but contracted them? The bustling town square in Gallatin, Missouri, where a hundred years ago Mervin Kelly worked in his father's hardware store and confronted the counterpoint between the present and the future, is now an assemblage of empty storefronts. The hardware store and old telephone exchange is now a vacant building, with a few telltale hanks of telephone wire dangling from a dilapidated corner mount. There are a variety of reasons for the decline of small-town America. But when all kinds of communications and entertainment are delivered to your home, there are fewer and fewer reasons to go into town and exchange greetings in person.

Information brings with it unintended consequences, too. Some technology journalists, notably the writer Nicholas Carr, have asked recently whether an increasing reliance on instant communications and Internet data is eroding our need, or ability, to think deeply. "What the Net seems to be doing is chipping away my capacity for concentration and contemplation," Carr writes. "My mind now expects to take in information the way the Net distributes it: in a swiftly moving stream of particles."[5] It is the dark side, in many respects, of Kelly's 1951 prediction, which has proven largely correct, that future networks would be "more like the biological systems of man's brain and nervous system." The tiny transistor, as Kelly saw it, would reduce dimensions and power consumption "so far that we are going to get into a new economic area, particularly in switching and local transmission, and other places that we can't even envision."[6] But we can envision it now.

The wash of information has risen so near to flood stage that Jeong Kim, the most recent president of Bell Labs, has suggested that the future of communications will be defined by an industry yet to be created— not the kind of business that simply delivers or searches out information, but one that manages the tide of information so that it doesn't drown us. At least in the communications industry, the greatest innovative challenge on the horizon, Kim says, is "to organize information in a way that allows you to live the way you want to live, to take time off with your kids without fear you're going to miss out on something."[7] The larger idea, then, is that electronic communication is a miraculous development, but it is also, in excess, a dehumanizing force. It proves Kelly's belief that even as new technology solves one problem, it creates others.

To follow this line of reasoning, the contemporary iterations of Bell Labs may have to solve some of the problems created by the solutions of the old Bell Labs.

THE PURPOSE OF INNOVATION is sometimes defined as new technology. But the point of innovation isn't really technology itself. The point of innovation is what new technology can do. "Better, or cheaper, or both"— Kelly's rule—is one way to think about this goal. In testimony before a U.S. Senate subcommittee in 1977, John Pierce gave a slightly more elaborate explanation. "The only really important thing about communication is how well it serves man," he said. "New gadgets or new technologies are important only when they really make good new things possible or good old things cheaper or better."[8] Put another way, a new technology can put more money in our pockets, and it can allow us to do things—call across the country, send email, write software, design skyscrapers, model pharmaceuticals—in ways that were never possible before. The results can be manifested in new products and civilizing comfort, as well as by economic growth. "The history of modernization is in essence a history of scientific and technological progress," Wen Jiabao, the premier of China, said recently. "Scientific discovery and technological inventions have brought about new civilizations, modern industries, and the rise and fall of nations." A recent report by the National Academy of Sciences argues that the United States, by consistently underinvesting in its education system and in scientific research over the past few decades, seems to have forgotten this lesson—a lesson that in many respects the country demonstrated for the rest of the world during the second half of the twentieth century. "While only four percent of the [U.S.] work force is composed of scientists and engineers," the National Academy report points out, "this group disproportionately creates jobs for the other 96 percent."[9]

One can only speculate about how Kelly, Pierce, Baker, and the rest would react to the most acclaimed American innovations of recent years—iPhones, say, or Google searches or Facebook. They would likely see them as vital, sophisticated tools for the information age. A more provocative question, however, is whether they would perceive them as paths to the future, as many economic commentators often do. Regrettably, the language that describes innovations often fails to distinguish between an innovative consumer product and an innovation that represents a leap in human knowledge and a new foundation (or "platform," as it is often described) for industry. In an effort to explain his motivations, Pierce once wrote in a memo, "Things should be done only when there is the possibility of a *substantial* gain, and this must be weighed against risk."[10] The italics were Pierce's own. Bell Labs' substantial innovations, John Mayo, the former Labs president, points out, "account for a large fraction of the jobs in this country and around the world. And they also account for a lot of the social status of the world."

Mayo and many of his former colleagues worry about where the foundation for the next generation's jobs will come from. Will they emanate from America, or from abroad? Are the next great leaps in energy research or biotechnology? Do we yet have the scientific base—akin to the "substantial gains" of transistors or lasers or optical fiber—on which to build that future economy? Or are we still living off the dividends from ideas that were nurtured, and risks that were taken, a half century ago?

EVEN AS MERVIN KELLY'S LIFE was drawing to a close, the models for innovation he had spent his career devising were changing. Kelly's philosophy is sometimes summed up as a belief that innovation occurs by the movement of ideas in one direction: first a fundamental scientific discovery, which is then developed into a product, which is then pushed into the market. The textbook example was the transistor. In truth, he believed big scientific advances could come from any engineers or scientists encountering interesting problems. Kelly had learned that when he managed the vacuum tube shop in lower Manhattan in the 1920s. His larger view of innovation, as a result, was that a great institution with the capacity for both research and development—a place where a "critical mass" of scientists could exchange all kinds of information and consult with one another for explanations—was the most fruitful way to organize what he called "creative technology." A corollary to his vision was that size and employee numbers were not the only crucial aspect. A large group of physicists, certainly, created a healthy flow of ideas. But Kelly believed the most valuable ideas arose when the large group of physicists bumped against other departments and disciplines, too. "It's the interaction between fundamental science and applied science, and the interface between many disciplines, that creates new ideas," explains Herwig Kogelnik, the laser scientist. This may indeed have been Kelly's greatest insight.

And yet his grand design was undone by time. In his memorial tribute to Kelly, John Pierce pointed out that Kelly never had the opportunity to change his views on research and development in the wake of evolving business circumstances. As a result, Pierce concluded, "Kelly may have overestimated the amount and quality of research that could in the future be expected from industry, and perhaps from the nation."[11] Pierce was probably correct. In succeeding decades, for instance, Bell Labs' own journey—as it moved from its monopoly status to Lucent and Alcatel-Lucent, shucking off employees and entire departments all the while—demonstrated that a large industrial laboratory had to change with political and legal regimes. It became increasingly difficult to fund basic research; instead, Bell Labs had to focus more on development and engineering. The Labs also needed a narrower focus on products and short- or medium-term goals. The new industrial lab had to succeed not only in engineering, but in business, too.

As industrial science was evolving, a very different model for innovation arose. From the 1970s on, a host of Silicon Valley entrepreneurs proved that new ideas didn't need to be attached to a large corporation to become world-altering technologies. A good idea could arise from a teacher or student at a school like Stanford, and the purveyor of that idea could then get funding from a venture capitalist on Sand Hill Road, the wide avenue that runs along the university's western boundary. In turn, the idea purveyor—now simply labeled an entrepreneur—could launch his or her technology through a small start-up company in a nearby town like Palo Alto or Cupertino or Mountain View. Not incidentally, in the process of backing a winner, everyone involved could get very, very rich. Bell Labs invariably lent some of its genetic material to this process—a number of the new ideas for computers or software relied on transistors or lasers or the Unix programming language, for instance. Eugene Kleiner, moreover, a founding partner at the premier venture capital firm Kleiner Perkins, was originally hired by Bill Shockley at his ill-fated semiconductor company. But the Silicon Valley process that Kleiner helped develop was a different innovation model from Bell Labs. It was not a factory of ideas; it was a geography of ideas. It was not one concentrated and powerful machine; it was the meshing of many interlocking small parts grouped physically near enough to one another so as to make an equally powerful machine. The Valley model, in fact, was soon so productive that it became a topic of study for sociologists and business professors. They soon bestowed upon the area the title of an "innovation hub."

Early on, Bell Labs executives were aware of the vitality of California.

At one point in the mid-1960s, for instance, Bill Baker and a New Jersey business consortium hired Frederick Terman, the Stanford engineering dean who had wooed Bill Shockley to Palo Alto in the mid-1950s. Terman is often credited as the father of Silicon Valley. (Shockley, by comparison, is sometimes called the Moses of Silicon Valley, since his failures prevented him from entering the Valley's promised land of wealth and influence.) The hope was that Terman might be able to map out an innovation hub for New Jersey, based in part around the technological excellence of Bell Labs. One seemingly insoluble problem was that New Jersey was too geographically diffuse for the Palo Alto model to work there. The universities that did exist, such as Princeton and Rutgers, were either too far away from one another or too theoretical in their scientific focus to act as fertile training grounds for East Coast entrepreneurs. What's more, while Bell Labs was dynamic—the scientists and engineers there shared ideas and knowledge with one another and, to a certain extent, with their friends in the academy and other industrial labs—it was not as dynamic as Silicon Valley.[12] In the Valley, engineers changed jobs constantly while companies formed and dissolved and then formed again, as if in constant agitation. Terman believed a new college, perhaps modeled after Caltech and deemed "Summit University," could help solve some of these problems. This new school—graduate students only—would provide to the region's telecommunication and pharmaceutical companies a steady stream of expert scientific and engineering talent. The problem was the cost. It turned out that funding the school would be expensive—Terman projected a start-up cost of $15 million, according to the management historian Stephen B. Adams. The pharmaceutical industries were not interested, which meant that Bell Labs would have to be the main backer.[13]

It was too difficult for Baker to justify the costs. And ultimately, Summit University, along with the Terman study, was shelved.

AT LEAST FOR THE past few decades, the venture economy has proven a more adaptable model for innovation than Mervin Kelly's. The products coming out of Silicon Valley—and to a lesser extent the Route 128 area outside of Boston—have evolved fluidly, from new applications in electronic hardware, to new applications in computer software, to new applications in biotechnology and clean energy. Perhaps the only thing lacking is that venture firms are averse, understandably, to funding an entrepreneur seeking out new and fundamental knowledge. Without any way to predict the difficulty of obtaining new knowledge, and without any tools to assess its market value, how could someone bet money on it? As one venture capitalist for Kleiner Perkins puts it, "We don't fund science experiments." In some respects, then, this leaves a gap. While it is frequently the case that new knowledge can arise from academia or a government laboratory and then secure venture capital afterward, it seems a more difficult proposition in Silicon Valley than it was long ago in New Jersey. The value of the old Bell Labs was its patience in searching out new and fundamental ideas, and its ability to use its immense engineering staff to develop and perfect those ideas. Some of the other great and now diminished industrial labs—General Electric, RCA, IBM—followed a similar tack, though with smaller staffs and less spectacular results.

John Pierce did not flatter himself so much as to think that success in basic or applied research—those big leaps in scientific knowledge—were necessarily more heroic than development. "You see, out of fourteen people in the Bell Laboratories," he once remarked, "only one is in the Research Department, and that's because pursuing an idea takes, I presume, fourteen times as much effort as having it."[14] Still, Pierce understood that the big new ideas—satellites, transistors, lasers, optical fibers, cellular telephony—could create an entirely new industry. "You may find a lot of controversy over how Bell Labs managed people," John Mayo, the former Bell Labs president, says. "But keep in mind, I don't think those managers saw it that way. They saw it as: How do you manage ideas? And that's very different from managing people. So if you hear something negative about how John Pierce managed people, I'd say, well, that's not surprising. Pierce wasn't about managing people. Pierce was about managing ideas. And you cannot manage ideas and manage people the same way. It just doesn't work. So if somebody tells you Pierce wasn't a great manager . . . you say, of what?"

Mayo and other Bell Labs veterans don't always call people like Pierce or Baker the Young Turks, the name this group of men, long ago, gave themselves. Sometimes they call them, without irony, the Giants. "Pierce did not let people get in the way of his pursuit of ideas," Mayo adds. "He did not compromise because it would make people feel good. He did his thing because he felt it was necessary to accomplish the development of ideas the way he wanted. He was excellent at that. And I loved those research people for that. They weren't about making people feel good. They were about motivating people—not to do the conventional thing, but to do the unconventional thing." To follow the progress of business now, Mayo adds, is to become accustomed to watching successful technology companies offer new engineers rich incentives for their work. Pierce and Bell Labs couldn't do that because they were funded like a public utility. But they also couldn't do that because it chafed against their belief in how innovations arise. "Incentives are fine," Mayo says, "but they produce incremental improvements in what's there. That's not what Pierce was about."

Mayo continues, "There are a lot of people that just don't see the kind of things that are going to happen or likely to happen. They would prefer to invest in incremental improvements, and to have wonderful picnics, and make this quarter's earnings without strain." In part, Mayo connects this to the "immense stress" associated with funding research on ideas that may destroy your business if the results make your current product obsolete. Those who study innovation know this as the *innovator's dilemma*, a term coined by the Harvard professor Clayton Christensen. "This is a very strong force," Mayo points out. "It's in me. And in everybody." Strangely enough, however, it may not have been in Mervin Kelly or in some of his disciples—perhaps because the monopoly, at least for a time, guaranteed that the phone company's business would remain sturdy even in the face of drastic technological upheaval. Kelly, for instance, who toiled for decades to improve and perfect the vacuum tube, effectively lobbied for a research program on the transistor that, when it succeeded, rendered his entire previous career in science irrelevant. And an array of other technologies at Bell Labs had a similar effect of discarding the old in favor of the new. As Mayo says, "When I came to Bell Labs, and if you told me we were going to put a billion [transistors] on a single chip of silicon, we're going to make glass so pure that you can shine light through it for hundreds of miles, or we're going to use a computer to listen and speak like people do, I would have said, 'You're out of your mind. We'll move in that direction, but that's too far.' But then, here we are."

"CAN WE LEARN SOMETHING FROM THE EXAMPLE OF BELL LABS?" John Pierce asked, in all capital letters, one day in late January 1997. At the age of eighty-six—five years before his death—he had sat down at his computer, at home in Palo Alto, to write a proposal for a book that explored what his old institution had taught him. He never wrote the book. But the problem Pierce wrestled with that day was how to decouple Bell Labs' success from its circumstances. "Bell Labs functioned in a world not ours," he noted. The links between government and business were different in that era; the monopoly was deemed acceptable as well as vital. And the compensation scale for its researchers and managers could never suffice in the modern economy. In Pierce's era, the top officer at Bell Labs made about twelve times that of the lowest-paid worker; in the late 1990s, it was more typical at large American firms for the CEO to make one hundred times the salary of the lowest-paid worker. Back in the 1940s and 1950s, moreover, smart and talented graduate students could never be wooed away from the Labs by the prospect of making millions. It wasn't even thinkable. You were in it for the adventure. "I don't think I was ever motivated by the notion of winning prizes, although I have a couple of dozen of them in the other room," Claude Shannon said late in life. "I was motivated more by curiosity. I was never motivated by the desire for money, financial gain. I wasn't trying to do something big so that I could get a bigger salary."[15]

It may be obvious, but it is nonetheless worth noting that in 1997 and in the present day there is little possibility—and, admittedly, little reason—for the return of a phone monopoly. Communications is a thriving, innovative industry, thanks in part to the fact that it has been built upon the foundation laid down by Bell Labs. "I've often said to my old friends that we were very lucky we got to work there, in an environment that I don't think will ever exist again," remarks Dick Frenkiel, who worked on the first generation of cellular technology. "It's hard to say something will never happen again. But with the monopoly gone, with the whole concept of monopoly essentially discredited, how could there ever be a place like that again?"[16] Still, to explore what we can learn from the example of Bell Labs isn't the same as pining for its return. Rather, it is to ask what aspects of Bell Labs made it succeed, and whether other organizations—or other governments, as they attempt to plan far into the future in energy, biotechnology, nanotechnology, information technology, and the like—can salvage what was valuable. Pierce, to put it simply, was asking himself: What about Bell Labs' formula was timeless? In his 1997 list, he thought it boiled down to four things:

A technically competent management all the way to the top.
Researchers didn't have to raise funds.
Research on a topic or system could be and was supported
    for years.
Research could be terminated without damning the researcher.[17]

It is an interesting list. But it is hard to see it as complete. The fact that the telephone engineers faced an unceasing stream of technical and logistical problems always urged them toward innovative solutions. Without question, the size of the staff at Bell Labs, and its interdisciplinary nature, were large factors in its success, too. So was the steadiness of the Labs' funding stream, guaranteed by the monthly bill paid by phone subscribers, which effectively allowed the organization to function much like a national laboratory. Bell Labs managers knew they could support projects—the undersea cable, for example, or cellular telephony—that might require decades of work. The funding stream also assured the managers that they could consistently support educational programs to improve the staff's expertise and capabilities. And as Morry Tanenbaum, the inventor of the silicon transistor, points out, Bell Labs' sense of mission—to plan the future of communications—also had an incalculable value that endured for sixty years. The mission was broad but also directed. Bell Labs' researchers, Tanenbaum notes, had a "circumscribed freedom" that proved to be liberating and practical at the same time.

And what about competition? It is now received wisdom that innovation and competitiveness are closely linked. Companies that are good at innovating are good at competing in the market; the uncompromising nature of the market, in turn, is a powerful force on companies to innovate. But Bell Labs' history demonstrates that the truth is actually far more complicated. It also suggests that we tend to misinterpret the value of markets. What seems more likely, as the science writer Steven Johnson has noted in a broad study of scientific innovations, is that creative environments that foster a rich exchange of ideas are far more important in eliciting important new insights than are the forces of competition.[18] Indeed, one might concede that market competition has been superb at giving consumers incremental and appealing improvements. But that does not mean it has been good at prompting huge advances (such as those at Bell Labs, as well as those that allowed for the creation of the Internet, for instance, or, even earlier, antibiotics). It's the latter types that pay to society the biggest and most lasting dividends. And it was almost always the latter types that Kelly and Pierce and Baker were striving for. It may be the case, too, that we not only mistake the potential for free market competition to prompt big breakthroughs. We may also misunderstand how the private sector produces the most promising innovations in any given year. For instance, a 2008 study titled "Where Do Innovations Come From?" concluded that partnerships among corporations, government laboratories, and federally funded university researchers has become increasingly essential to the U.S. innovation pipeline over the past several decades. In 2006, for instance, "77 of the 88 U.S. entities" that produced significant innovations were beneficiaries of federal funding.[19] Clearly, at least in regard to innovation, capitalism is more deeply intertwined with government than many of us realize.

There may be one other observation worth adding to Pierce's list. In recounting what he learned from Bell Labs, John Mayo, among other things, offers this: "We learned that the impossible is not impossible. We learned that if you think you can do something you may very well be able to do one thousand times better once you understand what's going on." It may be easy to overlook something crucial in what Mayo is saying. But it relates back to Bell Labs' origins—back when Frank Jewett and Harold Arnold set about creating the laboratory at Western Electric that ultimately became Bell Labs. The men built it upon the notion that by encouraging their staff to understand a technology, they could create advances that were not only useful but revolutionary. An industrial lab, Jewett explained, was a group of intelligent men "specially trained in a knowledge of the things and methods of science." As he saw it, a properly staffed and organized lab could avoid the mistakes of cut-and-try experimentation and in turn "bring to bear an aggregate of creative force on any particular problem which is infinitely greater than any force which can be conceived of as residing in the intellectual capacity of an individual."[20] The vacuum tube repeater, which allowed a phone call to reach from New York to San Francisco, was only the first great proof of this. And then many others followed.

A drive for understanding separated the great scientists and engineers of the twentieth century from their predecessors. And it separated their inventions and business successes, too. The thin, slow-moving physicist named Clinton Davisson, who was hired at the Western labs just before World War I, found that it was the only thing he cared about. And that, in turn, made a permanent impression on Davisson's best friend and office mate, the young and impetuous physicist from Missouri named Mervin Kelly. They were there to get ahead. But Kelly could see that they were only going to get ahead by understanding what they were doing.

COMPARING THE INFORMATION BUSINESS of Kelly's era to that of the present can be enlightening as well as tricky. How do the technology giants of today—companies like Apple, Microsoft, Google, or Facebook—measure up to Bell Labs? To be sure, there are similarities. All of these companies have carved out a near-monopoly status in various electronic hardware or computer software markets. All are sitting on enormous reserves of cash—tens of billions of dollars in some cases—that they could invest at will on research or new ideas. All of these companies seem intent on controlling, or at least dominating, our communications markets.

All of these companies meanwhile employ some of the finest engineers and computer scientists on the planet. And to house those employees, corporate executives have built citadels on expansive, grassy campuses—informal, creative environments that reward innovative thinking with financial rewards and (thanks to the easy proliferation of software) speedy product rollouts. Google has even picked up on an old Bell Labs tradition: It encourages workers to spend part of their time—up to 20 percent—on a project that captures their interest, just as Joel Engel did when he planned a cellular phone system in Holmdel's Black Box in the late 1960s.

Still, the contrasts between these organizations and Bell Labs are crucial. "This was a company that literally dumped technology on our country," the physics historian Michael Riordan has said of Bell Labs. "I don't think we'll see an organization with that kind of record ever again."[21] The expectation that, say, Google or Apple could behave like Bell Labs—that such companies could invest heavily in basic or applied research and then sprinkle the results freely around California—seems misplaced, if not naive. Such companies don't exist as part of a highly regulated national public trust. They exist as part of our international capital markets. They are superb at producing a specific and limited range of technology products. And at the end of the day, new scientific knowledge matters far less to them than the demands—for leadership, growth, and profits—of their customers, employees, and shareholders.

Perhaps information technology, then, is the wrong place to look for a new Bell Labs. We might do better to poke around in other parts of the economy. One place to consider is a complex of buildings set amid a 689-acre campus some thirty miles north of Washington, D.C. Known as Janelia Farm, the campus serves as an elite research center for the Howard Hughes Medical Institute. Janelia opened in 2006 with the intent of attacking the most basic biomedical research problems; it is patterned after Bell Labs and backed by a multibillion-dollar endowment. The primary goal is to understand consciousness and how the human brain processes information, but the approach to innovation is familiar: a close, interdisciplinary exchange of ideas between the world's brightest science researchers, all of whom are given ample funding and tremendous freedom. The directors of Janelia urge their researchers to take risks and to flirt with failure as they "explore the unknown." There are no classes to teach, no papers to grade, no federal grants to pursue. And while the scale of the research effort is smaller than at the Labs—Janelia is home to about three hundred researchers and a hundred visiting scholars—it's difficult not to conclude, just as Kelly might have, that it's very much an institute of creative technology. By early indications, too, the results at Janelia and Howard Hughes outshine the results of academics working within the existing structure for federally financed medical research.[22]

Also, there is another place to look for a new Bell Labs. Moving the global economy from one that runs on fossil fuels to one that runs on renewables is almost certainly the most difficult challenge—the wickedest problem—of the twenty-first century. Whether we succeed or fail at these efforts, which are only now in their infancy, will determine how dramatically our climate will change in the coming century. It will also determine the political and economic strength of nations that now rely on the production or consumption of oil, gas, and coal. As U.S. secretary of energy, Steven Chu, who won the Nobel Prize for his research at Bell Labs in the early 1980s, has proposed a number of research projects to spur clean energy innovation. Chu calls these projects "innovation hubs," which are effectively meant to function as miniature copies of his old employer. So far, the hubs focus on making breakthroughs in synthetic fuels, nuclear power, and energy efficiency. "I believe that to solve the energy problem," Chu said in 2009 at a U.S. Senate committee hearing, "the Department of Energy must strive to be the modern version of Bell Labs in energy research."[23]

In size and scope, the hubs are a modest first step. But on a number of counts, Bell Labs represents a useful model for energy innovation—a model that's arguably better than the Manhattan Project (for the first atomic bomb) or the Apollo program (for the first moon landing). Both aimed for a lofty but singular goal. By contrast, the creation of a clean-energy economy will be a process without end. It will involve the management of vast, sophisticated, interconnected systems, much like communications networks, that require great technological leaps forward as well as constant, incremental improvements. The Labs' research department was conceived upon the notion of constantly looking far ahead, toward the goal of big and risky breakthroughs. The search for clean, affordable energy undoubtedly requires such dramatic advances. Yet in 2012, a host of newly fashioned inventions (in solar, wind, and tidal power, among others) already await the ingenuity of engineers who are able to develop them into innovations that are cheaper and better than what we currently use.[24] And in this respect, Bell Labs' other dimension— the ability to exhaustively develop a product and get it ready for mass manufacturing and deployment—is perhaps even more crucial. To think long-term toward the revolutionary, and to simultaneously think near-term toward manufacturing, comprises the most vital of combinations.

The need for an energy quest, as it happens, might not surprise the founders of the Labs. In the spring of 1923, an editor at the *New York Times* wrote to Frank Jewett, soon to become Bell Labs' first president, and invited him to contribute to a symposium of ideas sponsored by the newspaper. Jewett agreed, and his four-hundred-word piece, appearing on the May 20, 1923, front page, set the tone for the edition. "Water, Energy Limited; Scientists Look to the Sun Next," the headline read. Jewett wrote, "It seems clear that a great, if not the greatest, present day need is the development of some new source of cheap utilizable energy." With the tools of "research and invention," Jewett urged scientists to figure out ways to take advantage of solar or tidal power, or "fuel from the luxuriant vegetable growths of the tropics"—a predecessor, most likely, of today's biofuels.[25]

The question the *Times* editor had posed to Jewett was, "What invention does the world need most?"

. . .

RALPH BOWN, the director of research who pondered the significance of the transistor as a snowstorm moved in on New Jersey on Christmas Eve 1947, would sometimes ask his colleagues: What was Bell Labs? As John Pierce recounted it, Bown would say: If we marched all the people out and destroyed the buildings and the equipment and the records, would Bell Laboratories be destroyed? Bown's answer was no, it would not. On the other hand, he would say that if the buildings, equipment, and records remained intact but the people were removed, Bell Laboratories would be destroyed. His obvious point was that Bell Labs was a human and not a material organization. Yet perhaps it was more complicated than that. For instance, Bown never explained whether the institution's success was a result of thousands of engineers and scientists working together, or of the few exemplars who towered above everyone else.

"Everybody has their own list," remarks Bob Lucky, the former Bell Labs executive who succeeded John Pierce as executive director of communications sciences. On Lucky's list are the mathematicians Claude Shannon, Steve Rice, and David Slepian. And the physicist Sol Buchsbaum, who rose to an executive position with Bell Labs in the 1970s. "Shockley of course," he adds. "Certainly Pierce and Baker." In Lucky's view, the exceptional individuals lent the institution its reputation of exceptionalism. "I just don't think they make people like the kind of people we had," Lucky says. "Not that nature doesn't make them, just that the environment doesn't make them. We had these people who were bigger than life back then. And we don't seem to have them anymore—though people might say Steve Jobs or Bill Gates." In Lucky's view, a list of Bell Labs' exemplars captures the essence of the organization. "They set the examples that permeated the whole place. They created the fame and were what other people aspired to be. They were the leaders, even if they weren't high up in management. If you knew them, you knew Bell Labs." While it's true that the handful of famous people overshadows tens of thousands of other people, he adds, if you take that handful away, "you don't have Bell Labs."

There is another way of answering Bown's question, however. Chuck Elmendorf, John Pierce's friend from Caltech who roomed with Pierce in New York City in the late 1930s, sees the institution of Bell Labs as far more important than what he calls "the great men." Elmendorf asks, "How do you capture the aspects of the institution that have nothing to do with these great names? What I'm troubled about is—even though these are great names . . . John Pierce was practically my brother; Barney Oliver, we were buddies; hell, I learned solid-state physics sitting on the couch in Bill Shockley's living room, because John Pierce sent me up there. I *knew* these guys. But they weren't Bell Laboratories." Nor, Elmendorf adds, were the other Nobel Prize winners. To him, the essence of Bell Labs was its immense and complete institutional capabilities—how it could develop anything from the tiniest element of a small electronic device to the grand plan for a national network; also, how it could develop people, turning callow college graduates into competent researchers and managers. As a result, it could solve the biggest of problems. "I worked with guys who made some tremendous contributions and you've never heard of them," he says. So maybe this argument—the individual versus the institution; the great men versus the yeomen; the famous versus the forgotten—is insoluble. Or maybe the argument is easily deflected. Perhaps the most significant thing was that Bell Labs had both kinds of people in profusion, and both kinds working together. And for the problems it was solving, both kinds were necessary.

Amazement is a common thread in conversations with Bell Labs veterans. Some of the amazement is in simple observations—for instance, in the speed and capabilities of our present-day information networks, as well as in the processing power and versatility of our cellular phones. Or it comes in the realization that connectivity in today's world is far more important than, say, fidelity. That phone calls often crackle or fade or echo is something that never would have been permissible thirty years ago, where perfect transmission and victory over noise were the ultimate goals. Some of the amazement, however, runs deeper. It resides in the fact that the world of Bell Labs is disappearing, and the contribution of its staff is mostly forgotten. "Frankly," Bob Lucky observes, "if you ask people on the street who invented the transistor, they don't have the foggiest idea."

John Pierce, the most eloquent of the Young Turks, seemed to have a deep respect for the destructive quality of new technology. Pierce carried this understanding with him from his youth until his death. Upon receiving the Japan Prize in the mid-1980s, he wrote, "However nostalgic I may be about the world of my childhood, it is gone, and so are the sorts of people who lived in it. Science and technology destroyed that world and replaced it with another." Typical of Pierce, he could sound bloodless in public about the process of change and innovation. But confidentially, some aspects of these social disruptions seemed to rankle him.

Pierce hoped privately that the work he and his colleagues did would someday be broadly recognized. This was not a late-in-life wish; rather, it was something he considered as early as the 1950s, before the launch of Echo, before he had even reached his fiftieth birthday. "In general we are no more sentimental about the relics of science and technology than Shakespeare's contemporaries were about his house and possessions," Pierce wrote in an unpublished essay from 1959. "What mementos will our heirs have of our romantic present to tell them that men created the things which they take for granted?" For someone sentimental enough to actually believe that the origins of technology were worth seeking out, Pierce noted, it might be possible to find the old laboratory that Mervin Kelly shared with his friend Clinton Davisson at the West Street labs in Manhattan. You might be able to locate the room in which Davisson worked, Pierce noted, even if the walls have been moved and the configuration of the laboratory had been changed. But in the case of then-recent inventions such as the maser, which Pierce already saw as a momentous development, "it will probably be impossible a few years from now even to find the movable partitions which surrounded the work. It is clear that we build for the day and not for the ages, and what we build has a community and functional rather than an individual character."[26]

There was no way around the conclusion. Pierce and his friends were making ideas and things that would either disappear in an instant, or would be absorbed into the ongoing project of civilization. He feared that any memories of the makers would perish, too. "I am afraid that there will be little tangible left in a later age," Pierce wrote of his world at Bell Labs, "to remind our heirs that we were men, rather than cogs in a machine."

Acknowledgments

To a certain extent, my interest in Bell Labs arose out of personal experience. I grew up in Berkeley Heights, New Jersey, just a few hundred yards from Bell Labs' Murray Hill campus. From an early age, I was familiar with every building at Murray Hill. I also knew—long before I understood its significance—that the transistor had been invented there. I never imagined my parents' decision to settle nearby would some-day enrich the texture of this book, but it did, and I thank them for that twist of fate. The more important point, of course, is the debt I owe my parents, Doreen and Bud Gertner, for their love and encouragement. They were supportive of my writing career from the beginning, and never wavered (though they always worried). My father, the true scientist in the family, died unexpectedly a week before I finished this manuscript. Had he gotten the opportunity, I imagine he would have been immensely relieved to see this book complete. I am finished now, Dad.

THIS BOOK EXISTS ONLY because so many people offered help along the way. Any list would need to begin with Sarah Burnes and Eamon Dolan, my agent and editor, respectively. When I walked into Sarah's office nine years ago to talk about possible book ideas, I hadn't imagined it would take this long. But she was willing to back an unwieldy topic, simply because it held my interest above all other ideas, and then was able to give the proposal shape and sensibility. All this—and her wise counsel—leaves me deeply in her debt. Eamon, meanwhile, gave the manuscript itself shape and sensibility. He maintained his good humor, as well as his faith in me, over the course of many years and several neglected deadlines. I am grateful. Others at Penguin Press also provided help. Scott Moyers guided the manuscript through its final stages and offered excellent advice; Emily Graff made sure this project glided over its production and organizational hurdles.

THE REPORTING OF THIS BOOK took me all over the country. In New Jersey, George Kupczak at the AT&T archives welcomed me to his offices year after year. At Alcatel-Lucent in Murray Hill, Ed Eckert, Peter Benedict, and Paul Ross were tremendously helpful. Also at Alcatel-Lucent, Gary Feldman and Jeong Kim talked with me about innovation and the future of telecommunications. At the Crawford Hill lab, Herwig Kogelnik gave me a long interview as well as a personal tour of the hill itself. Dozens of Bell Labs veterans welcomed me into their homes and allowed me to dredge up ancient history for hours on end. Several in particular—John Mayo, Morry Tanenbaum, Ian Ross, Chuck Elmendorf, Henry Pollak, Bob Lucky, and Dick Frenkiel—were especially generous with their time. Mike Noll, who long ago organized the archival papers of Bill Baker and John Pierce, was an indispensable ally in my research. Joan and Michael Frankel, the current residents of Mervin Kelly's old house in Short Hills, New Jersey, gave me a top-to-bottom tour of their home and allowed me to wander around the backyard tulip gardens reconstructing the past. Patricia Neering gathered some of the early periodical research for this book. Gerald Dolan remastered for me an obscure and inaudible Claude Shannon interview.

In Washington, D.C., the staff at the Library of Congress helped me make my way through the papers of Claude Shannon, Vannevar Bush, and Harald Friis; in Pasadena, California, the librarians at the Huntington Library helped me navigate through the papers of John Pierce. At Stanford, Leslie Berlin was a huge help in getting me started on the Shockley papers—an effort made even more efficient by the library's excellent staff. I'm also obliged to the staff of the Mudd Manuscript Library at Princeton; to the IEEE History Center; and to Harriet Zuckerman and the Columbia University Center for Oral History. And in Gallatin, Missouri, Dave Stark, a resourceful local historian, gave me a guided tour of the town square, introduced me around, and shared with me his own archival spadework. On the same trip, the kind women at the Daviess County Library and the Daviess County records office offered assistance and vital information.

As I worked on this book for the past five years, I also worked as a writer for the *New York Times Magazine*. My reporting for the *Times* often shaped my thinking about the innovative process and American technology. I'm especially grateful to Dean Robinson, my editor at the magazine for the better part of eight years, who somehow manages to make every story he touches better. Likewise, I'm grateful to the Sunday magazine's editor in chief during that time, Gerry Marzorati, and to the *Times* art directors, photo editors, fact-checkers, and copy editors who worked on my features. They consistently made those stories smarter, more attractive, and more accurate. As it happens, the first story I wrote on Bell Labs appeared not in the *Times* but in *Money* magazine in 2003. Two editors encouraged me to write that feature: Bob Safian and Denise Martin. I'm grateful to both for that opportunity, and also for the manner in which they've encouraged my career in the time since.

Several people read through this manuscript before publication. Any errors within the text—technical, perceptual, judgmental—are completely my own. I nevertheless owe a debt to Bob Lucky, Robert Gallager, and Ross Bassett, all of whom saved me from embarrassment and offered excellent suggestions that made the text clearer and more accurate. Similar thanks go to copy editor Roland Ottewell.

On a somewhat tangential note: Every part of this book was written to music. I'm thankful for the compositions of Brad Mehldau, Radiohead, and the National. Also, to Andras Schiff, for his rendition (both books) of J. S. Bach's "The Well-Tempered Clavier," and especially BWV 881.

.   .   .

MUCH TO MY SURPRISE, family and friends offered unfeigned interest in this project for years on end. Special thanks to my brother and sister, David Gertner and Patricia Stern, and to the rest of our kin in Charlotte: Marianne Gertner (and Andrew and Jack); and David Stern (and Allie and Lexie). The San Francisco branch of the family has been great, too—and never hesitated to put me up for the night during my swings through the Bay Area. Thanks to Laura Weinstein and Mark Lorenzen (and to Lily and Joe). And thanks, too, to our Beltway office: Joey and Adrienne Weinstein (and to Aaron). Also, I am especially grateful to my in-laws, Henry and Roz Weinstein, who have been such a warm and supportive presence over the years. I am blessed to have them in my life.

Friends have helped in ways large and small over the past five years. Dave Henehan read an early version of the manuscript during his lunch breaks in Hood River, Oregon; and during his visits east, he and Kyndaron Reinier made our days and nights immeasurably brighter. J. P. Olsen kept me both amused and sane; I can't imagine how I'd get by without his companionship, or without his demo tapes. Jim Kearns and Susan Panepento have offered unwavering friendship and have fed my family on every major Christian holiday since, like, forever. I'm grateful for the food, but really for the kindness. Mike Doheny and Paul McCormick have been trusted and valued friends. Peter Carbonara lent a sympathetic ear at the Millburn Diner. Also, a special thanks to Laurie Weber, who has graced my family with her friendship. Other debts go to our Columbus Day group—Jim and Susan, Adam, Danielle, Tom, Kathy, Sean, Cathy, April, John, and all the bazillion kids. They buoyed my flagging spirits every fall, wherever it was we traveled. The Wellfleet crew—Laurie W., Lauren, Tom, Hillary, Tony, and a kajillion kids—buoyed my spirits every August. I'm grateful, too, to old friends Tim and Jim Fullowan, Dave Allegra, Tony DePaolo, and Donna Romankow.

Thanks also go to Marty Weinraub for his good advice, and to Alex Levi, who helped me see why a new metaphor might help more than an old one. The journey was indeed better than the mountain.

. . .

FINALLY, this book is dedicated to my children, Emelia and Ben, and to my wife, Lisabeth. I discovered it is not easy to be a parent, husband, journalist, and author at the same time. So to Em and Ben, I'll say this: For my distractions; for my absorptions; for my weeks away from home during reporting trips; for all the times I should have closed the laptop to pay more or better attention; for the terrible and dull repetition of all those dinners I cooked—I beg forgiveness. Your patience and affection helped more than you'll ever know these past few years. As for Lizzie, my partner in life, the debts are, I'm afraid, too great to express. For you, for you only, words fail me.

Endnotes and Amplifications

INTRODUCTION: WICKED PROBLEMS

1   David A. Kaplan, *The Silicon Boys and Their Valley of Dreams* (New York: Harper-Perennial, 2000), p. 41.
2   Arthur C. Clarke, *Voice Across the Sea* (New York: Harper & Row, 1974), p. 152.

CHAPTER ONE: OIL DROPS

1   Robert Conot, *A Streak of Luck: The Life and Legend of Thomas Alva Edison* (New York: Seaview Books, 1979), p. 86: "He had the Victorian aversion to water, and throughout his life took at most one bath a week . . . [his] appearance was often accentuated by a pungent odor of things organic and inorganic."
2   Visitors can still see some of the items in the Edison stockroom firsthand, as I did, at Edison's laboratory complex—now a national historic park—in West Orange, New Jersey.
3   M. D. Fagen, ed., *A History of Engineering and Science in the Bell System: The Early Years (1875–1925)* (Bell Telephone Laboratories, 1975), pp. 69, 75.
4   Ronald W. Clark, *Edison: The Man Who Made the Future* (New York: G. P. Putnam, 1977), p. 10.
5   Daniel Kevles, *The Physicists: The History of a Scientific Community in Modern America* (Cambridge, MA: Harvard University Press, 2001), p. 8.
6   For details on Millikan's life and experiments, I've relied on his book *The Autobiography of Robert A. Millikan* (New York: Prentice-Hall, 1950); Robert H. Kargon's *The Rise of Robert Millikan: Portrait of a Life in American Science* (Ithaca, NY: Cornell University Press, 1982); the National Academy of Sciences biographical memoir "Robert Andrews Millikan: 1868–1953," by Lee DuBridge and Paul A. Epstein, 1959; and the National Academy of Sciences biographical memoir by Fletcher's son, Stephen F. Fletcher, "Harvey Fletcher: 1884–1981," 1992.
7   Fletcher, Millikan's graduate student, almost certainly suggested the switch from water to oil and helped Millikan create the testing apparatus. Only after Fletcher's funeral, in an autobiography that was deliberately left unpublished, Fletcher would assert that Millikan showed up at his apartment just before the publication of their results in the journal *Science* to inform Fletcher that he could not earn his PhD through a cowritten paper; Millikan would thus take full credit for the primary oil-drop

write-up, he suggested, and Fletcher could earn his doctorate on a different experiment they did together—one for which he could take the sole credit. "It was obvious that he wanted to be the sole author on the first paper," Fletcher recalled. The graduate student didn't like the idea but saw that he had no real alternative. Millikan had been gracious to him in every other respect; he had essentially nurtured Fletcher's talents and made his career in physics possible. And so Fletcher, a devout Mormon, decided he could not—would not—challenge his mentor until both were dead. For reference, see Fletcher's autobiography, now in the Brigham Young archives, or the National Academy of Sciences biographical memoir by Stephen F. Fletcher, "Harvey Fletcher: 1884–1981," 1992.

8 John Brooks, *Telephone: The First Hundred Years* (New York: Harper & Row, 1976), p. 102.

9 Bell and Gray raced to the patent office on the same morning—February 14, 1876. Various historical analyses seem to point in the direction that Gray was more deserving of the patent, and that Bell's lawyers used some of Gray's advances to support the claims of Bell's invention. Nevertheless, the patent was awarded to Bell, sparking several years of litigation.

10 Brooks, *Telephone*, p. 109.

11 Vail had been president of AT&T briefly in the 1880s and had, in the decades since, traveled the world as a businessman and bon vivant.

12 Leonard S. Reich, *The Making of American Industrial Research: Science and Business at GE and Bell, 1876–1926* (New York: Cambridge University Press, 2002), p. 151.

13 Brooks, *Telephone*, p. 144. Vail's full quote reads: "Society has never allowed that which is necessary to existence to be controlled by private interest."

14 Ibid., p. 132. In Vail's 1907 annual report for AT&T, he noted he "had 'no serious objection' to public control over telephone rates, 'provided it is independent, intelligent, considerate, thorough, and just.'"

15 Louis Galambos, "Theodore N. Vail and the Role of Innovation in the Modern Bell System," *Business History Review* 66, no. 1 (Spring 1992): 95–126.

16 Albert Bigelow Paine, *Theodore N. Vail: A Biography* (New York: Harper & Brothers, 1929), p. 238. "In 1908 the publicity department of the telephone company prepared a statement setting forth the close relationship between the American Telephone and Telegraph and the associated Bell companies." Curiously, according to Paine, some phone company executives worried even then that the slogan might be evidence that the Bell System "had the characteristics of a trust."

17 Frank B. Jewett, "Carty—The Engineer and the Man," *Bell Laboratories Record*, September 1930.

18 Millikan, *The Autobiography of Robert A. Millikan*, p. 117.

19 Fagen, ed., *A History of Engineering and Science in the Bell System*, p. 264. The exact cost in 1915 was $20.70 for three minutes.

CHAPTER TWO: WEST TO EAST

1 Oliver Buckley, "Frank Baldwin Jewett: 1879–1949," National Academy of Sciences, Biographical Memoir, 1952.

2 George Eberhardt, author interview.

3 The early history of research at Bell Labs is an interesting story in its own right. It is best chronicled in two definitive papers by Lillian Hoddeson, "The Emergence of Basic Research in the Bell Telephone System, 1875–1915," *Technology and Culture* 22 (1981): 512–44; and "The Roots of Solid-State Research at Bell Labs," *Physics Today*, March 1977. I've also relied on the Bell System's internal history, *A History of Engineering and Science in the Bell System: The Early Years (1875–1925)*, edited by M. D. Fagen (Bell Telephone Laboratories, 1975), as well as Leonard Reich's *The Making of American*

Industrial Research: Science and Business at GE and Bell, 1876–1926 (New York: Cambridge University Press, 2002).

4   M. J. Kelly, "Dr. C. J. Davisson," *Bell System Technical Journal*, October 1951; Mervin J. Kelly, "Clinton Joseph Davisson: 1881–1958," National Academy of Sciences, Biographical Memoir, 1962.

5   Claude S. Fischer, *America Calling: A Social History of the Telephone to 1940* (Berkeley: University of California Press, 1992), p. 50.

6   The Willis-Graham Act followed from an important event eight years before. In 1913, AT&T proffered a document called the Kingsbury Commitment, named after its vice president Nathan Kingsbury, that made certain concessions to the government in exchange for allowing AT&T's near-monopoly to remain intact. AT&T agreed that it would cease buying up local phone companies. Also, it would offer to the independent phone companies that still existed access to its long-distance system.

7   "Telephone Co. Forms Concern for Research," *New York Times*, December 18, 1924.

8   In a private November 9, 1925, memo, Frank Jewett noted, "Duality of interest in the physical equipment and the specialized personnel for its operation bespoke joint ownership of the A.T.&T. and Western Electric in this organization and payment of its operating expense in proportion to the work done for each company." Though some functions of AT&T's Department of D & R (development and research) were to be assumed by Bell Labs upon its establishment, the department was not formally folded into Bell Labs until the 1930s. AT&T archives.

9   Reich, *The Making of American Industrial Research*. By 1930, moreover, the Bell Labs budget had grown substantially. According to "Bell Telephone Laboratories: A Description of the Laboratory Research Organization of the Bell System" (booklet, December 1930), "At the present time [the Labs'] budget of such authorized expenditures is at the rate of nineteen million dollars per year, and the personnel engaged in those activities is approximately forty-six hundred." AT&T archives.

10   Harold Arnold, "Organizing Our Researches," *Bell Laboratories Record*, June 1926.

11   M. J. Kelly, "The Manufacture of Vacuum Tubes," *Bell Laboratories Record*, June 1926.

12   Interview of Katherine Kelly by Lillian Hoddeson, July 2, 1976, Niels Bohr Library & Archives, American Institute of Physics, College Park, MD; www.aip.org/history/ohilist.

13   Interview of Dean Woolridge by Lillian Hoddeson, August 21, 1976, Niels Bohr Library & Archives, American Institute of Physics, College Park, MD; www.aip.org/history/ohilist. As it happens, in the AT&T archives I came across Harvey Fletcher's notes on potential recruits of the era. His entry on Dean Woolridge reads, "evidences good judgment and has very good personality best man [M]illikan has had in many years."

14   Charles Townes, *How the Laser Happened* (New York: Oxford University Press, 1999), p. 17.

15   Michael Riordan and Lillian Hoddeson, *Crystal Fire: The Invention of the Transistor and the Birth of the Information Age* (New York: W. W. Norton, 1997), p. 51.

16   Interview of Walter Brattain by Alan Holden and W. J. King, 1964, Niels Bohr Library & Archives, American Institute of Physics, College Park, MD; www.aip.org/history/ohilist.

17   Townes, *How the Laser Happened*, p. 35.

CHAPTER THREE: SYSTEM

1   Interview of Karl Darrow by Henry Barton and W. J. King, April 2, 1964, and June 10, 1964, Niels Bohr Library & Archives, American Institute of Physics, College Park, MD; www.aip.org/history/ohilist.

2   Richard P. Feynman, *Six Easy Pieces: Essentials of Physics Explained by Its Most Brilliant Teacher* (Reading, MA: Helix/Addison-Wesley, 1995), pp. 33–34.

3   "News of the Month," *Bell Laboratories Record*, November 1935.
4   Interview of William B. Shockley by Lillian Hoddeson, September 10, 1974, Niels Bohr Library & Archives, American Institute of Physics, College Park, MD; www.aip.org/history/ohilist.
5   F. Barrows Colton, "Miracle Men of the Telephone," *National Geographic*, March 1947.
6   According to a rundown in the *Bell Laboratories Record* of January 1941, "Telephone conversations per day averaged 78,700,000, or nearly 5,000,000 more per day than in 1939."
7   N. R. Danielian, *A.T.&T.: The Story of Industrial Conquest* (New York: Vanguard Press, 1939).
8   Arno Penzias, author interview.
9   One of Shewhart's disciples was W. Edwards Deming, who brought quality control to Japan's automobile industry.
10   Mervin J. Kelly, "Testimony Before the Public Utilities Commission of the State of California, San Francisco, May 19–21, 1947; Los Angeles, May 12, 1947." AT&T archives.
11   Richard Rhodes, *The Making of the Atomic Bomb* (New York: Simon & Schuster, 1986), p. 261.
12   William Shockley, interviewed in 1969 by Jane Morgan for the Palo Alto 75th Anniversary; http://www.youtube.com/watch?v=LWGVuoisDbI.
13   William H. Shockley, diaries, May 27, 1912, and June 5, 1912. Shockley Collection, Stanford University.
14   Interview of Dean Woolridge by Lillian Hoddeson, August 21, 1976, Niels Bohr Library & Archives, American Institute of Physics, College Park, MD; www.aip.org/history/ohilist.
15   John L. Moll, "William Bradford Shockley: 1910–1989," National Academy of Sciences, Biographical Memoir, 1995.
16   Interview of Addison White by Lillian Hoddeson, September 30, 1976, Niels Bohr Library & Archives, American Institute of Physics, College Park, MD; www.aip.org/history/ohilist.
17   Interview of William B. Shockley by Lillian Hoddeson, AIP.
18   Walter Brown, author interview.
19   William Shockley, interviewed in 1969 by Jane Morgan.
20   Interview of Walter Brattain by Alan Holden and W. J. King, 1964, Niels Bohr Library & Archives, American Institute of Physics, College Park, MD; www.aip.org/history/ohilist.

CHAPTER FOUR: WAR
1   Interview of Dean Woolridge by Lillian Hoddeson, August 21, 1976, Niels Bohr Library & Archives, American Institute of Physics, College Park, MD; www.aip.org/history/ohilist.
2   The request for uranium research came from Vannevar Bush, one of Jewett's closest friends, who was then in charge of the National Defense Research Committee (NDRC), which later became the Office of Scientific Research and Development (OSRD). Bush would also be a mentor to Claude Shannon, the Bell Labs engineer-mathematician.
3   Vannevar Bush, letter to Frank B. Jewett, May 2, 1940, Bush Papers, Library of Congress. Bush wrote, "There is a matter on which I am decidedly puzzled, and it may or may not be of great importance. This is the matter of the fission of uranium."
4   Uranium in its natural form contains mostly U-238 and only trace amounts of U-235. Most modern-day fuel for nuclear reactors is composed of slightly "enriched"

uranium—that is, U-238 mixed with about 5 percent of the more volatile and fission-able U-235. Weapons-grade uranium contains much higher percentages of U-235. U.S. Nuclear Regulatory Commission, http://www.nrc.gov/materials/fuel-cycle-fac/ur-enrichment.html.

5   Interview of James Fisk by Lillian Hoddeson, June 24, 1976, Niels Bohr Library & Archives, American Institute of Physics, College Park, MD; www.aip.org/history/ohilist.

6   Charles Townes, *How the Laser Happened* (New York: Oxford University Press, 1999), p. 39.

7   Though Mervin Kelly used the words "scramble" and "unscramble" in his description of multiplexing and quartz filters, Bob Lucky points out that those terms erroneously suggest encrypted transmissions. It might be more precise to think of the telephone conversations of the era as "stacked" atop one another during the transmission process.

8   "Artificial Quartz Is Produced Here," *New York Times*, September 19, 1947.

9   James Gleick, *Genius: The Life and Science of Richard Feynman* (New York: Vintage, 1993), p. 85.

10 Lloyd Espenschied, "Memo: Visit by a Young Investigator of the War Department," January 25, 1944, 4:50 p.m. AT&T archives.

11 Buckley and Jewett traded memos on the Espenschied incident between late January 1944 and November 1944.

12 There is no evidence as to Mervin Kelly's personal feelings, yet Kelly was the first Bell Labs executive to hire Jews—for the Labs' efforts in radar and wartime development. Meanwhile, it took many more years before the Labs executives embraced the idea of hiring African Americans or promoting women into leadership positions. Those steps toward diversity—in the research department, at least—did not come until Bill Baker's tenure.

13 T. R. Kennedy Jr., "Theory of Radar: More Information on Radio Detection Device Is Made Public," *New York Times*, April 29, 1945.

14 William S. White, "Secrets of Radar Given to World," *New York Times*, August 15, 1945; Russell Owen, "Radar Promises Peacetime Miracles, Too," *New York Times Magazine*, August 19, 1945.

15 Robert Buderi, *The Invention That Changed the World* (New York: Touchstone/Simon & Schuster, 1997), p. 126.

16 Ibid., p. 29.

17 "Radar," *Time*, August 20, 1945.

18 M. D. Fagen, ed., *A History of Engineering and Science in the Bell System: National Service in War and Peace (1925–1975)* (Bell Telephone Laboratories, 1975), p. 114.

19 Buderi, *The Invention That Changed the World*, p. 49.

20 J. B. Fisk, H. D. Hagstrum, and P. L. Hartman, "The Magnetron as a Generator of Centimeter Waves," *Bell System Technical Journal*, April 1946.

21 "Facing the 70's: An Interview with President Fisk," *Bell Labs News*, December 11, 1970.

22 Bohr and Kelly probably became close in the early 1950s, when Kelly became involved in the "Atoms for Peace" program. Kelly's son-in-law, Robert Von Mehren, described to me their friendship. Some correspondence at the Niels Bohr archive in Copenhagen verifies this. "My wife and I look forward with great pleasure to meeting you and Mrs. Kelly at the Award Ceremony in October and to seeing you both later in Boston. We shall also be happy to come and visit you in your home in Short Hills from which we have so pleasant memories." Niels Bohr to M. J. Kelly, May 2, 1957. Courtesy Niels Bohr archive, Copenhagen.

23 Paul Hartman, letter to Homer Hagstrum, November 23, 1973. AT&T archives.

24   Homer Hagstrum, letter, December 1973. Fisk collection, AIP.

25   P. L. Hartman, letter to J. B. Fisk upon his retirement. Fisk Collection, AIP.

26   Fisk collection, AIP.

27   Philip M. Morse, *In at the Beginnings: A Physicist's Life* (Cambridge, MA: MIT Press, 1977), p. 184.

28   In the Stanford archives, where Shockley's papers are housed, there are usually various notebooks and diaries for the same year but for different aspects of his life. Some are spiral-bound; others are leather-bound; others still are two-inch-by-two-inch square-bound.

29   Occasionally, Shockley's trips involved hazardous secret missions. There was, for instance, a white-knuckle November 1942 trip that Philip Morse and Shockley took to London, on a special plane built for aquatic landings and takeoffs. In his memoir, *In at the Beginnings*, Morse recalled, "We went to Lisbon in a Pan American flying boat, with lengthy stops at Bermuda and again at the Azores to refuel and to wait for the sea to be calm enough so we could take off again. We finally landed in the bay at Lisbon, in the evening, after circling for a half hour waiting for the fishing boats to get out of the way. After a day's wait in Lisbon, we were flown at night to London in a blacked-out British transport plane, to avoid being spotted by German fighters from France" (p. 191).

30   Joel N. Shurkin, *Broken Genius: The Rise and Fall of William Shockley* (New York: Macmillan, 2006), pp. 77–78.

31   O. E. Buckley, memo, January 21, 1944. AT&T archives.

32   Mervin J. Kelly, "A First Record of Thoughts Concerning an Important Post-War Problem of the Bell Telephone Laboratories and Western Electric Company," memo, 29 pages, May 1, 1943. AT&T archives.

33   William Shockley, "The Invention of the Transistor—An Example of Creative-Failure Methodology," *Electrochemical Society Proceedings* 98, no. 1 (1998): 31.

CHAPTER FIVE: SOLID STATE

1   "Big Development in Murray Hill by Bell Telephone Laboratories," *Summit (NJ) Herald*, August 1, 1930.

2   O. E. Buckley to F. Jewett, January 16, 1939. AT&T archives.

3   O. E. Buckley to Dr. Jewett, May 17, 1938. AT&T archives.

4   In my discussion of the Murray Hill complex, I've drawn on numerous memos from the late 1930s and early 1940s, a June 1943 Bell System pamphlet by Franklin L. Hunt entitled "New Buildings of Bell Telephone Laboratories at Murray Hill," and a limited-edition photographic book on the design. I've also used my own visits for description.

5   Interview of Addison White by Lillian Hoddeson, September 30, 1976, Niels Bohr Library & Archives, American Institute of Physics, College Park, MD; www.aip.org/history/ohilist.

6   A box at the AT&T archives purports to hold Kelly's files; the box is mostly empty, containing his published speeches, some papers on an obscure patent dispute, and a disintegrating leather-bound log of his incoming mail during several years of the late 1930s. Kelly's long career at Bell Labs suggests that he would have amassed dozens of boxes of archive-worthy materials, as was the case with his predecessors, Oliver Buckley and Frank Jewett. I tried for several years to locate the Kelly papers, but to no avail. Kelly's wife and children are deceased; his son-in-law had only some published technical papers and a few random Polaroids of Kelly relaxing on his patio in Short Hills, New Jersey. Neither of Kelly's former schools—the Missouri School of Mines and the University of Chicago—were bequeathed any relevant archival materials, nor was Sandia Labs or any other government institution. It is possible (but unlikely) that some Kelly papers have been given to the AT&T archives but have not yet been cataloged. The archives has warehoused numerous crates and boxes of unexamined materials

that cannot be evaluated due to limited staffing. Most likely, however, the Kelly papers have been lost or destroyed. What remains exists in the collections of various contemporaries, such as Oliver Buckley, Frank Jewett, and Ralph Bown.

7   "Authorization for Work," Case No. 38139. I actually came across two distinct work authorizations for the transistor. Both authorizations are dated January 1, 1945, but one is unsigned. The other (signed) has an approval date of June 21, 1945. The $417,000 cost was illegible in the latter and is taken from the former. AT&T archives.

8   Brattain's notebooks from 1945 to 1948 are the ur-text of the transistor's genesis. I reviewed them at the AT&T archives.

9   William O. Baker, interviewed by Marcy Goldstein and Jeffrey L. Sturchio at AT&T Bell Laboratories, Murray Hill, New Jersey, May 23 and June 18, 1985. Chemical Heritage Foundation, Philadelphia, Oral History Transcript #0013. Quoted by permission.

10   J. H. Scaff, R. O. Grisdale, and J. K. Galt, "A History of the Chemical Laboratories, Part I," internal Bell Laboratories document, p. 6. AT&T archives.

11   Frank B. Jewett, "Research Methods," *Bell Laboratories Record*, July 1928.

12   Mervin J. Kelly, Testimony Before the Public Utilities Commission of the State of California, Los Angeles, May 12, 1947. AT&T archives.

13   For my explanation of semiconductors, I'm particularly indebted to several sources: G. L. Pearson and W. H. Brattain, "History of Semiconductor Research," *Proceedings of the I.R.E.*, December 1955, Bell Telephone System Monograph 2538; K. K. Darrow, "Electronic Conductors," *Bell Laboratories Record*, May 1955; J. H. Scaff, "The Role of Metallurgy in the Technology of Electronic Materials," *Metallurgical Transactions*, March 1970; and Ernest Braun and Stuart Macdonald, *Revolution in Miniature: The History and Impact of Semiconductor Electronics* (Cambridge: Cambridge University Press, 1982).

14   This episode is described in "The Story of the Transistor," a 1958 Bell Labs pamphlet, and is explored in deeper detail in both Walter Brattain's AIP oral history with A. N. Holden and W. J. King and an AIP oral history interview of Russell Ohl by Lillian Hoddeson.

15   Scaff, "The Role of Metallurgy in the Technology of Electronic Materials."

16   Ibid.

17   Morgan Sparks, "Semiconductor Research," *Bell Laboratories Record*, June 1958, p. 193.

18   Michael F. Wolff, "The R&D 'Bootleggers': Inventing Against Odds," *IEEE Spectrum*, July 1975.

19   My rendition of the chronology of events leading up to the invention is drawn from a number of sources. Among the documents recounting the path to the transistor's invention, a few have proven more essential and more frequently quoted in the science history literature than others. These are Walter Brattain's two lengthy oral histories (one with Holden and King; the other with Weiner); an official and unpublished internal history, commissioned by the Labs in 1948 and written as a Memorandum for Record on December 27, 1949, by W. S. Gorton; William Shockley's long and detailed article "The Invention of the Transistor—An Example of Creative-Failure Methodology," *Electrochemical Society Proceedings* 98, no. 1 (1998); personal accounts written by Brattain ("Genesis of the Transistor," *Physics Teacher*, March 1968) and Ralph Bown ("The Transistor as an Industrial Research Episode," *Scientific Monthly*, January 1955); and J. H. Scaff's "The Role of Metallurgy in the Technology of Electronic Materials," *Metallurgical Transactions*, March 1970. Also essential was Charles Weiner, "How the Transistor Emerged," *IEEE Spectrum*, January 1973. I consulted the 1956 Nobel Prize speeches of Bardeen, Brattain, and Shockley, as well as a special issue of the *Bell Laboratories Record* on the transistor's tenth anniversary (June 1958), especially for Morgan Sparks's article "Semiconductor Research." Shockley's "Transistor Physics," *American Scientist*, January 1954, and Brinkman, Haggan, and Troutman's "A History

of the Invention of the Transistor and Where It Will Lead Us," *IEEE Journal of Solid-State Circuits* 32, no. 12 (December 1997), added insightful points. I took a close look at the actual first transistor, now on display at Alcatel-Lucent headquarters in Murray Hill. Also, the oral histories of Lillian Hoddeson (especially Russell Ohl, Gerald Pearson, Addison White, and John Bardeen) and Harriet Zuckerman's oral histories of Bardeen, Brattain, and Shockley proved vital, as did the private Bell Labs memos and promotional and explanatory literature published by the Bell System that I reviewed at the AT&T archives.

20 Interview of Walter Brattain by Alan Holden and W. J. King, 1964, Niels Bohr Library & Archives, American Institute of Physics, College Park, MD; www.aip.org/history/ohilist.

21 Author interviews; see also Brinkman et al., "A History of the Invention of the Transistor and Where It Will Lead Us."

22 Interview of James Fisk by Lillian Hoddeson, June 24, 1976, Niels Bohr Library & Archives, American Institute of Physics, College Park, MD; www.aip.org/history/ohilist.

23 Shockley, "The Invention of the Transistor—An Example of Creative-Failure Methodology."

24 Walter Brattain, interviewed by Harriet Zuckerman, 1964, as part of her larger study of the research careers of Nobel laureates in the sciences, Columbia Oral History Research Office. See Harriet Zuckerman, *Scientific Elite: Nobel Laureates in the United States* (New York: Free Press, 1977), and the enlarged edition (New Brunswick, NJ: Transaction Press, 1996). Quoted by permission of Dr. Zuckerman.

25 Interview of Walter Brattain by Alan Holden and W. J. King, AIP.

26 Walter Brown, author interview.

27 William Shockley, "The Invention of the Transistor—An Example of Creative-Failure Methodology," draft version, July 31, 1973. AT&T archives.

CHAPTER SIX: HOUSE OF MAGIC

1 William Shockley, "The Invention of the Transistor—An Example of Creative-Failure Methodology," *Electrochemical Society Proceedings* 98, no. 1 (1998).

2 Interview of Walter Brattain by Alan Holden and W. J. King, 1964, Niels Bohr Library & Archives, American Institute of Physics, College Park, MD; www.aip.org/history/ohilist.

3 Ibid.

4 Walter Brattain, interviewed by Harriet Zuckerman, 1964, as part of her larger study of the research careers of Nobel laureates in the sciences, Columbia Oral History Research Office. See Harriet Zuckerman, *Scientific Elite: Nobel Laureates in the United States* (New York: Free Press, 1977), and the enlarged edition (New Brunswick, NJ: Transaction Press, 1996).

5 William Keefauver, former Bell Labs patent attorney, author interview.

6 In Shockley's recollection, it was Harvey Fletcher, not Ralph Bown, who issued this challenge.

7 Walter Brattain, interviewed by Harriet Zuckerman, 1964. By permission of Dr. Zuckerman.

8 Bell Telephone Laboratories, "Memorandum for File: Terminology for Semiconductor Triodes," May 28, 1948. AT&T archives.

9 Ralph Bown, "Memorandum to Those Working on Surface States Phenomena," July 16, 1948. AT&T archives.

10 Memorandum, "BTL Confidential," May 27, 1948. AT&T archives.

11 J. H. Scaff, "The Role of Metallurgy in the Technology of Electronic Materials," *Metallurgical Transactions* (March 1970).

12 W. Shockley, letter to Robert Gibney, June 29, 1948. AT&T archives.

13 "Press Release from Bell Telephone Laboratories: A.M. Papers of Thursday, July 1, 1948." AT&T archives.

14 J. Bardeen and W. H. Brattain, "A Semi-Conductor Triode." AT&T archives.

15 Quoted in Shirley Thomas, *Men of Space*, vol. 4 (Philadelphia: Chilton Books, 1962), p. 175.

16 William Shockley, interviewed by John L. Gregory on the twenty-fifth anniversary of the transistor, April 1972.

17 Shockley, "The Invention of the Transistor—An Example of Creative-Failure Methodology."

18 Interview of Walter Brattain by Charles Weiner, May 28, 1974, Niels Bohr Library & Archives, American Institute of Physics, College Park, MD; www.aip.org/history/ohilist.

19 Shockley, "The Invention of the Transistor—An Example of Creative-Failure Methodology."

20 Bell Labs document, March 23, 1953. AT&T archives.

21 Phil Anderson, "Physics at Bell Labs, 1949–1984," unpublished. Given to the author by Anderson.

22 Interview of Addison White by Lillian Hoddeson, September 30, 1976, Niels Bohr Library & Archives, American Institute of Physics, College Park, MD; www.aip.org/history/ohilist.

23 Shockley, "The Invention of the Transistor—An Example of Creative-Failure Methodology."

24 Richard Rhodes, *The Making of the Atomic Bomb* (New York: Simon & Schuster, 1986), p. 655. Rhodes notes that the Trinity core could not have been larger "than a small orange." While he put the weight at eleven pounds, other descriptions of the plutonium core have ranged higher—up to fourteen pounds, and the size of a softball.

25 Michael Riordan and Lillian Hoddeson, *Crystal Fire: The Invention of the Transistor and the Birth of the Information Age* (New York: W. W. Norton, 1997), p. 164.

26 Letter, Jay W. Forrester to Ralph Bown, July 22, 1948. AT&T archives.

27 Letter, Ralph Bown to Jay W. Forrester, July 26, 1948. AT&T archives.

28 A discovery of similar import occurred in the early 1960s at Bell Labs, when Arno Penzias and Robert Wilson discovered the cosmic microwave radiation that remained from the Big Bang. The men later won the Nobel Prize for their work. Jansky would have almost certainly been considered for the prize, too, had he not suffered an early death, in 1950, at the age of forty-five. After Jansky's death, a lingering and insoluble disagreement arose over Bell Labs' suggestion that Jansky move on from his discovery and redirect his research to more practical matters. Jansky's brother, C. M. Jansky, has made the case that Bell management's orders not only were shortsighted but were implemented against Jansky's wishes. Jansky's supervisor Harald Friis has disagreed.

29 *Oxford English Dictionary*, 2nd ed. See "innovate" and "innovation," vol. 8, pp. 997–98. Usually, for science mandarins such as Frank Jewett, the chairman of the Labs and head of the National Academy of Sciences, the descriptive language for innovation in the World War II era, and immediately thereafter, was "ingenuity," "invention," and "development." Though it may well have been used earlier, the first reference I came across to the word "innovate" in the Bell Labs literature was in a 1958 speech by Jack Morton and in the 1959 *Bell Laboratories Record* touting the Labs' involvement in the military's Distant Early Warning (DEW) line. For further reference, see chapter 9, endnote 4.

30 Ernest Braun and Stuart Macdonald, *Revolution in Miniature: The History and Impact of Semiconductor Electronics* (Cambridge: Cambridge University Press, 1982), p. 5.

31 Ralph Bown, "The Transistor as an Industrial Research Episode," *Scientific Monthly*, January 1955.

32 Jack A. Morton, "From Research to Technology," *International Science and Technology*, May 1964.

33   Jack A. Morton, "The Innovation Process," date unknown. AT&T archives.
34   Eugene I. Gordon, "Morton's Legacy," *New Jersey Council News*, 1991; also author interviews.
35   Michael Wolff, "The R&D 'Bootleggers': Inventing Against the Odds," *IEEE Spectrum*, July 1975.
36   "Kelly's favorite room" is from an author interview with Robert Von Mehren, Kelly's son-in-law; the description of the "stately room" comes from my visit to the actual house in Short Hills.
37   Interview of Walter Brattain by Charles Weiner, AIP.
38   Ralph Bown had already explained the technical reasons for spreading the invention around. An internal Bell Labs memo written a decade after the transistor became a commercial product noted that the large semiconductor industry, "with its center of gravity outside the Bell System," was the deliberate result of a policy that the Labs' managers settled on in the months after the invention. By involving engineers around the world in the evolution of the device—making it better, cheaper, more reliable—the hope was that everyone would profit from the advances, especially the Bell System.
39   Ralph Bown, letter to M. J. Kelly, Baur au Lac Hotel, Zurich, Switzerland, August 20, 1948. AT&T archives.
40   Oliver Buckley to M. J. Kelly, Savoy Hotel, London, England, September 15, 1948. AT&T archives.
41   M. J. Kelly, "Remarks Before Bell System Lecturer's Conference," October 2, 1951. AT&T archives.
42   Milton Silverman, "Ma Bell's House of Magic," *Saturday Evening Post*, July 1947. The publication of the feature sparked an internal correspondence between Bell Labs executives, all of whom viewed the story critically.
43   William G. Pfann, "Some Remarks on the Discovery of Zone Melting," May 3, 1965. AT&T archives.
44   Mervin J. Kelly, "Semiconductor Electronics: A New Technology—A New Industry," 1958. AT&T archives.

CHAPTER SEVEN: THE INFORMATIONIST
1   Claude Shannon, Kyoto Prize acceptance speech, 1985. Shannon Collection, Library of Congress. See also Anthony Liversidge, "Profile of Claude Shannon," in *Claude Elwood Shannon Collected Papers*, reprinted (in a slightly different form) from *Omni* magazine, August 1987. Used by permission of Liversidge.
2   Vannevar Bush, letter to Professor E. B. Wilson, Harvard School of Public Health, December 15, 1938. "As he appeared to have great promise, and moreover appeared to be a decidedly unconventional type of youngster, I made it possible for him to go on with his studies." Shannon Collection, Library of Congress.
3   Vannevar Bush, recommendation for Claude Shannon for a National Research Fellowship, undated, circa late 1939. Shannon Collection, Library of Congress.
4   Erico Marui Guizzo, "The Essential Message: Claude Shannon and the Making of Information Theory" (master's thesis, MIT, 2003).
5   Len Kleinrock, a former student of Shannon's, author interview.
6   Liversidge, "Profile of Claude Shannon."
7   Claude Shannon, letter to Dr. V. Bush, December 13, 1939. Shannon Collection, Library of Congress.
8   Liversidge, "Profile of Claude Shannon." Biographical facts relating to Shannon's father are in a personal letter Shannon wrote, October 20, 1981, to Ms. Shari Bukowski: "[My father] was born in Oxford, New Jersey in 1862, came to Ovid, Michigan when very young and was raised and graduated there. He was a traveling salesman for a period

and came to Gaylord shortly after 1900. There he bought a furniture and undertaking business, and, having confidence in Gaylord's future, built the Shannon Block and Post Office building on Main Street." Shannon Collection, Library of Congress.

9   Robert McEliece, *Claude Shannon, Father of the Information Age*, directed and written by Doug Ramsey, produced by Ramsey and Mike Weber; http://www.youtube.com/ watch?v=z2Whj_nL-x8. In a profile of Shannon in his *Collected Papers* (New York: IEEE Press/John Wiley & Sons, 1993), the book's editors, N. J. A. Sloane and Aaron D. Wyner, refer to H. H. Goldstine's book *The Computer from Pascal to Von Neumann*, which described the thesis as "one of the most important master's theses ever written . . . a landmark in that it helped to change digital circuit design from an art to a science."

10 Claude Shannon, letter to Dr. V. Bush, December 13, 1939. Shannon papers, Library of Congress.

11 Norma (Levor) Barzman, author interview.

12 Claude Shannon, letter to Dr. V. Bush, March 8, 1940. Shannon Collection, Library of Congress.

13 Claude Shannon, letter to Dr. V. Bush, February 16, 1939. "Off and on I have been working on an analysis of some of the fundamental properties of general systems for the transmission of intelligence [*sic*], including telephony, radio, television, telegraphy, etc." Shannon, *Collected Papers*, p. 455.

14 Claude Shannon, oral history conducted in July 1982 by Robert Price, IEEE History Center, New Brunswick, NJ.

15 Claude Shannon, interview with Robert Price, December 20, 1983. Shannon said: ". . . a lot of information theory . . . I had worked out before, during the five years between 1940 and 1945. Much of that work I did at home." Shannon's penciled manuscripts: author's review, Shannon Collection, Library of Congress.

16 Betty Shannon, author interview.

17 T. C. Fry, oral history, unpublished. Interviewed by H. O. Pollak and D. M. La Porte. Author's copy, p. 4.

18 Ibid., p. 55.

19 Ibid., p. 33.

20 Henry Pollak, author interview.

21 M. D. Fagen, ed., *A History of Engineering and Science in the Bell System: National Service in War and Peace (1925–1975)* (Bell Telephone Laboratories, 1975), p. 172.

22 Claude Shannon, oral history with Robert Price, IEEE.

23 David Hagelbarger, author interview.

24 David Kahn, a historian of cryptography, has noted, "Shannon's insight, his great contribution to cryptology, lay in pointing out that redundancy furnishes the ground for cryptanalysis." Essentially, Shannon's analysis had "given an explanation for the constancy of letter frequency . . . he has thus made possible, for the first time, a fundamental understanding of the process of cryptogram solution." David Kahn, *The Codebreakers: The Comprehensive History of Secret Communications from Ancient Times to the Internet* (New York: Scribner, 1996), p. 748.

25 Like many innovations in the Bell System, the precise history of PCM includes competing claims. AT&T has maintained, somewhat unpersuasively, that it was an early, albeit unrecognized invention of P. M. Rainey of Western Electric in 1926. But most accounts credit the invention to A. H. Reeves of International Telephone and Telegraph in 1938. Shannon's primary contribution to the technology is in B. M. Oliver, J. R. Pierce, and C. E. Shannon, "The Philosophy of PCM," in *Collected Papers*.

26 An internal history of Bell Laboratories described two major reasons for PCM in slightly more technical terms: First, its traffic capacity could be high. The pulses and

codes of a PCM system "can be very short in duration," allowing for the "multiplexing" of other signals (audio as well as video) in the intervals between. Also, it would be easier to preserve the quality of a digital signal, since "the uniform pulse coded samples may be exactly reconstructed at regenerators along a line so that there need be no accumulation of impairments with distance, as there is in all analog systems."

27 Claude Shannon, oral history with Robert Price, IEEE.
28 "We had an internal distribution system," recalls a Bell Labs mathematician, Brock McMillan, who worked in the office next to Shannon's at Murray Hill at the time. "So people in the math department could read things before publication. Shannon was not particularly talkative. He talked to us about his theories maybe a little bit. But that 1948 paper caught me almost stone cold." Most of the Bell researchers, with the exception of Shannon's friend Barney Oliver, experienced a similar sense of shock. John Pierce likened the surprise he felt upon encountering his friend's ideas to the dropping of a powerful explosive.
29 Shannon, Kyoto Prize speech.
30 Robert Lucky, *Silicon Dreams: Information, Man, and Machine* (New York: St. Martin's Press, 1991).
31 "Information Theory," from Shannon's article on the topic in *Encyclopedia Britannica*, 14th ed., 1968.
32 John Robinson Pierce, quoted in M. Mitchell Waldrop, "Claude Shannon: Reluctant Father of the Digital Age," *Technology Review*, January 7, 2002. For my explication of information theory, I owe a debt to a number of people who have written on Shannon's theory: John Pierce, Warren Weaver, David Kahn, Robert Gallager, Bob Lucky, Neil Sloane, Aaron Wyner, John Horgan, Wiliam Poundstone, and Erico Marui Guizzo. Before me, Pierce, Poundstone, and Guizzo also made the insightful point juxtaposing the reduction of redundancy in cryptanalysis with the addition of redundancy in coding. Finally, Shannon was especially eloquent on the genesis and meaning of his own work in his speeches throughout the 1950s, which were rich with metaphors to explain his ideas; in his interviews with Robert Price and Anthony Liversidge; and (later still) in his Kyoto prize speech in 1985.
33 Richard Hamming and David Slepian are most often credited for their work on these codes.
34 Waldrop, "Claude Shannon: Reluctant Father of the Digital Age." As Bob Gallager at MIT points out, communications engineers (Fano included) had long understood that errors could be reduced by increasing the power of a transmission and/or increasing the channel capacity—that is, the "bandwidth" through which transmissions moved. Shannon's great breakthrough was in showing that they could reduce errors without doing either.
35 Shannon, *Collected Papers*; Lucky, *Silicon Dreams*, p. 37.
36 Claude Shannon, letter to Francis Bello at *Fortune*. Shannon Collection, Library of Congress.
37 Hamming complained about this in an unpublished oral history.
38 The supervisor was John Pierce, whose work is detailed later in this book. The anecdote comes from Henry Landau, a former Bell Labs mathematician. Author interview.
39 William Shockley, interviewed by Harriet Zuckerman, 1964, as part of her larger study of the research careers of Nobel laureates in the sciences, Columbia Oral History Research Office. See Harriet Zuckerman, *Scientific Elite: Nobel Laureates in the United States* (New York: Free Press, 1977), and the enlarged edition (New Brunswick, NJ: Transaction Press, 1996). Quoted by permission of Dr. Zuckerman.
40 John R. Pierce, "Mervin Joe Kelly: 1894–1971," National Academy of Sciences, Biographical Memoirs, 1975.
41 C. Chapin Cutler, oral history, IEEE. This point was also made to me in an interview with William Keefauver, a former patent attorney at the Labs.

42 Liversidge, "Profile of Claude Shannon." When asked, "Can Bell Labs take credit to some extent for your achievement?" Shannon answered, "I think so. If I had been in another company, more aimed at a particular goal, I wouldn't have had the freedom to work that way."

43 Walter Brattain conceded this in a 1963 oral history interview with Harriet Zuckerman, Columbia University Oral History Research Office.

44 Toby Berger, *Claude Shannon, Father of the Information Age*, directed and written by Doug Ramsey, produced by Ramsey and Mike Weber; http://www.youtube.com/watch?v=z2Whj_nL-x8.

CHAPTER EIGHT: MAN AND MACHINE

1 Shannon Collection, Library of Congress.

2 Claude E. Shannon, "Programming a Computer for Playing Chess," March 9, 1949, *Collected Papers*, edited by N. J. A. Sloane and Aaron D. Wyner (New York: IEEE Press/John Wiley & Sons, 1993).

3 Claude Shannon, Kyoto Prize acceptance speech, 1985. Shannon Collection, Library of Congress.

4 It was actually built of Meccano parts, rather than Erector Set parts, purchased at the Hudson Dobson store in New York City.

5 Brock McMillan, author interview.

6 William Keefauver, former Bell Labs patent attorney, author interview.

7 "Mouse with a Memory," *Time*, May 19, 1952.

8 Claude Shannon, letter to Miss Irene Angus, August 8, 1952. Shannon Collection, Library of Congress.

9 Mervin Kelly to Ralph Bown, March 24, 1950. Shannon Collection, Library of Congress.

10 Walter B. Smith to Dr. M. J. Kelly, May 4, 1951. Shannon papers, Library of Congress.

11 Francis Bello, *Fortune*, December 1953.

12 Claude E. Shannon, "Game Playing Machines," 1955, in *Collected Papers*.

13 THROBAC stands for "THrifty ROman numeral BAckward-looking Computer." Shannon later admitted that the name (and to some extent the calculator itself) was inspired by the era's computer acronyms—ENIAC, UNIVAC, and so on. Unlike some of Shannon's other machines, which are now owned by the MIT Museum, his original "ultimate machine," with the mechanical hand, has been lost, according to his wife, Betty.

14 Shannon, "Game Playing Machines," 1955, *Collected Papers*.

15 It was unclear, fifty years after the fact, which game Slepian was singling out. It may have been Shannon's version of the complex game Hex.

16 Robert Lucky, *Silicon Dreams: Information, Man, and Machine* (New York: St. Martin's Press, 1989), p. 51.

17 Claude Shannon to H. W. Bode, October 3, 1956. Shannon Collection, Library of Congress.

18 Claude Shannon, oral history conducted in July 1982 by Robert Price, IEEE History Center, New Brunswick, NJ..

19 Shannon Collection, Library of Congress.

20 "I hear via my daughter and your son that you've bought a Volkswagen microbus and are going to California." Edward Moore, letter to Claude Shannon, February 26, 1957. Shannon Collection, Library of Congress.

21 Shannon's work on cryptography with the NSA probably did not continue beyond the 1960s. Even in the 1980s, though, he was forbidden to speak of it—a hallmark of the secrecy obligations that involve any type of work with the NSA. In response to a Freedom of Information Act request by this book's author, the NSA responded that "Dr. Shannon was never an affiliate of NSA. He was, however, involved as a consultant to

NSA as a member of various committees." Documents demonstrate that while Shannon was still at Bell Labs he was at least involved with the Mathematics Panel of the National Security Agency Scientific Advisory Board (NSASAB). When asked by Bob Price during an oral history many years later about his National Security Agency contributions, Shannon replied, "I was a consultant. I probably should . . . I don't know . . . I think, I don't know that I have any . . . even though this was a long time ago, I'd better not talk about . . ." Claude Shannon, oral history with Robert Price, IEEE.

22 Claude Shannon, letter to James R. Newman, April 9, 1958. Shannon Collection, Library of Congress.
23 Bob Lucky, author interview.

CHAPTER NINE: FORMULA
1   "Dr. Kelly Visits Technical Groups in Europe," *Bell Laboratories Record*, June 1950.
2   Mervin Kelly, letter to Ralph Bown, March 24, 1950. Shannon Collection, Library of Congress.
3   Mervin J. Kelly, "The Bell Telephone Laboratories—An Example of an Institute of Creative Technology," March 23, 1950. AT&T archives.
4   It is by all means possible that a Bell Labs scientist used the word "innovation" before 1958. However, in a review of thousands of Bell Labs documents and hundreds of speeches from the 1950s, the first instance I came across was in a speech by Jack Morton, "Some Thoughts About the Future," at the tenth anniversary of the transistor, June 17, 1958. "In the past," Morton remarked, "American industry has led the world because we have used existing knowledge with ingenuity, energy and innovation." Perhaps simultaneously, John R. Pierce began using the word, too. "Innovation in Technology," an essay by Pierce, appeared in the September 1958 issue of *Scientific American*. Moreover, a draft of a paper of Pierce's later titled "Myths of Creation," dated September 10, 1958, was originally titled "On Innovation in Science."
5   Ernest Braun and Stuart Macdonald, *Revolution in Miniature: The History and Impact of Semiconductor Electronics* (Cambridge: Cambridge University Press, 1982), p. 33.
6   John Rowell, author interview.
7   Some of the most acclaimed Bell Labs scientists would often stand amazed by the talents of their technical assistants. For Morris Tanenbaum, for instance, it was Ernie Buehler, who was adept at growing crystals.
8   Phil Anderson, author interview.
9   Sara Silver, "With Uncertain Future, Bell Labs Turns to Commerce," *Wall Street Journal*, August 21, 2006.
10 Mervin J. Kelly, *Bell Telephone Magazine*, Summer 1953.
11 William T. Golden, "Conversations with Drs. Oppenheimer, Robert Bacher, and Charles Lauritsen," Memorandum for the file, December 21, 1950. "We commented on the volume of his [Kelly's] work and they said that he seems to be working himself to death although he does not look as badly they said as he did during the war." Golden's search for a presidential science advisor involved numerous interviews with Kelly and his peers—Oppenheimer, DuBridge, and so forth. These interviews are chronicled in his contemporaneous memoranda, available at http://archives.aaas.org/golden.
12 Interview of Katherine Kelly by Lillian Hoddeson, July 2, 1976, Niels Bohr Library & Archives, American Institute of Physics, College Park, MD; www.aip.org/history/ohilist.
13 Author visit to 2 Widemere Terrace, Short Hills, New Jersey.
14 Joe Parisi, the Kellys' former gardener, author interview.
15 Interview of Katherine Kelly by Lillian Hoddeson, AIP.
16 Eugene Gordon, author interview.

17 William T. Golden, Memorandum for the file, December 19, 1950. "[H]e is next in line to become President upon the expected retirement of Dr. Oliver Buckley in about a year and a half. It was completely clear to me, in fact, Buckley said so, that Kelly is the man to succeed him."

18 William T. Golden, oral history, 1989, conducted by Niel M. Johnson. Truman Library; http://www.trumanlibrary.org/oralhist/goldenw.htm.

19 John R. Pierce, "Mervin Joe Kelly: 1894–1971," National Academy of Sciences, Biographical Memoir, 1975.

20 Ibid.

21 William T. Golden, Executive Office of the President, Memorandum for the file, December 4, 1950: "Kelly spends about one quarter of his time on the United States Air Force research and development matters . . . another one quarter of his time on various Government matters . . . [t]he remaining half of his time is devoted to Bell Tel Laboratories." American Association for the Advancement of Science, History and Archives.

22 William T. Golden, oral history, 1989.

23 William T. Golden, memoranda. Golden wrote on December 19, 1950, "It is clear that Kelly would be a most excellent man for scientific adviser to the President in view of the practicality and directness of his approach, his comprehensive knowledge of the military services, and his wide experience in various phases of research and development matters. . . . It is also clear that he does not want the job, that he is most certainly an empire builder, that it would be a sacrifice to him in terms of his status at the Bell Tel. Labs where he is next in line to become president upon the expected retirement of Dr. Oliver Buckley in about a year and a half."

24 Anthony Lewis, "A.T.&T. Settles Antitrust Case; Shares Patents," *New York Times*, January 25, 1956.

25 Leroy Wilson, letter to David E. Lilienthal, Chairman, U.S. Atomic Energy Commission, July 1, 1949. "It is my understanding that you have discussed with Mr. Clark, the Attorney General. . . . I also understand that you intend to acquaint the President with the situation." AT&T archives.

26 Necah Stewart Furman, *Sandia National Laboratories: The Postwar Decade* (Albuquerque: University of New Mexico Press, 1990).

27 "Pertinent Information—Sandia," AEC memo, May 17, 1949. AT&T archives.

28 Furman, *Sandia National Laboratories: The Postwar Decade.*

29 Harry Truman, letter to Leroy A. Wilson, May 13, 1949. "I am writing a similar note direct to Dr. O. E. Buckley," Truman noted. AT&T archives.

30 David E. Lilienthal, letter to Mervin J. Kelly, July 13, 1949. AT&T archives.

31 "Nike Added to Nation's Defense Arsenal," *Bell Laboratories Record*, February 1954.

32 Mervin Kelly, "The Contribution of Industrial Research to National Security," paper presented at the annual meeting of the American Association for the Advancement of Science, Boston, December 29, 1953. AT&T archives.

33 These innovations included a radio technique called "tropospheric scatter" and "over-the-horizon" radio transmission.

34 Remarks by Dr. M. J. Kelly, "Before Bell System Lecturers' Conference," October 2, 1951. AT&T archives.

CHAPTER TEN: SILICON

1 Francis Bello, "The Year of the Transistor," *Fortune*, March 1953.

2 Ernest Braun and Stuart Macdonald sum up the military's reasons concisely: "The military in the early fifties was not primarily concerned with price. Instead it was more concerned with such matters as availability, whether performance could match military specifications, and whether semiconductor devices would be reliable. . . . One considerable attraction of the new devices was the reduction in weight they offered,

both in the components themselves and in the batteries of power supplies required." Ernest Braun and Stuart Macdonald, *Revolution in Miniature: The History and Impact of Semiconductor Electronics* (Cambridge: Cambridge University Press, 1982), p. 70.

3   The automatic routing device, a fabulously complex piece of machinery, was known as the "card translator"; it used both a phototransistor and a current-amplifying transistor.

4   Mervin J. Kelly, "The First Five Years of the Transistor," *Bell Telephone Magazine*, Summer 1953.

5   Remarks by Dr. M. J. Kelly, "Before Bell System Lecturers' Conference," October 2, 1951. AT&T archives.

6   Long before the transistor came along, there is some historical evidence that if a Bell Labs innovation seemed a threat to company revenue it could be suppressed by upper management at AT&T. One instance, described by technology historian Mark Clark, is the Labs' work on magnetic recorders during the 1930s. While it seems likely that the Bell engineers in that period developed the best magnetic recording devices in the world, Clark concludes that AT&T "management feared that the availability of a recording device would make customers less willing to use the telephone system." Recording devices, for instance, by providing a permanent record of phone conversations, could threaten a loss of privacy. They might also make managers less likely to use the phone system to conduct business and sensitive negotiations. The development work on magnetic recorders was discontinued after World War II. I came across no similar examples after the war—indeed almost the opposite seemed to hold true. Innovations that should have been discontinued (such as the waveguide and Picturephone) were not; meanwhile, innovations that seemed a conceivable threat to wireline service, such as cellular telephony, were funded.

7   "The Bell Is Ringing," *Time*, May 29, 1964.

8   There were some good reasons to favor germanium over silicon, too. For instance, electrons move faster within a germanium crystal than within silicon. Moreover, some of the essential advantages of silicon—that it could grow a surface oxide layer that makes it ideal for etching and fabrication—were not yet clear.

9   Morris Tanenbaum, author interview.

10 "DuPont Says It Can Make Pure Silicon, May Find Use in Transistors," *Wall Street Journal*, May 11, 1953.

11 Calvin S. Fuller, interviewed by James J. Bohning, Vero Beach, Florida, April 29, 1986, Chemical Heritage Foundation, Philadelphia, Oral History Transcript 0020. Used by permission of the Chemical Heritage Foundation.

12 As Fuller went on to explain in his oral history, the taint within the germanium crystals being grown at Bell Labs "was a curious property of germanium that nobody seemed to understand called 'thermal conversion.' . . . I found out that it had to do with the way people were handling the crystals when they etched and washed them. If one took very great pains to make very pure water, better than conductivity type, and then looked for this thermal effect in the germanium crystal, it did not happen. That is, if you were good enough to use extremely pure conditions of preparation, the crystal was so sensitive that if you went into a laboratory and grabbed the doorknob and then happened to lightly touch the crystal, it would convert; that is it would change type from n to p on subsequent heating above about 500 degrees C. With the highly purified water as a post-etchant rinse the effect would not occur." Ibid.

13 D. M. Chapin, C. S. Fuller, and G. L. Pearson, "The Bell Solar Battery," *Bell Laboratories Record*, July 1955.

14 Interview of G. L. Pearson by Lillian Hoddeson, August 23, 1976, Niels Bohr Library & Archives, American Institute of Physics, College Park, MD; www.aip.org/history/ohilist.

15 Chapin, Fuller, and Pearson, "The Bell Solar Battery."

16 Calvin S. Fuller, interview by James J. Bohning.
17 Ibid.
18 According to the front page of the *New York Times*.
19 Interview of Gerald Pearson by Lillian Hoddeson, AIP.
20 John Perlin, *Space to Earth: The Story of Solar Electricity* (Ann Arbor, MI: Aatec Publications, 1999), p. 36. Perlin also points out that in 1956 Chapin put the cost of a one-watt cell at $286.
21 Remarks by Dr. M. J. Kelly, "Before Bell System Lecturers' Conference," October 2, 1951.
22 Eugene O'Neill, ed., *A History of Engineering and Science in the Bell System: Transmission Technology (1925–1975)* (New York: AT&T Bell Laboratories, 1985), p. 187.
23 "Radio Phone Line Links East, West," *New York Times*, August 18, 1951.

CHAPTER ELEVEN: EMPIRE
1   I have borrowed the quote, with apologies, as it appears in *Voice Across the Sea*, a history of transatlantic communications by Arthur C. Clarke (New York: Harper & Row, 1974), p. 39. The British royal astronomer (or astronomer royal, as Clarke called him) was Sir George Airy.
2   Mervin J. Kelly, "Oliver Ellsworth Buckley: 1887–1959," National Academy of Sciences, Biographical Memoir, 1964, p. 14.
3   Clarke, *Voice Across the Sea*, p. 141.
4   M. J. Kelly, Sir Gordon Radley, G. W. Gilman, and R. J. Halsey, "A Transatlantic Telephone Cable," Bell Telephone System technical publications monograph, 1954–1955.
5   Kelly, "Oliver Ellsworth Buckley," p. 15.
6   Kelly, Radley, Gilman, and Halsey, "A Transatlantic Telephone Cable."
7   In August 1954, for instance, the Bell System closed down a twenty-five-year test on phone poles—buried halfway in the "adobe-type" soil—in Limon, Colorado. "Limon was chosen as the site for the tests because of its location in a typically dry, 'dust-bowl' climate noted for its severe exposure conditions." *Bell Laboratories Record*, August 1954, p. 290. This issue of the *Record* also noted that there was a phone pole test in Gulfport, Mississippi, that had been ongoing since 1925, making it even older than the Colorado investigation.
8   Eugene O'Neill, ed., *A History of Engineering and Science in the Bell System: Transmission Technology (1925–1975)* (AT&T Bell Laboratories, 1985), p. 342.
9   In reacting to a journal paper Kelly had written about Bell Labs, for instance, Shockley, citing deficiencies with laboratory supplies available to the scientists, chided Kelly. According to Shockley, Kelly had displayed "a glaring lack of knowledge . . . a tone expressed . . . which implies a lack of knowledge of shortcomings that is surprising." Shockley Collection, Stanford University.
10 Ian Ross, author interview.
11 In her book *The Man Behind the Microchip: Robert Noyce and the Invention of Silicon Valley* (New York: Oxford University Press, 2005), Leslie Berlin offers additional details on the "intense" courtship between Terman and Shockley. As Berlin notes, "Terman envisioned a symbiotic relationship in which technically oriented companies would support advanced research at Stanford while at the same time benefiting from a supply of well-educated graduates and professors interested in consulting work" (p. 57).
12 T. R. Reid, *The Chip: How Two Americans Invented the Microchip and Launched a Revolution* (New York: Random House, 2001), p. 87.
13 John Pierce, "Mervin Joe Kelly: 1894–1971," National Academy of Sciences, Biographical Memoir, 1975.
14 William Shockley, letter to Mervin Kelly, November 20, 1956. Shockley Collection, Stanford University.

15 "What the Anti-Trust Suit Consent Decree Means," January 31, 1956. Issued by the Public Relations Department, American Telephone and Telegraph Company, "For Administrative People of AT&T General Departments." No author. AT&T archives.

16 Eileen Shanahan, "Company Replies—'Astonished,' It Says, Warns a Break-up Would Raise Rates," *New York Times*, November 21, 1974.

17 Peter Temin, with Louis Galambos, *The Fall of the Bell System* (New York: Cambridge University Press, 1987), p. 15.

18 Mervin J. Kelly, "Semiconductor Electronics: A New Technology—A New Industry," June 17, 1958.

19 William H. Whyte Jr., *The Organization Man* (New York: Doubleday Anchor, 1957), p. 230.

20 Francis Bello, "The World's Greatest Industrial Laboratory," *Fortune*, November 1958.

21 Robert Gallager, MIT, author interview.

22 Max Mathews, author interview.

CHAPTER TWELVE: AN INSTIGATOR
1 John R. Pierce, "The Wings of the Wind," in *Science, Art, and Communication* (New York: Clarkson Potter, 1968), p. 17.

2 John R. Pierce, interviewed by Harriet Lyle, Pasadena, California, April 16, 23, 27, 1979. Oral History Project, California Institute of Technology Archives. Used with permission.

3 William C. Jakes, author interview.

4 Steven M. Spencer, "Dial 'S' for Satellite," *Saturday Evening Post*, January 14, 1961.

5 Henry Landau, author interview.

6 John Pierce, oral history conducted by Andy Goldstein, IEEE History Center, New Brunswick, NJ, August 19–21, 1992. "My father was a good man. He was a quiet man, not brilliant, not dumb. My mother had a sharper mind than he did."

7 John R. Pierce, *My Career as an Engineer: An Autobiographical Sketch* (University of Tokyo, 1988), p. 54.

8 John Pierce, oral history conducted by Andy Goldstein.

9 Ibid.

10 The instructor was S. S. Mackeown, whom Pierce also credited for getting him a job at Bell Laboratories.

11 Pierce was an active and effective self-promoter. His profile in the *New Yorker*, which focused on his communication satellite work, was published a few years after Pierce and his deputy, Rudi Kompfner, sent a long, unsolicited letter (August 12, 1959) to the editor of the magazine, suggesting the editor assign someone to cover the events leading up to the Echo satellite launch. "We have little doubt that the news media will dish out the usual exaggerated, distorted and unbalanced accounts, regardless of the accuracy of the data supplied to them," Kompfner wrote. He then suggested that a reporter from the august publication, by spending significant amounts of time with the Bell Labs team, could produce an account of the satellite experiment "which will be all that the usual newspaper accounts are not, something written by grown-ups for grown-ups, and which may in due course become a document of our time." An editor at the magazine (Eugene Kinkead, letter to Rudi Kompfner, September 9, 1959) responded positively a few weeks later: "We may well want to look into these and, if we decide they're our dish, you should be hearing before long from one of our writing staff." AT&T archives.

12 Chuck Elmendorf, author interview.

13 John R. Pierce, interview by Harriet Lyle. Used by permission.

14 Pierce himself later acknowledged this to Elmendorf, who was amused by the concession.

15 Pierce, *My Career as an Engineer: An Autobiographical Sketch*.

16   The close colleague was William O. Baker, who made the observation at Pierce's retirement.

17   John Pierce, oral history conducted by Andy Goldstein.

18   Calvin Tomkins, "Woomera Has It!," *New Yorker*, September 21, 1963.

19   John Pierce, oral history conducted by Andy Goldstein.

20   Pierce's various explanations for his derivation of the word "transistor" are fairly technical and not wholly clarifying. In his interview with Lincoln Barnett, for instance, he said, "There are already resistors and thermistors in the Bell System . . . there is a term 'conductance,' and vacuum tubes have 'transconductance.' Well, there's a term 'resistance.' The transistor is supposed to be the dual of the vacuum tube, so if you had transconductance, why shouldn't you have transresistance? So I thought we should call it the transistor."

21   "An interview of Dr. J. R. Pierce by Mr. Lincoln Barnett for the American Telephone & Telegraph Company," February 13, 1963. AT&T archives.

22   W. S. Shockley, daily planner, December 31, 1947. Shockley files, Stanford University. Reviewed by author.

23   Clarke dedicated *Voice Across the Sea*, his book on overseas communications, "To John Pierce, who suggested this book and bullied me into writing it."

24   Clarke made an extended visit to Bell Labs in the late 1950s. In *My Career as an Engineer*, Pierce writes, "A great deal of effort went into generating samples representing speech. At Bell Labs, Max Mathews, John Kelly, Peter Denes, Jim Flanagan and others were pioneers in such work. As a part of this work, a computer sang 'A Bicycle Built for Two' ('Daisy, Daisy, give me your answer true . . .'). Arthur Clarke heard this, and the computer Hal sang the song in the movie *2001*."

25   As it happened, Pierce was also commanding a team in the lab next door to Jim Fisk during the Second World War. While Fisk was doing his wartime work on magnetrons for radar, Pierce was working on a related technology known as klystrons. One night, Pierce filled Fisk's magnetrons with tiny scraps of paper laced with thyme oil, so that when Fisk's group gave supervisors a demonstration of the magnetrons the next day, the first step of the demonstration—blowing cold air through the devices—resulted in a blizzard of perfumed confetti.

26   H. L. McDowell, "The Traveling-Wave Tube Goes to Work," *Bell Laboratories Record*, June 1960.

27   Walter Brown, author interview.

28   J. R. Pierce and H. D. Hagstrum, "Report on Visit to British Electronics Laboratories." Pierce Collection, Stanford Special Collections archive.

29   T. R. Kennedy Jr., "New Tube Expands Radio Possibilities," *New York Times*, July 6, 1946.

30   Ernest Braun and Stuart Macdonald, *Revolution in Miniature: The History and Impact of Semiconductor Electronics* (Cambridge: Cambridge University Press, 1982).

31   "I wanted to write for *Astounding Science Fiction*; I had found I could sell factual articles, and I didn't want to annoy the Publications Department or have them annoyed with me by sending all this junk through for release." "An Interview of Dr. J. R. Pierce by Mr. Lincoln Barnett for the American Telephone & Telegraph Company."

32   Dennis Flanagan, managing editor, *Scientific American*, letter to John Pierce, March 3, 1949. Pierce Collection, Huntington Library.

33   Pierce was rejected, repeatedly, by most of the major magazines: *Harper's*, *Time*, the *Atlantic*, the *New York Times Magazine*, and so forth.

34   Pierce, *My Career as an Engineer*.

35   John Pierce, oral history conducted by Andy Goldstein.

36   J. R. Pierce, "Visit with H. G. Wells." Pierce typed up his handwritten notes from the meeting in November 1944 on July 8, 1971—"a transcription of an account that Mr. Pierce wrote shortly afterward." Pierce Collection, Huntington Library.

CHAPTER THIRTEEN: ON CRAWFORD HILL

1   Simon Flexner quoted in Ron Chernow, *Titan: The Life of John D. Rockefeller, Sr.* (New York: Random House, 1998), p. 474.

2   Charles H. Townes, Patent No. 2,879,439, March 24, 1959. The device emits radiation when the molecules drop from a high-energy state to a lower-energy state. Townes's device was known as a two-level maser; in subsequent years, physicists developed a three-level maser and replaced the gas component with solid-state materials. Many of these improvements were attributed to a team led by the Harvard physicist (and later Nobel laureate) Nicolaas Bloembergen.

3   The scientists at the Labs most closely associated with the maser work were H. E. D. Scovil and R. W. DeGrasse. Nicolaas Bloembergen (see above) had come up with a workable idea for a solid-state maser slightly ahead of the Bell Labs team. In a long private memo (September 14, 1956) Rudi Kompfner detailed how "a rumor" reached the Bell Labs team of Bloembergen's invention; what followed was an invitation by Kompfner to Bloembergen to advise Bell Labs on their maser work. The precipitating reason seems to have been the avoidance of a patent infringement suit at a later date. Bloembergen declined a formal consultancy with the Labs but agreed to an "understanding" with Bell Labs regarding a future patent license.

4   H. E. D. Scovil, "A Three-Level Solid-State Maser," *Bell Laboratories Record,* July 1958.

5   Calvin Tomkins, "Woomera Has It!," *New Yorker,* September 21, 1963.

6   John R. Pierce, *My Career as an Engineer: An Autobiographical Sketch* (University of Tokyo, 1988).

7   Ibid.

8   "An Interview of Dr. J. R. Pierce by Mr. Lincoln Barnett for the American Telephone & Telegraph Company," February 13, 1963. AT&T archives.

9   Chuck Elmendorf, author interview.

10   John R. Pierce, memo to Mr. J. A. Morton, April 14, 1959. AT&T archives.

11   Advent, after burning up more than $170 million, according to Pierce, never got off the ground. As it happened, though, the first communications satellite test—at least in a technical sense—was Project Score, launched by the Army Signal Corps in December 1958; it was an Atlas rocket, outfitted with a short recorded message by President Eisenhower that was transmitted to earth as it orbited. Score burned up upon a planned reentry the following month.

12   John R. Pierce, *My Career as an Engineer.*

13   Estill I. Green, memo to Mervin Kelly, July 25, 1958. The memo suggested that a satellite system "at best . . . would probably yield barely acceptable transmission performance as contrasted with the excellent performance obtainable with submarine cable. . . . The satellite system would be a major undertaking. With so many uncertainties in the picture no accurate estimate of research and development cost can be had but comparison with other large system developments would suggest a minimum cost of $20,000,000 and a possibility of a considerably higher figure." The memo also predicted, incorrectly, that the "propaganda value" (i.e., the public relations boost) from a communications satellite would be minimal.

14   The term was coined years after Kelly's reign, by Harvard professor Clayton Christensen.

15   J. R. Pierce and R. Kompfner, letter to M. J. Kelly, 8 October 1958. In Kompfner's handwriting at the top it reads: "Copy given to J.R.P. who showed it to W.O.B. Not Sent." J.R.P. is Pierce; W.O.B. is William (Bill) Baker. AT&T archives.

16   J. R. Pierce and R. Kompfner, "Proposal for Research Toward Satellite Communication," January 6, 1959. AT&T archives.

17   Hugh L. Dryden, Deputy Administrator, NASA, letter to Mr. R. Kompfner, January 13,

1959. "At the present time it is our plan to launch a 100-foot balloon sometime in the early fall." AT&T archives.

18 Bill Jakes and Mary Jakes, author interview. Also see P. B. Findley, "The Laboratories in Monmouth County," *Bell Laboratories Record*, March 1950.

19 In fact it was named Crawford's Hill, though it was almost always referred to as Crawford Hill.

20 Costs for the antennas from "Project Echo Meeting" memo, by W. C. Jakes, August 20, 1959. AT&T archives

21 W. C. Jakes, "Project Echo Conference Report," June 16, 1959. AT&T archives.

22 "News Release, Project Echo Payload and Experiment," Release No. 60-186, National Aeronautics and Space Administration.

23 Mary Jakes, author interview.

24 Calvin Tomkins, "Woomera Has It!," *New Yorker*, September 21, 1963.

25 Bill Jakes, author interview.

26 "A Different Drummer," *Time*, August 22, 1960.

27 "BALLOON SATELLITE ORBITS; RELAYED MESSAGE HERALDS NEW COMMUNICATIONS ERA" was the headline of a front-page story in the following day's *New York Times*. The issue also contained a "Man in the News" profile of Pierce.

28 "Project Echo: Chronology 8/12/60–4/24/62," undated memo. AT&T archives.

29 Steven M. Spencer, "Dial 'S' for Satellite," *Saturday Evening Post*, January 14, 1961.

30 John R. Pierce, "Telstar: A History," SMEC Vintage Electrics 1990; www.smecc.org/john_pierce1.htm.

31 Lawrence Lessing, "Laying the Great Cable in Space," *Fortune*, July 1961.

32 "Interview with Frederick R. Kappel, President of AT&T," *U.S. News & World Report*, July 10, 1961.

33 The meetings for what was called the "Active Satellite Program" were begun in the spring of 1960. Pierce as well as Kompfner were attendees. Memoranda by L. C. Tillotson, March 8, 1960, and April 11, 1960. AT&T archives.

34 Eugene O'Neill, oral history conducted in 2001 by David Hochfelder, IEEE History Center, New Brunswick, NJ.

35 Ibid.

36 J. R. Pierce, *The Beginnings of Satellite Communications* (San Francisco: San Francisco Press, 1968).

37 John W. Finney, "Nation Sees Tests: Experimental Device Launched by Space Agency for AT&T," *New York Times*, July 11, 1962.

38 Richard Kostelanetz, *Master Minds: Portraits of Contemporary American Artists and Intellectuals* (New York: Macmillan, 1969).

39 John Pierce, oral history conducted by Andy Goldstein, IEEE History Center, New Brunswick, NJ, August 19–21, 1992.

40 A deeper discussion of the political tensions between AT&T and NASA, along with a broader overview of the history of modern satellites, navigation, and tracking, can be found in Helen Gavaghan's *Something New Under the Sun: Satellites and the Beginning of the Space Age* (New York: Copernicus/Springer-Verlag, 1998).

41 Pierce was not of the belief, however, that all Bell Labs work was automatically superior to the competition. In active satellites, for instance, he admired the work done by Harold Rosen at Hughes Aircraft, which in 1963 and 1964 put up a series of high (22,300-mile) geosynchronous satellites known as Syncom.

42 For instance, W. S. Shockley—Journal, Tuesday, 4 October 1955. Shockley archives, Stanford University. Reviewed by author.

43 Pierce was often a visitor, according to interviews with Shannon's wife, Mary. Atypically for Shannon, he would actually write back to Pierce. A March 8, 1957, letter

from Shannon to Pierce reads, "We are all excited here about a) our new house on the Mystic Lake, a tremendous Thomas Jefferson Type establishment with built in swimming, boating, skating, skiing and arboretum, b) a year (next year) in California at the Behavioral Institute in Palo Alto, of all places, c) a new Microbus we are outfitting for camping to make the trip this summer to California. Why don't you come up to Boston sometime and see these things (also us) and make sidetrips to the M.I.T. labs., etc?" Pierce Collection, Huntington Library.

44   John Pierce, oral history conducted by Andy Goldstein.
45   John R. Pierce, March 9, 1961, letters to Copland and Bernstein. Pierce Collection, Huntington Library.
46   Cronkite interviewed Pierce for a January 29, 1967, show, "The Communications Explosion." Transcript in the Pierce Collection, Huntington Library.
47   John R. Pierce, letter to Harald Friis, June 6, 1957. Pierce Collection, Huntington Library.
48   "Death of a Satellite," *Bell Laboratories Record*, July 1968.
49   Bill Jakes, author interview.

CHAPTER FOURTEEN: FUTURES, REAL AND IMAGINED
1   An internal AT&T memo noted that one of the main reasons for participating in the New York show was "to take advantage of the opportunity to create, at a reasonable cost, a favorable image of the Bell System in the minds of millions of people." AT&T archives.
2   Some of the Bell System exhibits in Seattle, as well as the company's strenuous public relations efforts, are captured in the blithe Bell-funded short film "Century 21 Calling," posted at http://www.youtube.com/watch?v=S4Iu3JEsoQY&feature=PlayList& p=15491C1FCC325BC4&playnext_from=PL&playnext=1&index=16.
3   John R. Pierce, letter to W. E. Kock, August 20, 1956. Pierce Collection, Huntington Library.
4   "Study of Picturephone Service, 1964 World's Fair," American Telephone and Telegraph Company, Business Research Division, June 1965. AT&T archives.
5   From the start of telephone service in the 1880s, it was understood that each phone could not be connected to every other phone and that therefore a central switching office would be essential; a formula for figuring out the number of interconnections—$n(n-1) \div 2$—demonstrates why. Even a small group of n callers—100, say—necessitates nearly 4,950 possible combinations, meaning 4,950 wires. But it increases exponentially after that. A phone system of 10,000 subscribers, for instance, where each subscriber had a line to each other, would necessitate nearly 50 million lines.
6   The first call was made through ESS No. 1 in Succasunna, New Jersey, on May 27, 1965.
7   "First Call Made Through Electronic Central Office in Succasunna," *195 News Bulletin*, May 28, 1965. AT&T archives.
8   Arthur Albiston, memo, "Guidelines for Talking and Writing About #1 ESS," from "Public Relations Program for the Introduction of No. 1 ESS," November 30, 1964. AT&T archives.
9   Robert Alden, "A Shift to All-Electronic Phones Begun in Biggest Step Since Dial," *New York Times*, September 20, 1964.
10   James B. Fisk, text of statement submitted to the U.S. House of Representatives Subcommittee on Science, Research and Development, December 11, 1963.
11   Julius P. Molnar, "BTL's Response to Competitive Thrust: Talk to Western Electric Spring Conference," May 19, 1971. AT&T archives.
12   Bill Fleckenstein, author interview.

13   Ed David, author interview.

14   These comments were repeated, almost verbatim, in several author interviews, including those with Ian Ross and Henry Pollak.

15   Irwin Dorros, author interview.

16   Ian Ross, author interview.

17   Helen M. Baker, *Turkeys: Common Sense Theories, Practical Management, Incubation and Brooding in Detail, Feeding Directions, Feeding Formulas,* 2nd rev. ed. (Baltimore, 1933).

18   William O. Baker, interviewed by Marcy Goldstein and Jeffrey L. Sturchio, Chemical Heritage Foundation oral history, May 23 and June 18, 1985. Used by permission.

19   On the Baker farm, at least, the diseases affecting their turkeys were coccidiosis and blackhead.

20   William O. Baker, interviewed by Marcy Goldstein and Jeffrey L. Sturchio.

21   Helen M. Baker, *Turkeys.*

22   William O. Baker, "Materials: The Grand Alliance of Engineering and Science," Dedication of Center for Materials Science and Engineering, MIT, October 1, 1965.

23   William O. Baker, "James Brown Fisk," American Philosophical Society, 1988.

24   Dean Gillette, memo, "Replacement of Lead Sheath by Polyethylene and Copper Wire by Optical Fibers," February 14, 1979. Baker Collection, Princeton University.

25   Calvin Tomkins, "Woomera Has It!," *New Yorker,* September 21, 1963.

26   Ken Reese, "World War II: The Rubber Shortage," *Today's Chemist at Work,* June 1992; also see William O. Baker, interviewed by Marcy Goldstein and Jeffrey L. Sturchio.

27   Baker apparently knew a great deal about Pierce's family life—his handwritten notes taken during meetings and phone calls with Pierce through the 1960s follow the course of Pierce's drawn-out and painful divorce, during which Baker served as a sounding board and informal counsel. Baker Collection, Princeton University.

28   Baker Collection, Princeton University.

29   Bill Baker, telegram to William Shockley, November 1, 1956. Shockley Collection, Stanford University.

30   William O. Baker, letter to James B. Fisk, August 21, 1963. Baker Collection, Princeton University.

31   Will Lepkowski, interview with Dr. Frederick Seitz, March 2006. From "Final Report: The William O. Baker Papers and Biography," booklet. Baker Collection, Princeton University.

32   William O. Baker, National Reconnaissance Office Oral History, conducted by R. Cargill Hall, May 7, 1996.

33   William O. Baker, Q&A, *Chemical and Engineering News,* November 25, 1996.

34   There exists a lack of clarity regarding the commission's official members and evolution. A photo provided by Baker to an interviewer from the National Reconnaissance Office shows the members "circa 1957" as being Hendrik Bode, Oliver Selfridge, Nathan Rochester, David Huffman, Luis Alvarez, Richard Garwin, William Friedman, William Baker, and John Pierce. In *The Puzzle Palace* (New York: Penguin, 1983), author James Bamford's exposé of the National Security Agency, he cites Claude Shannon and Andrew M. Gleason as members but not Nathan Rochester (p. 429).

35   "Charge to a Study of Foreign Intelligence Gathering," May 13, 1957. The quote I've included reflects some of the inserted text that Baker had noted in the margins of his draft. AT&T archives.

36   Bamford, *The Puzzle Palace,* pp. 427–29.

37   *New Scientist,* January 30, 1975, p. 274.

38   Baker's handwritten notes. Baker Collection, Princeton University.

39   William O. Baker, National Reconnaissance Office Oral History. "I was sitting around

the Cabinet room during the Cuban missile crisis of October 1962," Baker would recall, and ultimately served as a "legman," going back and forth between the White House and State Department, relaying information about the Russian fleets steaming toward Cuba with their cargo.

40   Clark Clifford, "Serving the President," *New Yorker*, May 6, 13, 20, 1991. According to both Baker and Clifford, the two men were eating lunch together in the White House mess when they were informed that Kennedy had been shot in Dallas.

41   Louis Tordella, letter to William O. Baker, January 15, 1959. Baker Collection, Princeton University.

CHAPTER FIFTEEN: MISTAKES

1   Ian Ross, author interview.

2   William O. Baker, "Transistor Making and Mind Stretching," speech given at the transistor's twenty-fifth anniversary, Western Electric, Allentown, Pennsylvania, December 5, 1972. AT&T archives.

3   William O. Baker, Murray Hill, New Jersey, June 21, 1966. In his closing remarks at a two-day Bell Labs symposium entitled "The Human Use of Computing Machines," Baker said, "Thus our nation, and to some degree mankind altogether, have become trustees for one of the heroic capabilities so far evolved from science and engineering—that of the giant computing machine and its associated networks of interactions and man-machine complementarities."

4   Jack Kilby, oral history with Michael Wolff, December 2, 1975, IEEE Global History Network. For a deeper discussion of how Morton and Bell Labs failed to perceive the integrated circuit, see T. R. Reid's *The Chip*, and Michael Riordan, "How Bell Labs Missed the Microchip," *IEEE Spectrum*, December 2006.

5   William O. Baker, letter to Jim Fisk, July 11, 1958. Baker Collection, Princeton University.

6   In Kilby's device, the different parts of the circuit—transistors and resistors and capacitors—were connected by wires rather than "etched" connections in the silicon, as was the case with Noyce's design.

7   Arthur L. Schawlow, *Optics and Laser Spectroscopy, Bell Telephone Laboratories, 1951– 1961, and Stanford University Since 1961*, an oral history conducted in 1996 by Suzanne B. Riess, Regional Oral History Office, Bancroft Library, University of California, Berkeley, 1998.

8   The first laser patent—which happens to be filed casually in a box buried in the AT&T archives—resulted in a ferocious and long-standing litigation. Gordon Gould, a former colleague of Charles Townes, laid a competing and (ultimately) partially successful claim to the laser.

9   Thanks to Herwig Kogelnik for pointing this out. The Schawlow and Townes March 1960 patent—No. 2,929,922—is titled "Maser and Maser Communications System." In later years, it would become popular to say that the laser, during its early incarnation, was a solution looking for a problem—a quip apparently made by a colleague of Ted Maiman's at Hughes Aircraft. In his bestselling book *The Black Swan* (New York: Random House, 2007), for instance, Nassim Nicholas Taleb asserts that the laser had "no real purpose." The evidence contradicts this.

10   Joan Lisa Bromberg, *The Laser in America, 1950–1970* (Cambridge, MA: MIT Press, 1991), p. 92.

11   The reason that a laser is so much more suited to carrying information than ordinary light is that it is highly ordered (coherent) and monochromatic.

12   "Research Breakthroughs in Optical Masers and Superconductors," *Bell Laboratories Record*, March 1961. In technical terms, Javan and colleagues "impressed a telephone conversation on a maser signal . . . by using an electro-optical device—the Kerr Cell."

13   Herwig Kogelnik, author interview.

14   An undated explanatory flyer about the Bell laser research explained, "When Picture-phone service becomes common, when high-speed data communication between computers is more widespread, and when all of today's communications services have expanded, present message carrying capacities may not be enough." AT&T archives.

15   Stewart Miller, "Communication by Laser," *Scientific American*, January 1966.

16   C. G. B. Garrett, "The Optical Maser," *Electrical Engineering* (April 1961): 248–51.

17   The waveguide pipes, as Jeff Hecht points out in *City of Light* (New York: Oxford University Press, 1999), his history of optical communications, were also sensitive to temperature changes and vibrations, not to mention seismic activity.

18   One possible solution to keeping a light beam focused within the pipe seemed to be the insertion of gas "lenses" inside the tubes that would continually refocus the light wave as it moved along. Still, waveguide transmissions might still be ruined by fluc-tuations in air temperature, ground vibrations, and tremors (see above). The problem appeared to be vexing.

19   Hecht, *City of Light*, p. 111.

20   Tingye Li, author interview. In Li's view, Kao deserves enormous credit for three separate aspects that hastened the introduction of fiber optics. The first was proposing fused silica as the material for the glass strands. "The second thing was that [Kao] went ahead and measured a very low loss to the material. And the third was, he went around the world to get people to work on this. He came to Bell Labs, and he went to Corning."

21   Ira Jacobs, author interview.

22   Memo, "Work Forecast," S. E. Miller to R. Kompfner and J. R. Pierce, July 18, 1969. As Stew Miller put it to his bosses, "Our current mainstream effort is centered around laser transmission systems. Primary effort has been on the transmission media, the following being of prime interest: 1) Glass-lens (beam) waveguides . . . ; 2) Gas-lens waveguides; 3) Glass fiber waveguides." The memo goes on to detail the Labs' efforts at all three and acknowledges its work on fiber was lagging behind the work at Corning, which had been described to the Bell Labs' team through "a private communication."

23   William O. Baker, "Testimony of William O. Baker," May 31, 1966. F.C.C. Docket No. 16258.

24   Memo, R. Kompfner to J. K. Galt, November 30, 1970. AT&T archives.

25   Unix was created primarily by Ken Thompson, Dennis Ritchie, Brian Kernighan, Douglas McIlroy, and Joe Ossanna.

26   Willard Boyle and George Smith were credited as inventors of the CCD device and were awarded the 2009 Nobel Prize—to be shared with Charles Kao, the early cham-pion of fiber optics. Immediately after Boyle and Smith won the award, Eugene Gor-don, another Bell Labs veteran, challenged the credit they received and claimed that some of the ideas for the device originated with him. And as for making the device usable in a camera? That credit would probably go to Michael Tompsett, another Bell Labs engineer, who worked on developing the actual invention with Carlo Sequin and Ed Zimany. A series of articles in *Spectrum*, the magazine of the IEEE (the Institute of Electrical and Electronics Engineers), ably summarized the dispute: http://spectrum .ieee.org/tech-talk/semiconductors/devices/nobel-controversy-former-bell-labs-employee-says-he-invented-the-ccd-imager. Without question, the unsettled contro-versy highlights the inadequacy of the Nobel awards in singling out a small number of individuals for industrial advances where authorship and credit could be more broadly dispersed. The transistor award that went to Shockley, Bardeen, and Brattain—but not to any of the material scientists who fabricated the silicon and germanium—is arguably another case in point.

27   William O. Baker, National Reconnaissance Office Oral History, conducted by R. Car-gill Hall, May 7, 1996. There is some confusion as to the date that Baker brought the CCD to Washington. He claims in this interview, probably mistakenly, that it was 1963. Almost certainly it was several years later.

28 In missing the invention of the integrated circuit, however, Bell Labs and AT&T were not at all precluded from using the idea. Patent agreements guaranteed that the Labs would have access to the technology—though not the glory.

29 By several estimates, the Picturephone cost nearly half a billion dollars to develop.

30 J. P. Molnar, memo, "PICTUREPHONE Program," August 19, 1966. Baker Collection, Princeton University.

31 "Exploratory Study of the Market Potential for Picturephone Service," American Telephone & Telegraph Company, Business Research Division, December 1967. AT&T archives.

32 James B. Fisk, speech given at the Midwest Research Institute, Kansas City, Missouri, May 22, 1968. Whether or not the technological hunch behind the Picturephone was correct, Bell Labs executives could rightly point to the device as a significant feat of engineering, one that indirectly yielded not only the CCD device but a variety of new applications in integrated circuits and (during the manufacturing process) ruby lasers.

33 To some extent, there was an expectation that in preparing the Bell System for Picturephones, Bell Labs was also laying the groundwork for high-speed network services for businesses. "The switching and transmission capacity provided for Picturephone service will also enable us to move into flexible high-speed message data services which customers could use as easily as they place an ordinary call today," Ray Ralston, the Picturephone's engineering chief, remarked (*195 Magazine*, November–December 1967). Indeed, the Picturephone was able to connect with a computer and thereby provide rudimentary data to its user.

34 Julius Molnar, "Technical Program of Bell Laboratories: Talk to Department Heads," March 2, 1972. AT&T archives.

35 One of the more curious things about Picturephones emerged from research in John Pierce's communications science department. Tests showed that it was easier to lie to someone during a Picturephone conversation than in an ordinary telephone conversation. The researchers concluded that when not distracted by a visual image, a caller attends more closely to the content and veracity of the exchange.

36 I'm indebted to Bob Lucky for this comparison. More precisely, Metcalfe's law suggests that the value of a network grows in proportion to the square of its number of users.

37 John R. Pierce, "Rudolf Kompfner: 1909–1977," National Academy of Sciences, Biographical Memoir, 1983.

CHAPTER SIXTEEN: COMPETITION

1 The album of black-and-white photographs is part of the Shockley Collection, housed at the Stanford University archives. In addition to Kelly, Fisk, Shockley, Bown, Baker, Bardeen, Brattain, Molnar, and Pierce, other men at the party included Cal Fuller and Gerald Pearson, who would collaborate nine years later on the solar silicon cell. On the dais with Fisk was Harald Friis, the Danish-born antenna wizard who ran the Holmdel lab and served as a mentor to John Pierce and his Caltech friend Chuck Elmendorf. Friis had been instrumental in work leading to the construction of the microwave long-distance network.

2 Joe Baker, author interview.

3 Peter Temin with Louis Galambos, *The Fall of the Bell System: A Study in Prices and Politics* (New York: Cambridge University Press, 1987), p. 10.

4 Steve Coll, *The Deal of the Century: The Breakup of AT&T* (New York: Atheneum, 1986), p. 59.

5 The post-1956 legal tangles between AT&T, the U.S. Department of Justice, and the FCC are compelling but highly involved; moreover, because those legal tangles were often unrelated (or indirectly related) to the science and engineering of Bell Labs, they are not fully detailed here. Some of the competition to the Bell System dates back to

the 1950s and 1960s, to cases involving two independent companies and their devices, Hush-a-Phone and Carterfone. Hush-a-Phone was an attachment to a handset's mouthpiece that allowed customers to talk or whisper into the phone with more confidentiality; Carterfone was a device that allowed mobile radios to be tied into the landline network. AT&T contended that both could conceivably harm the network, an assertion that was ultimately proven unfounded. Eventually, based on court appeals and FCC decisions, both devices were allowed. A number of books offer a useful presentation of the issues, players, and implications. See *The Deal of the Century*, above, for instance, or *The Fall of the Bell System*, also above. A concise and valuable summary is also given in the introduction to the three-volume set *Decision to Divest: Major Documents in U.S. v. AT&T, 1974–1984*, edited by Christopher H. Sterling, Jill F. Kasle, and Katherine T. Glakas (Washington, DC: Communications Press, 1986).

6   MCI was started by the entrepreneur Jack Goeken. McGowan joined in the late 1960s and soon became its singular, driving force. Goeken was forced out in the early 1970s.

7   The 1971 FCC decision was known as *Specialized Common Carrier*. It followed on the heels of several earlier decisions that opened the door for MCI. These were *Above 890* (1959), which allowed for private companies to build "point-to-point" microwave systems, such as those that might connect one corporate branch to another; *Carterfone* (1968), which allowed non-AT&T equipment to be connected to the network; and *Microwave Communications Inc.* (1969), which allowed MCI to build a point-to-point microwave service between Chicago and St. Louis.

8   Complaint, November 20, 1974, *United States of America v. American Telephone and Telegraph Company; Western Electric Company, Inc.; and Bell Telephone Laboratories, Inc.*, pp. 11–14. As published in *Decision to Divest: Major Documents in U.S. v. AT&T, 1974–1984*. MCI also initiated its own private antitrust suit against AT&T.

9   John deButts, letter to John Pierce, March 31, 1975. DeButts had seen an article that Pierce wrote in the *Money Manager*. Pierce Collection, Huntington Library.

10 Mervin J. Kelly, memo, "A First Record of Thoughts Concerning an Important Post War Problem of the Bell Telephone Laboratories and Western Electric Company," May 1, 1943. AT&T archives.

11 *Bell Labs News*, "Special Report: An Interview with W. O. Baker," May 1973.

12 Rudi Kompfner, letter to Mr. J. R. Pierce, April 13, 1970. AT&T archives.

13 I. Hayashi and Mort Panish developed the first breakthrough room-temperature semiconductor laser at Bell Labs. A Russian team made a simultaneous discovery. Shortly thereafter, in 1971, Herwig Kogelnik and C. V. Shank invented the first distributed feedback laser.

14 Morton B. Panish, "Heterostructure Injection Lasers," *Bell Laboratories Record*, November 1971.

15 Along with the new lasers, a group of Bell Labs scientists also discovered that they could make light-emitting diodes into a satisfactory source for light-pulse communications. These were tiny semiconductor sandwiches, too, but were slightly different than lasers (the light was not as bright, for instance). The upside was that the diodes promised to last longer. The LEDs, too, could be modulated—that is, impressed with a communications signal. They blinked at rates of up to 100 million times per second.

16 Bernard C. DeLoach Jr., "On the Way: Lasers for Telecommunications," *Bell Laboratories Record*, April 1975. The ends of the crystal laser were polished to a mirrorlike finish so that some of the energy emitted from the laser could be sent back to stimulate even more radiated light.

17 Rudi Kompfner, handwritten note on memo from Dean Gillette, October 20, 1971. AT&T archives.

18 A good overview of the process for making fiber during that era is given in the *Bell System Technical Journal*, July–August 1978.

19   Eugene O'Neill, ed., *A History of Engineering and Science in the Bell System: Transmission Technology (1925–1975)* (AT&T Bell Laboratories, 1985), p. 665.
20   Ira Jacobs, "Lightwaves System Development: Looking Back and Ahead," *Optics & Photonics News*, February 1995.
21   John R. Pierce, letter to Dr. W. O. Baker, November 23, 1976. Pierce Collection, Huntington Library.
22   O'Neill, *A History of Engineering and Science in the Bell System: Transmission Technology (1925–1975)*, p. 401. In 1929, "commercial service to ships on the high seas was inaugurated . . . the first ship equipped was the *SS Leviathan*, the largest ship in the world at the time, and the first shore stations were at Ocean Gate and Forked River, New Jersey."
23   W. R. Young, "Advanced Mobile Phone Service: Introduction, Background, and Objectives," *Bell System Technical Journal*, January 1979.
24   D. A. (Donald) Quarles, "Organization of Mobile Radio Development," memo to M. J. Kelly, April 16, 1945. AT&T archives.
25   William C. Jakes, ed., *Microwave Mobile Communications* (New York: IEEE Press, 1993).
26   D. H. Ring, "Mobile Telephony—Wide Area Coverage—Case 20564," December 11, 1947. Though W. R. Young is not credited as an author of the nine-page memo, he is credited in the text: "As pointed out by W. R. Young in his report . . . the best general arrangement of frequency assignments for the minimum interference and with a minimum number of frequencies is a hexagonal layout in which each station is surrounded by six equidistant adjacent stations." AT&T archives.
27   "Hearing Before Federal Communications Commission on Opening of UHF Band to Television, Docket No. 8976: Statement of Dr. O. E. Buckley, President of Bell Telephone Laboratories, Inc." The statement is undated, but a companion letter to Buckley, written by John Gepson (October 26, 1950), summarized the FCC proceedings: "Hearings on the Bell Laboratories petition . . . commenced on June 5, 1950, and continued for five days thereafter. . . . The witnesses were Dr. Buckley, Mr. Gilman, Mr. Hanselman and Mr. Ryan." AT&T archives.
28   Richard H. Frenkiel, "Creating Cellular: A History of the AMPS Project (1971–1983)," *IEEE Communications Magazine*, September 2010.
29   "An interview of Dr. J. R. Pierce by Mr. Lincoln Barnett for the American Telephone & Telegraph Company," February 13, 1963. AT&T archives. Pierce said, "And people—sometimes I think of science as really bringing the one thing—suppressing the individual—those two things—it suppresses him through television, but it frees him through the telephone, if you wish."
30   Also in 1964, the Bell System began offering "Improved Mobile Telephone Service," or IMTS. The improvements over the existing system were technical—you could dial a number directly, for instance. But the system's capacity was still limited.
31   "High Capacity Mobile Telephone System: Summary," unsigned memo, September 14, 1965, 9 pages. AT&T archives.
32   C. H. Elmendorf, letter to Mr. D. Gillette, March 13, 1967. AT&T archives.

CHAPTER SEVENTEEN: APART

1   H. J. Wallis, "The Holmdel Laboratories," *Bell Laboratories Record*, April 1961. Saarinen designed Holmdel between 1957 and 1959. He died before it was completed. The building was intended to have reflective glass, which was not yet available when it opened. It was meanwhile covered in gray glass.
2   Ibid. As Saarinen put it, "Emerging from concentration in laboratory or office, the individual will come upon the sweeping, uninterrupted views of gently rolling hills and formal planting and of the winter-garden interior court. At such moments of relaxation, walking down the periphery corridor, he can feel refreshed by the encounter with these views and really appreciate them." *Bell Laboratories Record*, April 1961.

3 Richard H. Frenkiel, *Cellular Dreams and Cordless Nightmares: Life at Bell Laboratories in Interesting Times*, 2009. Downloaded from http://www.winlab.rutgers.edu/~frenkiel/ dreams.

4 Richard Frenkiel, author interview.

5 Frenkiel was also given an important 1960 Bell Labs paper on cellular service by W. D. (Deming) Lewis, H. J. Schulte, and W. A. Cornell.

6 Joel Engel, author interview.

7 "Notes on Cellular Mobile Telephone Service," AT&T internal memo (unsigned), November 27, 1979, distributed to F. H. Blecher, R. H. Frenkiel, J. L. Troe, and J. T. Walker. "We have spent approximately 100 million dollars developing and demonstrating the feasibility of cellular technology capable of nationwide application in serving the public need." In 2010 dollars, the cost would therefore have been about $300 million. What's more, the actual costs of developing cellular were undoubtedly higher, since service was not approved until several years later. AT&T archives.

8 As Frenkiel notes in his book *Cellular Dreams and Cordless Nightmares*, the ESS used "stored program control." In essence it functioned like a computer. Unlike the crossbar switches of earlier generations, it did not have "fixed" logic. It could be programmed for a variety of functions.

9 Jakes focused, in part, on studying what's known as "diversity." It involves what happens to transmission and reception when multiple antennas are spaced closely together.

10 William C. Jakes, ed., *Microwave Mobile Communications* (New York: IEEE Press, 1993), p. 11.

11 "Nike Zeus," *Bell Laboratories Record*, March 1963. The *Record* puts the size of the horseshoe-shaped island at 819 acres, "up to one-half of a mile wide and about 2½ miles long. It has a tropical marine climate which is fairly constant most of the year, since it is only about 8 degrees above the equator. Annual rainfall averages 102 inches; both the relative humidity and temperature average a steady 82."

12 Gerry DiPiazza, author interview.

13 DiPiazza worked on something known as "discriminatory radar." The technology was meant to identify an enemy warhead amid an obscuring cloud of tiny foil decoys.

14 Within a few years, Motorola would propose a competing system and would develop the first portable, handheld cell phone. The Motorola handset, developed by Martin Cooper, is a good example of how technological leaps are often perceived reductively. The handset invention demonstrated that cellular receivers could be portable and handheld, as many people at Bell Labs—John Pierce especially—had long imagined. But without the development of the cellular system, Cooper's important advances would have had little impact.

15 These were not transmission "trunks" connecting intercity routes, nor were they transmission "loops" connecting subscribers to local switching offices. Rather, they were something of a hybrid: "metropolitan trunks," for what Ira Jacobs calls "backbone routing."

16 In Atlanta, Bell Labs would also, by agreement, test Corning's fiber.

17 Ira Jacobs, author interview.

18 W. O. Baker, letter to Dr. John Pierce, January 24, 1977. Pierce Collection, Huntington Library.

19 In addition to a demonstration of the cellular *market*, the FCC also wanted a deeper demonstration of cellular *technology*. For this, Bell Labs had set up a working cellular system in Newark, New Jersey. There were no customers, but there was a working cellular system, again being run by Gerry DiPiazza, who drove around in a retrofitted trailer home to sharpen his understanding of urban noise and cellular signals. "Most of the research was done in the middle of the night," DiPiazza says, "because the traffic was so bad in the urban center." Sometimes DiPiazza would find himself in

a rough neighborhood at 2 a.m. with a street gang banging on his car. The Newark system was especially helpful in exploring how to create statistical rules—based on environment and topography, for instance—for start-up cellular systems. At the same time, DiPiazza and his colleagues sought to work out mathematical algorithms so that computer software could decide when, and how, to switch a signal from one cell to another.

20 *Decision to Divest: Major Documents in U.S. v. AT&T, 1974–1984,* edited by Christopher H. Sterling, Jill F. Kasle, and Katherine T. Glakas (Washington, DC: Communications Press, 1986), pp. 1–15.
21 "Behind AT&T's Change at the Top," *Business Week,* November 6, 1978.
22 Judge Harold Greene, "Address, Consumer Federation of America," February 17, 1984. AT&T archives.
23 In my discussion of vertical and horizontal integration, I am indebted to *The Fall of the Bell System: A Study in Prices and Politics,* by Peter Temin with Louis Galambos (New York: Cambridge University Press, 1987).
24 Ibid., p. 252.
25 "Bell Labs at Fifty," *Bell Telephone Magazine,* January–February 1975, p. 14.
26 Andrew Pollack, "Two Settlements May Widen the Pressures on Competition," *New York Times,* January 9, 1982.
27 Christopher Byron, "Stalking New Markets," *Time,* January 25, 1982.
28 "Bell Labs: Imagination Inc.," *Time,* January 25, 1982.
29 Judge Harold Greene, "Address, Consumer Federation of America," February 17, 1984.
30 Peter F. Drucker, "Beyond the Bell Breakup," *Public Interest,* Fall 1984.

CHAPTER EIGHTEEN: AFTERLIVES
1 M. J. Kelly, letter to William O. Baker, February 9, 1959. Baker Collection, Princeton University.
2 Kelly's résumé, updated in December 1965, lists his posts as a director at Prudential Insurance and as a director at the optical company Bausch & Lomb (1959–63) and vacuum tube maker Tung-Sol Electric (1959–64).
3 Interview of Katherine Kelly by Lillian Hoddeson, July 2, 1976, Niels Bohr Library & Archives, American Institute of Physics, College Park, MD; www.aip.org/history/ohilist.
4 John Pierce, oral history conducted by Andy Goldstein, IEEE History Center, New Brunswick, NJ, August 19–21, 1992.
5 William Baker, letter to John Pierce, January 29, 1973. Pierce Collection, Huntington Library.
6 James Fisk, letter to John Pierce, February 1, 1973. Pierce Collection, Huntington Library.
7 A. O. Beckman, letter to Dr. William Shockley, September 3, 1955. In addition to the $30,000 per annum (about $237,000 in 2009 dollars) the letter also stated that "Beckman agrees to grant to Shockley an option to purchase 4000 shares of Beckman stock under its existing restricted stock option plan." Shockley Collection, Stanford University.
8 "John Bardeen, Walter Brattain, William Shockley, interviewed at the Sheraton-Park Hotel in Washington, D.C.," conducted by John L. Gregory, April 24, 1972. AT&T archives.
9 Moore's law describes how the number of transistors on computer chips doubles every two years or so. See http://www.intel.com/technology/mooreslaw.
10 Gordon Moore, "Solid-State Physicist," *Time,* March 29, 1999.
11 "Is Quality of U.S. Population Declining? Interview with a Nobel Prize–Winning Scientist," *U.S. News & World Report,* November 22, 1965.

12   John L. Moll, "William Bradford Shockley: 1910–1989," National Academy of Sciences, Biographical Memoirs, 1995.

13   Shockley's sensitivity had apparently been evident from a very young age, back when his father had recorded in his journal that "Billy always gets angry because he is thwarted or denied something." Shockley Collection, Stanford University.

14   Victor Cohn, *New York Post*, April 25, 1968, p. 79.

15   William Shockley, "Proposed Research to Reduce Racial Aspects of the Environment-Hereditary Uncertainty," April 24, 1968. Shockley Collection, Stanford University.

16   W. Shockley, memo to F. D. Leamer and J. A. Morton, "Reduction of Consulting Time for Bell Telephone Laboratories," April 26, 1968. Shockley Collection, Stanford University.

17   Rae Goodell, *The Visible Scientists* (Boston: Little, Brown, 1977), p. 192.

18   Baker Collection, Princeton University. The note is from October 14, 1981. Baker wrote, "I said I can't think of any source of funds."

19   Syl Jones, "Playboy Interview: William Shockley, a Candid Conversation with the Nobel Prize Winner—in Physics—About His Theories on Black Inferiority and His Donation of Sperm for a 'Super Baby,'" *Playboy*, August 1980.

20   "Shockley Runs to Air Theories on Genetics," Associated Press, from the *Camarillo (CA) News*, May 27, 1982.

21   "John Bardeen, Walter Brattain, William Shockley, interviewed at the Sheraton-Park Hotel in Washington, D.C.," conducted by John L. Gregory, April 24, 1972. AT&T archives.

22   Anthony Liversidge, "Profile of Claude Shannon," in *Claude Elwood Shannon Collected Papers*, reprinted (in a slightly different form) from *Omni* magazine, August 1987. Quoted by permission of Liversidge.

23   Whether it was a sign of his loosening up or his lifelong disregard for social norms, during the 1960s Shannon also had a brief reunion with his first wife, Norma. According to Barzman's autobiography, the two met in a hotel in Boston, where Barzman's daughter (by her second husband) was attending college. "Why did you leave me?" Shannon asked her angrily. Eventually, he relaxed and told her about his life: "I have a nice wife, wonderful kids. I teach, do research. I have a collection of twenty-three cars. I tinker." Later, Barzman relates, "We went upstairs to my room and made love." Norma Barzman, *The Red and the Blacklist: The Intimate Memoir of a Hollywood Expatriate* (New York: Nation Books, 2004).

24   "A Conversation with Claude Shannon," Robert Price and Claude Shannon, December 20, 1983. Shannon papers, Library of Congress.

25   Baker was particularly aware of the late-in-life financial travails of Lee De Forest, the inventor of the vacuum tube. Baker told his colleagues that he never wanted such a fate to befall someone like Shannon.

26   Shannon also believed a smart gambler could take advantage of certain inefficiencies in gaming systems, an idea that led him to travel to various Nevada casinos around this time with a mathematician and investor named Ed Thorp. Their intent was to beat the house in roulette and cards. Many of these exploits are detailed in William Poundstone's book *Fortune's Formula: The Untold Story of the Scientific Betting System That Beat the Casinos and Wall Street* (New York: Hill & Wang, 2005).

27   Arthur Lewbel, author interview.

28   Claude Shannon, "Scientific Aspects of Juggling," in *Collected Papers*, edited by N. J. A. Sloane and Aaron D. Wyner (New York: IEEE Press, 1993).

29   Arthur Lewbel, "A Personal Tribute to Claude Shannon," http://www2.bc.edu/~lewbel/Shannon.html.

30   Shannon still rarely answered letters or spoke with reporters. The exception was when he received a letter about chess or juggling, which he would often answer promptly, even if the correspondent was a high school student.

31   John Horgan, "Claude E. Shannon: Unicyclist, Juggler and Father of Information Theory," *Scientific American*, January 1990.
32   Rudi Kompfner, note to John Pierce, June 1, 1977. Pierce Collection, Huntington Library.
33   John Pierce, letter to Rudi Kompfner, July 5, 1977. Pierce Collection, Huntington Library.
34   Edward E. David Jr., Max V. Mathews, and A. Michael Noll, "John Robinson Pierce: 1910–2002," National Academy of Sciences, Biographical Memoir, 2004.
35   As Pierce remarked, the scale has "frequency ratios 3:5:7:9 rather than the frequency ratios of the major tetrad, 4:5:6:8; the new scale uses tones with odd partials only, and a 'tritive' with a frequency ratio 3 replaces the octave of frequency ratio 2." John R. Pierce interview by Harriet Lyle, Pasadena, California, April 16, 23, 27, 1979. Oral History Project. California Institute of Technology archives.
36   John Pierce, *My Career as an Engineer: An Autobiographical Sketch* (University of Tokyo, 1988).
37   Program notes to the Pierce concert. Pierce Collection, Stanford University archives.
38   William O. Baker, letter to Clark Clifford, April 22, 1991. Baker Collection, Princeton University.
39   On the contrary, the FCC seemed to have clear objectives, even if they weren't to Baker's liking. The simply stated goal of the Telecommunications Act of 1996, for instance, was "to let anyone enter any communications business—to let any communications business compete in any market against any other"; http://transition.fcc.gov/telecom.html.
40   William O. Baker, interview with Michael Noll.

CHAPTER NINETEEN: INHERITANCE
1   John S. Mayo, "Evolution of the Intelligent Telecommunications Network," *Science* 215 (February 12, 1982).
2   John R. Pierce, letter to A. Michael Noll, June 18, 1986.
3   Christopher Byron, "Stalking New Markets," *Time*, January 25, 1982.
4   Andrew M. Odlyzko, "The Decline of Unfettered Research, revised version," October 4, 1995, http://www.dtc.umn.edu/~odlyzko/doc/complete.html.
5   Today's AT&T is a different organization than the AT&T that existed in the mid-1990s. As the Southern Bell Corporation—originally one of the regional operating companies—grew in wealth and subscribers, it acquired its former parent company in 2005. Ultimately, it adopted its name as well.
6   Ron Insana, "Can He Save Lucent? Henry Schacht Comes Out of Retirement to Stop the Bleeding," *Money*, October 2001.
7   Kenneth Chang, "Panel Says Bell Labs Scientist Faked Discoveries in Physics," *New York Times*, September 26, 2002.
8   William Speed Weed, "The Way We Live Now: 10-13-02: Questions for Paul Ginsparg," *New York Times Magazine*, October 13, 2002.
9   Sara Silver, "With Its Future Now Uncertain, Bell Labs Turns to Commerce," *Wall Street Journal*, August 21, 2006.
10   Geoff Brumfiel, "Bell Labs Bottoms Out," *Nature*, August 20, 2008.

CHAPTER TWENTY: ECHOES
1   Brian Hayes, *Infrastructure: A Field Guide to the Industrial Landscape* (New York: W. W. Norton, 2005), p. 283.
2   The Crawford Hill lab has a curious history. As Saarinen's Black Box was being planned for the Holmdel site in the late 1950s, about sixty Bell Labs researchers in Holmdel—who had been working in the modest woodframe labs slated for replacement—splintered off. Led by Rudi Kompfner, they succeeded in getting funding from Bill

Baker to build a small, three-story lab at the base of nearby Crawford Hill, the site of the Echo satellite experiment, a few miles away. The researchers at the new lab focused on microwave and lightwave communications.

3 Press release, Intel Corporation, October 19, 2010.

4 Hendrik Hertzberg,"Open Secrets," *New Yorker*, August 2, 2010. Hertzberg's column summarized many points in the multipart *Washington Post* story "Top Secret America," by Dana Priest and William M. Arkin; http://projects.washingtonpost.com/top-secret-america.

5 Nicholas Carr, "Is Google Making Us Stupid?" *Atlantic*, July–August 2008.

6 M. J. Kelly, "Remarks Before Bell System Lecturer's Conference," October 2, 1951.

7 Jon Gertner, "Mad Scientist," *Fast Company*, February 2008.

8 John Pierce, Testimony, Subcommittee on Communication of the Senate Commerce Committee, March 22, 1977. Pierce Collection, Huntington Library.

9 "Rising Above the Gathering Storm, Revisited: Rapidly Approaching Category 5," National Academy of Sciences, 2010. The figure citing that 4 percent of the nation's workforce creates jobs for the other 96 percent is credited to the National Science Board, *Science and Engineering Indicators 2010*.

10 John Pierce, memo, February 16, 1968, in response to a question about why the Bell system got things done profitably and without disaster. Pierce Collection, Huntington Library.

11 John R. Pierce, "Mervin Joe Kelly: 1894–1971, National Academy of Sciences, Biographical Memoir, 1975.

12 Among academics, the widely accepted notion that Bell Labs was exclusively a "closed" innovation model—whereby every innovation is started and developed within the firm, as opposed to a more modern "open" innovation model, whereby a parent company depends collaboratively on customers and outside innovators for ideas—deserves more consideration. Without question, AT&T depended upon its vertical integration for innovations. But Bell Labs' staffers were open to a stream of outside ideas through regular contacts with other industrial labs, military contractors, academic affiliates, patent licensing partners, and participants in weekly Bell Labs symposiums. Managers also acquired superior technologies from outside when warranted. Examples of the latter are the vacuum tube repeater (from Lee De Forest), the three-level maser (from Harvard professor Nicolaas Bloembergen), and optical fiber (from Corning). A more typical example: During John Pierce's first week of work at the Labs he was directed to visit Philo Farnsworth's television shop in Manhattan to see if there was anything useful to license for the telephone company. Nothing that day caught Pierce's interest.

13 For a thorough examination of Terman's work on the New Jersey innovation hub, see Stephen B. Adams, "Stanford University and Frederick Terman's Blueprint for Innovation in the Knowledge Economy," in Sally H. Clarke, Naomi R. Lamoreaux, and Steven W. Usselman, eds., *The Challenge of Remaining Innovative: Insights from Twentieth Century American Business* (Stanford, CA: Stanford Business Books, 2009).

14 "An Interview of Dr. J. R. Pierce by Mr. Lincoln Barnett for the American Telephone & Telegraph Company," February 13, 1963. AT&T archives.

15 Claude Shannon, oral history conducted in July 1982 by Robert Price, IEEE History Center, New Brunswick, NJ.

16 Dick Frenkiel also made the case, as did many others I interviewed, that present telecommunications and Internet costs would likely be higher had the Bell System remained intact. Nevertheless, it is difficult to broadly contend that telecommunications is, at present, "cheap." Some technologies (Gmail, Skype) undoubtedly are; others (monthly iPhone plans with unlimited data) arguably are not.

17 John R. Pierce, letter, January 24, 1997. Pierce Collection, Huntington Library.

18 See also Steven Johnson, "Innovation: It Isn't a Matter of Left or Right," *New York Times*, October 30, 2010.

19 Fred Block and Matthew Keller, "Where Do Innovations Come From?" University of California–Davis, 2008. The authors looked at the provenance of the top 100 innovations (a list compiled annually by *R&D Magazine*) in the years 1970 to 2006. The authors noted that "the U.S. federal government's role in fostering innovation—both in terms of organizational auspices and funding—across the U.S. economy has significantly expanded in the last several decades." In 2006, "77 of the 88 U.S. entities that produced award-winning innovations were beneficiaries of federal funding."

20 F. B. Jewett, "Modern Research Organizations and the American Patent System" (New York: Bell Telephone Laboratories Incorporated, 1932).

21 "Missed Calls: AT&T Inventions Fueled Tech Boom, and Its Own Fall," *Wall Street Journal*, February 2, 2005.

22 Janelia is part of the larger efforts of the Howard Hughes Medical Institute. A recent paper on the performance of HHMI researchers, as compared with grant recipients of the National Institutes of Health, concluded they had "higher levels of breakthrough innovation." From Pierre Azoulay, Joshua S. Graff Zivin, and Gustavo Manso, "Incentives and Creativity: Evidence from the Academic Life Sciences," December 30, 2010.

23 Steven Chu, statement in front of the U.S. Senate Committee on Appropriations, May 19, 2009.

24 Jon Gertner, "Capitalism to the Rescue," *New York Times Magazine*, October 3, 2008.

25 *New York Times*, May 20, 1923; see also Lester Markel, letter to Frank B. Jewett, March 31, 1923. AT&T archives.

26 John Pierce, "Can Science Do Without Sentiment?," unpublished, March 12, 1959. Pierce Collection, Huntington Library.

Sources

T he brunt of the work for this book depended upon my personal interviews with Bell Labs veterans; a review of oral histories of Bell
Labs scientists and engineers; and my perusal of documents at the Alcatel-
Lucent archives, the AT&T Corporate archives, the Library of Congress
(for Claude Shannon and Harald Friis), the Huntington Library (for John
Pierce), the Mudd Manuscript Library at Princeton University (for William Baker), and the Stanford University Archives (for Pierce and Bill
Shockley). My research was centered around AT&T's archives, where over
the course of several years George Kupczak made available to me many
hundreds of archival boxes and what were, ultimately, tens of thousands
of document pages. Many of these papers were private or internal correspondence and official reports. But on several occasions I also had the
good luck to come across an odd and valuable nugget, such as several
hundred pages of sworn testimony made by Mervin Kelly in 1947 before
California's Public Utilities Commission, which served as an insightful
guide to Kelly's thinking and workplace responsibilities. At other times,
George carted out for me old crossbar switches, relays, vacuum tubes, and
silicon cells. Also, George gave me free rein to read through the complete,
fifty-nine-year collection of the *Bell Laboratories Record*, the monthly
magazine of Bell Labs, as well as his collection of the *Bell System Technical
Journal* and *WE* (the in-house magazine of Western Electric).

A large number of Bell Labs veterans, whose names are detailed on the pages that follow, gave me their time, usually several hours, and sometimes much more, so that I could interview them. Some of these veterans—Phil Anderson, Bill Fleckenstein, Mannfred Schroeder, Richard Frenkiel, and Irwin Dorros—also allowed me to read personal, unpublished memoirs that proved especially helpful. A few of the men I interviewed, including David Slepian, Ira Jacobs, Max Mathews, and Mannfred Schroeder, died during the researching of the book. I consider myself fortunate to have had the opportunity to speak with them before their passing and remain grateful for their time and consideration.

Also, a word about the oral histories listed in the following pages, which I've quoted from with the permission of the American Institute of Physics, AT&T, the California Institute of Technology, the Institute of Electrical and Electronics Engineers (IEEE), the Chemical Heritage Foundation, and Harriet Zuckerman of the Columbia University Oral History Project. I am grateful to all of these institutions and historians for giving me access to this work. But one oral historian in particular—Lillian Hoddeson—deserves special recognition. Her dogged efforts in the 1970s to track down the origins of solid-state physics, and her insightful interviews of Shockley, Fisk, Bardeen, Pearson, Ohl, and a number of other Bell Labs researchers, have been a resource for a generation of writers. These interviews now reside with the American Institute of Physics. I'm particularly lucky to have been directed to her interview of Katherine Kelly, Mervin Kelly's widow, in the AIP archives, which helped add flavor to the life of a man who left few clues, and several mysteries, behind.

IN THIS BOOK I've tried to describe a group of Bell Labs researchers whose lives have never been the subject of an ensemble narrative; I've also made it my priority to contextualize their achievements within the social, scientific, and commercial world of Bell Labs. What I have not done, however, is go as deep into the details of various innovations as some other historians. Fortunately, for readers more interested in, say, communications sciences or solid-state physics, a number of authors have done so. Their books have enriched my own reporting and are worth searching out in their own right. All of these texts are included in the bibliography on the pages that follow. A few are worth an additional plug here.

The definitive scientific and technological history of the Bell System is a seven-volume set, published by AT&T between 1975 and 1982, that comprises about forty-eight hundred pages. Though the writing is technical and uneven (and in places far too favorably disposed to its parent company), the set is illuminating and indispensable. Four of the volumes in particular—on the Bell System's early years; on transmission technology; on switching history; and on wartime service—were especially useful to me. A companion, eighth volume in the series entitled *Engineering and Operations in the Bell System* also proved helpful, especially in its methodical unraveling of the phone system's technical and organizational complexity.

These books should not be thought of as the final word, however. In the science literature about Bell Labs and its era, more accessible—and more independent-minded—histories abound. Daniel Kevles's *The Physicists* is a valuable book that explains the rise and influence of physics in America; Leonard Reich's *The Making of American Industrial Research* is an expert summation of the early rise of scientific research at GE and Bell Labs. Robert Buderi's *The Invention That Changed the World* is essential reading for anyone interested in the development of wartime radar systems. For those interested in a deeper history of the transistor and semiconductor electronics, meanwhile, two books in particular are worth seeking out. *Revolution in Miniature*, by Ernest Braun and Stuart Macdonald, is a thoughtful exploration of the invention and its implications. *Crystal Fire*, by Michael Riordan and Lillian Hoddeson, is an authoritative account of the transistor's birth and development. Two other excellent books—*The Chip*, by T. R. Reid, which focuses mostly on Jack Kilby; and *The Man Behind the Microchip*, by Leslie Berlin, which explores the life of Robert Noyce—explain how the transistor gave way to an even more powerful invention, the integrated circuit.

Claude Shannon is not yet the subject of a full biography. Yet his encyclopedia-sized *Collected Papers*, edited by Neil Sloane and Aaron Wyner, includes all of Shannon's seminal (and well-written) papers on

Boolean algebra, cryptography, information, chess, and juggling. It also contains a wide-ranging and valuable interview of Shannon, done originally for *Omni* magazine, by Anthony Liversidge. (Another valuable interview of Shannon, which I relied upon but is unfortunately not included in the book, was conducted by Robert Price for the IEEE.) For a better understanding of how Shannon's work fits into the science of cryptography, there is David Kahn's monumental book *The Codebreakers*. For a better sense of how Shannon's work formed the basis of communications engineering, there is Bob Lucky's *Silicon Dreams* and Robert Gallager's *Principles of Digital Communication*. James Gleick's *The Information*, which was published as this book was in the final stages of editing, struck me as essential reading for anyone interested in how Shannon's work fits into the evolution of our information-based society. And for those wishing to get more of a flavor for Shannon's life and thought process, I suggest watching the engaging thirty-minute documentary "Claude Shannon: Father of the Information Age," which is available for viewing on YouTube. I likewise suggest Googling "The Essential Message," an unpublished MIT thesis on Shannon by Erico Marui Guizzo, which is an insightful look into how he created information theory. Finally, no list of books on Shannon would be complete without *Fortune's Formula*, William Poundstone's entertaining narrative of how in the 1960s Shannon and his friend Ed Thorp tried to beat the Las Vegas casinos and Wall Street.

John Pierce wrote many technical books that are still available through used booksellers; with Mike Noll, he also wrote *Signals*, a useful and accessible introduction to the science of communications. Alas, one of Pierce's most captivating pieces of writing was *My Career as an Engineer*, an autobiography (unpublished in the United States) that he wrote—very quickly, I'm sure—for a small Japanese publisher upon receiving the Japan Prize in 1985. It is not available to a general readership, but I had the good fortune to get my hands on a copy thanks to Mike Noll. Another book useful for understanding Pierce and the developments that led to the early communications satellites is Arthur C. Clarke's *Voice Across the Sea*. Also, to get a better understanding of the laser and fiber optics work that went on under Pierce's deputy Rudi Kompfner, I recommend *The Laser in America*, Joan Lisa Bromberg's history of the device, as well as Jeff Hecht's exhaustive examination of the invention and development of fiber optics, *City of Light*.

Finally, there are a multitude of books on telephone history and AT&T that informed my conclusions. Among the best are Claude Fischer's *America Calling*, which documents the telephone's early social history, and John Brooks's *Telephone*, which deftly explains the business decisions that shaped AT&T's first hundred years. Two books on AT&T's breakup, meanwhile, though differing in their perspective, are especially well-written and compelling: *The Fall of the Bell System*, by Peter Temin, and *The Deal of the Century*, by Steve Coll. And two other books on the Bell System are worth noting as well. Sonny Kleinfield's *The Biggest Company on Earth* presents a vivid snapshot of AT&T just before the breakup, and Jeremy Bernstein's *Three Degrees Above Zero* captures life at Bell Laboratories in the midst of the empire's early-1980s transition.

Other books, along with many magazine and journal articles that proved useful, are listed in the selected bibliography below.

—J.G.

INTERVIEWS

Rod Alferness	Robert Dynes
Phil Anderson	George Eberhardt
Joe Baker	Chuck Elmendorf
Norma Barzman	Joel Engel
Walter Brown	Alan English
Alan Chynoweth	Gary Feldman
Steven Chu	Bill Fleckenstein
Edward E. David	Dick Frenkiel
Gerry DiPiazza	Robert Gallager
Phil DiPiazza	Ted Geballe
Irwin Dorros	Randy Giles

SOURCES

Eugene Gordon

Robert Gunther-Mohr

David Hagelbarger

Ira Jacobs

Bill Jakes

Mary Jakes

William Keefauver

Jeong Kim

Leonard Kleinrock

Herwig Kogelnik

Henry Landau

Arthur Lewbel

Tingye Li

Sandy Liebsman

Bob Lucky

John MacChesney

Max Mathews

John Mayo

Brock McMillan

Debasis Mitra

Cherry Murray

Michael Noll

Doug Osheroff

Joe Parisi

Arno Penzias

Henry Pollak

Ian Ross

John Rowell

Mannfred Schroeder

Betty Shannon

David Slepian

Neil Sloane

Dave Stark

Morris Tanenbaum

Robert Von Mehren

SOURCES

SELECTED ORAL HISTORIES

William O. Baker (Chemical Heritage)

William O. Baker (National Reconnaissance Office [NRO])

John Bardeen (American Institute of Physics [AIP])

John Bardeen (Harriet Zuckerman/Columbia University)

Nicolaas Bloembergen (AIP)

Walter Brattain (Alan Holden and W. J. King/AIP)

Walter Brattain (Charles Weiner/AIP)

Walter Brattain (Harriet Zuckerman)

C. Chapin Cutler (Institute of Electrical and Electronics Engineers [IEEE])

Karl Darrow (AIP)

James Fisk (AIP)

Thornton Fry (AT&T)

Cal Fuller (Chemical Heritage)

William Golden

Eugene Gordon (IEEE)

Richard Hamming (AT&T)

Conyers Herring (AIP)

Alan Holden (AIP)

Charles Kao (AIP)

Katherine Kelly (AIP)

Jack Kilby (Charles Babbage Institute)

Robert Lucky (IEEE)

John Mayo (IEEE)

Stanley Morgan (AIP)

Foster Nix (AIP)

Russell Ohl (AIP)

Barney Oliver (Hewlett-Packard)

Eugene O'Neill (IEEE)

Gerald Pearson (AIP)

John Pierce (Caltech)

SOURCES

John Pierce (AT&T)

John Pierce (IEEE)

Ian Ross (IEEE)

Arthur Schawlow (Stanford University)

Claude Shannon (IEEE)

William Shockley (AIP)

William Shockley (Harriet Zuckerman)

Gordon Teal (AIP)

Charles Townes (AIP)

Addison White (AIP)

Dean Woolridge (AIP)

Selected Bibliography

Adams, Stephen, and Orville Bustler. *Manufacturing the Future: A History of Western Electric*. Cambridge: Cambridge University Press, 1999.

Anderson, John B. *Digital Transmission Engineering*. 2nd ed. Piscataway, NJ: IEEE Press/ John Wiley & Sons, 2005.

Baldwin, Neil. *Edison: Inventing the Century*. New York: Hyperion, 1995.

Bardeen, John. "Semiconductor Research Leading to the Point Contact Transistor." Nobel Prize Lecture, December 11, 1956; http://130.242.18.21/nobel_prizes/physics/laureates/1956/bardeen-lecture.html.

Barzman, Norma. *The Red and the Blacklist: The Intimate Memoir of a Hollywood Expatriate*. New York: Nation Books, 2004.

Bell Laboratories Record, 1925–1986, vols. 1–64. Published by Bell Telephone Laboratories. Warren, NJ: AT&T Archives.

Bello, Francis. "The World's Greatest Industrial Laboratory." *Fortune*, November 1958.

Berlin, Leslie. *The Man Behind the Microchip: Robert Noyce and the Invention of Silicon Valley*. New York: Oxford University Press, 2005.

Bernstein, Jeremy. *Three Degrees Above Zero*. New York: Charles Scribner's Sons, 1984.

Block, Fred, and Matthew Keller. "Where Do Innovations Come From: Transformations in the U.S. National Innovation System, 1970–2006" (July 2008). Information Technology & Innovation Foundation; www.itif.org.

Bown, Ralph. "The Transistor as an Industrial Research Episode." *Scientific Monthly* 80, no. 1 (January 1955).

Braun, Ernest, and Stuart Macdonald. *Revolution in Miniature: The History and Impact of Semiconductor Electronics*. Cambridge: Cambridge University Press, 1982.

Brinkman, William, et al. "A History of the Invention of the Transistor and Where It Will Lead Us." *IEEE Journal of Solid State Circuits* 32, no. 12 (December 1997).

Bromberg, Joan Lisa. *The Laser in America, 1950–1970*. Cambridge, MA: MIT Press, 1991.

Brooks, John. *Telephone: The First Hundred Years; The Wondrous Invention That Changed a World and Spawned a Corporate Giant*. New York: Harper & Row, 1976.

Buckley, Oliver E. "Bell Laboratories in the War." *Bell Telephone Magazine*, Winter 1944.

Buderi, Robert. *The Invention That Changed the World: How a Small Group of Radar Pioneers Won the Second World War and Launched a Technological Revolution*. New York: Touchstone/Simon & Schuster, 1996.

Christensen, Clayton. *The Innovator's Dilemma: The Revolutionary Book That Will Change the Way You Do Business*. New York: HarperBusiness, 2003.

Clark, Mark. "Suppressing Innovation: Bell Laboratories and Magnetic Recording." *Technology and Culture* 34, no. 3 (July 1993): 516–38.

Clark, Ronald W. *Edison: The Man Who Made the Future*. New York: G. P. Putnam, 1977.

Clarke, Arthur C. *Voice Across the Sea*. New York: Harper & Row, 1974.

Clarke, Sally H., Naomi R. Lamoreaux, and Steven W. Usselman, eds. *The Challenge of Remaining Innovative: Insights from Twentieth-Century American Business*. Stanford, CA: Stanford Business Books, 2009.

Clogston, A. M. "The Scientific Basis of Solid-State Technology, with Case Histories." From *Physics in Perspective*. National Academy of Sciences, 1972.

Coll, Steve. *The Deal of the Century: The Breakup of AT&T*. New York: Atheneum, 1986.

Colton, F. Barrows. "Miracle Men of the Telephone." *National Geographic*, March 1947.

Conot, Robert. *A Streak of Luck: The Life and Legend of Thomas Alva Edison*. New York: Seaview Books, 1979.

Danielian, N. R. *A.T.&T.: The Story of Industrial Conquest*. New York: Vanguard Press, 1939.

DuBridge, Lee, and Paul A. Epstein. "Robert Andrews Millikan: 1868–1953." National Academy of Sciences, 1959.

Fagen, M. D., ed. *A History of Engineering and Science in the Bell System: The Early Years (1875–1925)*. Bell Telephone Laboratories, 1975.

———. *A History of Engineering and Science in the Bell System: National Service in War and Peace (1925–1975)*. Bell Telephone Laboratories, 1975.

Feynman, Richard. *Six Easy Pieces: Essentials of Physics Explained by Its Most Brilliant Teacher*. Reading, MA: Helix/Addison-Wesley, 1995.

Fischer, Claude. *America Calling: A Social History of the Telephone to 1940*. Berkeley: University of California Press, 1992.

Fletcher, Stephen F. "Harvey Fletcher: 1884–1981." National Academy of Sciences, 1992.

Galambos, Louis. "Theodore N. Vail and the Role of Innovation in the Modern Bell System." *Business History Review* 66, no. 1 (Spring 1992): 95–126.

Gallager, Robert G. "Claude E. Shannon: A Retrospective on His Life, Work, and Impact." *IEEE Transactions on Information Theory* 47, no. 7 (November 2001).

Gavaghan, Helen. *Something New Under the Sun: Satellites and the Beginning of the Space Age*. New York: Copernicus/Springer-Verlag, 1998.

Gertner, Jon. "Mad Scientist: Can Legendary Bell Labs—and Its Struggling Parent, Alcatel-Lucent—Be Saved by a 'Crazy Risk Taker' Who's Betting That Innovation Can Be Captured in a Mathematical Formula?" *Fast Company*, February 2008.

———. "The Lost World: What Did We Gain—and Lose—in the Telecom Revolution? Searching for Answers in the Fate of Lucent and Bell Labs." *Money*, March 2003.

Goodell, Rae. *The Visible Scientists*. Boston: Little, Brown, 1977.

Guizzo, Enrico Marui. "The Essential Message: Claude Shannon and the Making of Information Theory," master's thesis, MIT, 2003.

Hamilton, Loren Henry. *A Missouri Boyhood*. Self-published, 1983. Daviess County Library, Gallatin, MO.

Hayes, Brian. *Infrastructure: A Field Guide to the Industrial Landscape*. New York: W. W. Norton, 2005.

Hecht, Jeff. *City of Light: The Story of Fiber Optics*. Revised and expanded edition. New York: Oxford University Press, 1999.

Hoddeson, Lillian Hartmann. "The Roots of Solid-State Research at Bell Labs." *Physics Today*, March 1977.

Hoddeson, Lillian, and Vicki Daitch. *True Genius: The Life and Science of John Bardeen*. Washington, DC: Joseph Henry Press, 2002.

Hughes, Thomas P. *American Genesis: A Century of Invention and Technological Enthusiasm*. New York: Viking, 1989.

Jacobs, Ira. "Lightwave System Developments: Looking Back and Ahead." *Optics & Photonics News*, February 1995.

Jakes, William C., ed. *Microwave Mobile Communications*. New York: IEEE Press, 1993. First published in 1974 by John Wiley & Sons.

Kahaner, Larry. *On the Line: How MCI Took On AT&T—and Won!* New York: Warner Books, 1986.

Kahn, David. *The Codebreakers: The Comprehensive History of Secret Communication from Ancient Times to the Internet*. New York: Scribner, 1996.

Kaplan, David A. *The Silicon Boys and Their Valley of Dreams*. New York: HarperPerennial, 2000.

Kargon, Robert H. *The Rise of Robert Millikan: Portrait of a Life in American Science*. Ithaca, NY: Cornell University Press, 1982.

Kelly, Mervin. "The American Engineer." *The Bridge of Eta Kappa Nu*, September 1943.

Kevles, Daniel. *The Physicists: The History of a Scientific Community in Modern America*. Cambridge, MA: Harvard University Press, 2001.

Kleinfield, Sonny. *The Biggest Company on Earth: A Profile of AT&T*. New York: Holt, Rinehart and Winston, 1981.

Lewis, Tom. *Empire of the Air: The Men Who Made Radio*. New York: HarperPerennial, 1993.

Lucky, Robert. *Silicon Dreams: Information, Man, and Machine*. New York: St. Martin's Press, 1991.

Millikan, Robert. *The Autobiography of Robert A. Millikan*. New York: Prentice-Hall, 1950.

Millman, S., ed. *A History of Engineering and Science in the Bell System: Physical Sciences (1925–1980)*. AT&T Bell Laboratories, 1983.

———. *A History of Engineering and Science in the Bell System: Communications Sciences (1925–1980)*. AT&T Bell Laboratories, 1984.

Morse, Philip M. *In at the Beginnings: A Physicist's Life*. Cambridge, MA: MIT Press, 1977.

Morton, Jack. "From Research to Technology." *International Science and Technology*, May 1964.

O'Neill, Eugene, ed. *A History of Engineering and Science in the Bell System: Transmission Technology (1925–1975)*. AT&T Bell Laboratorics, 1985.

Paine, Albert Bigelow. *Theodore N. Vail: A Biography*. New York: Harper & Brothers, 1929.

Pierce, John R. *Science, Art, and Communication*. New York: Clarkson Potter, 1968.

———. *My Career as an Engineer: An Autobiographical Sketch*. University of Tokyo, 1988.

Pierce, John, and Michael Noll. *Signals: The Science of Telecommunications*. New York: Scientific American Library, 1990.

Poundstone, William. *Fortune's Formula: The Untold Story of the Scientific Betting System That Beat the Casinos and Wall Street*. New York: Hill & Wang, 2005.

Reich, Leonard S. *The Making of American Industrial Research: Science and Business at GE and Bell, 1876–1926*. New York: Cambridge University Press, 2002.

Reid, Loren. *Hurry Home Wednesday: Growing Up in a Small Missouri Town, 1905–1921*. Columbia: University of Missouri Press, 1978.

Reid, T. R. *The Chip: How Two Americans Invented the Microchip and Launched a Revolution*. Paperback edition. New York: Random House, 2001. First published in 1985 by Simon & Schuster.

Rey, R. F., technical ed. *Engineering and Operations in the Bell System*. Murray Hill, NJ: AT&T Bell Laboratories, 1984.

Rhodes, Richard. *The Making of the Atomic Bomb*. New York: Simon & Schuster, 1986.

Riordan, Michael. "How Bell Labs Missed the Microchip. *IEEE Spectrum*, December 2006.

Riordan, Michael, and Lillian Hoddeson. *Crystal Fire: The Invention of the Transistor and the Birth of the Information Age*. New York: W. W. Norton, 1997.

Saxenian, Annalee. *Regional Advantage: Culture and Competition in Silicon Valley and Route 128*. Cambridge, MA: Harvard University Press, 1996.

Scaff, J. H. "The Role of Metallurgy in the Technology of Electronic Materials." *Metallurgical Transactions* 1 (March 1970).

Schindler, G. E., ed., with A. E. Joel et al. *A History of Engineering and Science in the Bell System: Switching Technology (1925–1975)*. Bell Telephone Laboratories, 1982.

Shannon, Claude Elwood. *Collected Papers*. Edited by N. J. A. Sloane and Aaron D. Wyner. New York: IEEE Press/John Wiley & Sons, 1993.

Shockley, William. "The Invention of the Transistor—An Example of Creative-Failure Methodology." *Electrochemical Society Proceedings* 98, no. 1 (1998).

———. "Transistor Physics." *American Scientist* 42, no. 1 (January 1954).

———. "Crystals, Electronics, and Man's Conquest of Nature." Undated, circa 1958.

Shurkin, Joel. *Broken Genius: The Rise and Fall of William Shockley, Creator of the Electronic Age*. New York: Macmillan, 2006.

Smith, George David. *The Anatomy of a Business Strategy: Bell, Western Electric, and the Origins of the American Telephone Industry*. Baltimore: Johns Hopkins University Press, 1986.

Smits, F. M., ed. *A History of Engineering and Science in the Bell System: Electronics Technology (1925–1975)*. AT&T Bell Laboratories, 1985.

Temin, Peter, with Louis Galambos. *The Fall of the Bell System: A Study in Prices and Politics*. New York: Cambridge University Press, 1987.

Tompkins, Dave. *How to Wreck a Nice Beach: The Vocoder from World War II to Hip-Hop*. Brooklyn, NY: Melville House/Chicago: Stop Smiling Books, 2005.

Townes, Charles. *How the Laser Happened: Adventures of a Scientist*. New York: Oxford University Press, 1999.

Von Auw, Alvin. *Heritage and Destiny: Reflections on the Bell System in Transition*. New York: Praeger, 1983.

Weiner, Charles. "How the Transistor Emerged." *IEEE Spectrum* 10, no. 1 (January 1973).

Whyte, William H., Jr. *The Organization Man*. New York: Anchor, 1957.

Wolfe, Tom. "The Tinkerings of Robert Noyce." *Esquire*, December 1983.

Zachary, G. Pascal. *Endless Frontier: Vannevar Bush, Engineer of the American Century*. Cambridge, MA: MIT Press, 1999.

Index